Architecting a Satisficing Strategy for Designing Complex Systems

Lin Guo · Janet K. Allen · Farrokh Mistree

Architecting a Satisficing Strategy for Designing Complex Systems

Springer

Lin Guo
Department of Industrial Engineering
South Dakota School of Mines
and Technology
Rapid City, SD, USA

Janet K. Allen
School of Industrial and Systems
Engineering
University of Oklahoma
Norman, OK, USA

Farrokh Mistree
School of Aerospace and Mechanical
Engineering
University of Oklahoma
Norman, OK, USA

ISBN 978-3-032-04576-8 ISBN 978-3-032-04577-5 (eBook)
https://doi.org/10.1007/978-3-032-04577-5

This Springer imprint is published by the registered company Springer Nature Switzerland AG
The registered company address is: Gewerbestrasse 11, 6330 Cham, Switzerland

Foreword

I have spent my career guiding engineers—both hardware and software—who architect satisficing strategies that help them make sound decisions despite incomplete information. The authors (Guo, Allen and Mistree) pursue the same aim in *Architecting a Satisficing Strategy for Designing Complex Systems*. They remind readers that real designers rarely face tidy closed form optimization problems. Designers explore shifting landscapes of stakeholder expectations, evolving models, and uncertainties that resist collapse into single point estimates, especially when those systems span cyber, physical, and social dimensions. The authors respond by presenting a disciplined approach that prizes adaptability over illusory perfection and portrays design as a continuous process of architecting satisficing strategies rather than a single search for an optimal solution.

The authors organize nine chapters around five illustrative problems: water resources, supply chains, manufacturing, thermal systems, and social interventions. Each case reflects an evolving cyber-physical-social system in which technical performance intertwines with human behavior and institutional policy. The authors demonstrate how designers architect satisficing strategies in every domain and treat complexity as an opportunity for insight. They draw on behavioral economics, artificial intelligence, and systems thinking to help readers recognize why such multidimensional contexts require flexible reasoning styles.

The narrative aligns with several initiatives at the National Institute of Standards and Technology, where colleagues and I promote standards for model interoperability, knowledge representation, and trustworthy artificial intelligence. The authors emphasize transparency in a way that supports these efforts and keeps their approach auditable in high consequence applications.

I expect three trends to amplify the importance of architecting satisficing strategies for cyber-physical-social systems. First, engineers will harness generative artificial intelligence systems, along with traditional knowledge-based frameworks, as design copilots and will generate alternatives faster than human teams can evaluate them. Satisficing frameworks will help teams steer those tools toward solutions that honor diverse stakeholder values. Second, developers will build digital engineering pipelines that fuse real time operational data with design intent and will update design decisions continuously rather than in stage gate fashion. Third, policy makers and community advocates will press engineers to confront ethical, environmental, and social factors as they call for sustainable and resilient infrastructure. Satisficing methods will enable engineers to negotiate those multidimensional tensions.

The authors invite engineers, scientists, and educators to extend these ideas. I urge our community to collaborate in building the theories, tools, and demonstrations needed to architect satisficing strategies for evolving cyber-physical-social systems. We should refine ontologies, validate human AI workflows, and prepare designers who steward complex systems responsibly. I look forward to collaborating on this effort and welcome readers to join the conversation and experiments that the authors encourage. Let us work together to architect satisficing strategies that keep human values at the center of every evolving cyber-physical-social system!

June 2025

Ram D. Sriram
Fellow AAIA, ACM, AIMBE, ASME, AAAS, IEEE, IET, INCOSE, SMA, SME, WAS
Honorary Member IISE
Senior Member: AAAI
Chief, Software and Systems Division, Information Technology Laboratory
National Institute of Standards and Technology
Gaithersburg, MD, USA

Preface

Why Must We Architect *Satisficing* Strategies?

We confront shifting information, watch technology evolve, and listen as stakeholders express new needs, so we ask: *Why must we architect satisficing strategies?* Heraclitus answered centuries ago when he noted that no one steps into the same river twice; we see the same restless flow in the complex systems that we shape today. We invite designers to walk with us through the Design Evolution Cycle as they architect *satisficing* strategies that stay robust while they face shifting information, evolving technology, and changing needs. We draw on the wisdom of Herbert A. Simon, Albert Einstein, and Genichi Taguchi. Simon reminds us that decision makers gain practical value when they pursue solutions that *satisfice* instead of chasing an elusive optimum. Einstein warns us that mathematical certainty often drifts away from reality. Taguchi guides us toward designs that retain *robustness* in the face of variation. We draw on these viewpoints to replace a quest for perfection with a disciplined journey toward solutions that are good enough, adaptable, and resilient.

What We Set Out to Achieve

We set out to share a way that designers, researchers, and decision makers can refine their models and solutions as fresh evidence arrives. We use the Design Evolution Cycle as our guide for architecting satisficing strategies. In the chapters that follow we walk through formulation, approximation, exploration, and evaluation and show how each revision strengthens robustness and adaptability. We ground the discussion in five test problems: water resources, supply chains, manufacturing, thermal systems, and social interventions, and we show readers how they can carry the same reasoning to any domain shaped by change and uncertainty.

How We Organized This Monograph

We organize this monograph by moving from motivation to method and then from demonstration to reflection. We guide readers through a sequence of chapters that build evidence for architecting satisficing strategies.

- In Chap. 1 we establish why designers need a satisficing strategy and identify gaps that appear when they rely on strict optimization.
- In Chap. 2 we compare optimizing and satisficing mindsets, introduce five illustrative problems, and show how designers reshape early problem formulation with a satisficing perspective.
- In Chap. 3 we present the Design Evolution Cycle in detail and describe four methods that designers may use to realize each stage of the cycle.
- In Chaps. 4–7 we apply the cycle to domain problems so that readers see how people create satisficing strategies in practice.
- In Chap. 8 we assess the results and show how designers improve resilience to uncertainty and support diverse solution sets.
- In Chap. 9 we look ahead and discuss how designers may architect Cyber Physical Social Systems that harness generative AI for decision support in a world where change is the order of the day. We link the Design Evolution Cycle to Industry 5.0, public policy, and sustainable development.

We arrange the chapters to form a narrative arc that begins with motivation, equips readers with tools, and ends with opportunities for future exploration. We encourage readers to enter at any point yet follow the full journey to see how each insight supports architecting satisficing strategies for tomorrow's complex systems.

Who May Benefit

George Bernard Shaw observed, "Some men see things as they are and ask why. Others dream things that never were and ask why not." Robert F. Kennedy echoed that challenge. We encourage graduate students and researchers in engineering, operations research, data science, and artificial intelligence to adopt that spirit and use the Design Evolution Cycle to ask, "Why not?" as they architect satisficing strategies for the next generation of complex systems. Wayne Gretzky advised, "I skate to where the puck is going to be, not where it has been." Practicing engineers, system architects, AI developers, and policy analysts may apply the cycle to design for tomorrow's needs rather than yesterday's routines. Richard Riley stated, "We are preparing students for jobs and technologies that do not yet exist, to solve problems we do not even know are problems yet." By turning to the Design Evolution Cycle, all of these individuals participate in architecting satisficing strategies that keep complex systems resilient amid continuous change.

Expressions of Appreciation

We honor the late Prof. Owen F. Hughes, who invited Farrokh to work with him as a postdoctoral researcher at the University of New South Wales between 1974 and 1980. Professor Hughes guided our early exploration of decision making in engineering design. His collaboration with Farrokh and Bert Bras produced the Adaptive Linear Programming algorithm, documented in "The Compromise Decision Support Problem and the Adaptive Linear Programming Algorithm," in *Structural Optimization: Status and Promise*, edited by M. P. Kamat, AIAA, 1993 pp. 247–286. Lin Guo later refined this algorithm; we present her formulation refinements in Chap. 4 and show in Chaps. 5 and 6 how she applied the improved ALP within the Design Evolution Cycle.

We remember the late Douglas F. Muster and thank Warren F. Smith and Bert Bras. In 1990 we collaborated with them to publish "Decision Based Design: A Contemporary Paradigm for Ship Design" in the *Transactions of the Society of Naval Architects and Marine Engineers*. In that paper we placed the decision maker at the center of engineering design, introduced the Compromise Decision Support Problem construct, and argued that designers benefit when they pursue good enough solutions through iterative model refinement. We formalize these ideas in Chap. 3, apply them across domains in Chaps. 4–7, and assess their impact in Chap. 8.

We congratulate Lin Guo, who received the 2021 Provost's Outstanding Dissertation Award in Engineering and Science for her work on adaptive satisficing methods; we have expanded her dissertation into this monograph. We supported her doctoral studies with funds from the L. A. Comp Chair and the John and Mary Moore Chair at the University of Oklahoma.

Looking Forward

Simon urges us to *satisfice* rather than chase perfection, Einstein reminds us to respect model limits, and Taguchi guides us toward robustness. We draw inspiration from Heraclitus's flowing river, Shaw's and Kennedy's call to ask, "Why not?", Gretzky's advice to skate to where the puck will be, and Riley's warning about unknown problems, and we invite every reader to join us in architecting *satisficing* strategies for evolving Cyber Physical Social Systems.

June 2025

Janet K. Allen
John and Mary Moore Chair
and Professor
School of Industrial and Systems
Engineering
University of Oklahoma
Norman, OK, USA
janet.allen@ou.edu

Farrokh Mistree
L. A. Comp Chair and Professor,
School of Aerospace and Mechanical
Engineering
University of Oklahoma
Norman, OK, USA
farrokh.mistree@ou.edu

Contents

Chapter 1
Designing Complex Systems Using Satisficing Strategy

The Motivation for Writing This Monograph

Imagine a network of dams spread across a vast region. Each month, the water released from each dam must be meticulously planned to prevent floods and droughts while meeting the diverse needs of various users—humans, wildlife, and industries. Traditional optimization methods strive for a single best solution, but as Herbert A. Simon suggests, in complex, uncertain environments, optimality is often an illusion. Instead, we adopt a satisficing strategy, seeking solutions that are "good enough" rather than perfect, so that designers can remain adaptable and resilient when conditions become unpredictable. By using this pragmatic approach, we ensure that the system continues to function effectively even when new information emerges and conditions change.

We illustrate the evolving nature of the modern complex system, underscoring the necessity for designers to adopt a dynamic and robust design strategy. In this monograph, we present the design evolution framework as a computational tool. We embrace the concept of 'satisficing,' finding solutions that are good enough to meet our criteria and that we can refine over time to remain effective under changing conditions. This approach is particularly beneficial for managing complex systems with multiple goals, uncertainties, and dynamic environments. This approach is particularly beneficial for managing complex systems with multiple goals, uncertainties, and dynamic environments.

The motivation for adopting a satisficing strategy in design evolution is rooted in the need to pragmatically manage complexity. As Genichi Taguchi emphasized, robust design minimizes the effects of variation and ensures that solutions remain viable under uncertain conditions. Unlike optimizing strategies, where designers seek the single optimal solution, satisficing acknowledges the limitations of mathematical models in capturing real-world complexity—a perspective aligned with Albert Einstein's insight that mathematical certainty often diverges from reality. By integrating satisficing strategies with the Design Evolution Cycle, designers can iteratively refine their solutions, ensuring robustness, adaptability, and sustainability in dynamic environments.

L. Guo et al., *Architecting a Satisficing Strategy for Designing Complex Systems*,
https://doi.org/10.1007/978-3-032-04577-5_1

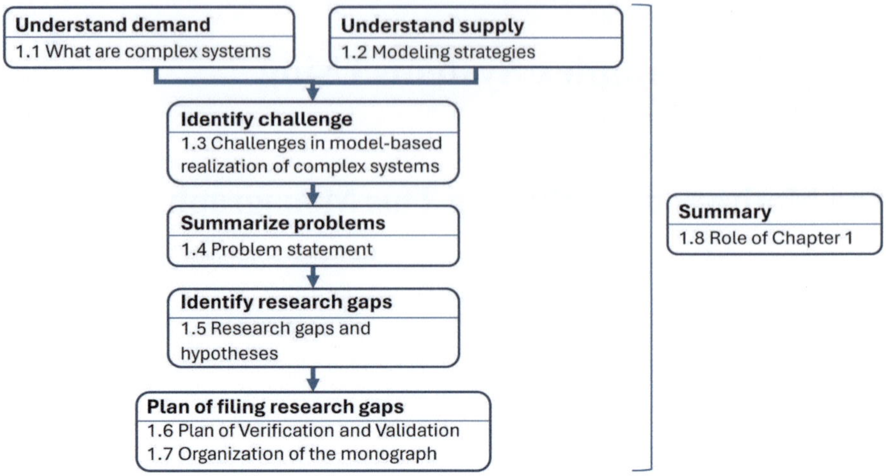

Fig. 1.1 Organization of this chapter

As proposed in this monograph, the design evolution cycle is a computational framework that is used to implement a satisficing strategy. With the design evolution cycle, a designer can iteratively refine solutions and models, adapting them to new information and changing circumstances. This ensures that the design remains viable and effective over time.

By adopting a satisficing strategy for a design evolution process, designers can better manage complex systems, make informed decisions, and in the long term, achieve both satisfactory and sustainable results. However, uncertainty and variability are inherent in water resource management, supply chain logistics, manufacturing, engineering design, and social systems.

1.1 Organization of This Chapter

In this chapter, the context is established, and the "why question" is answered: "Why do we need design evolution? We provide the motivation and purpose of design evolution in managing complex systems. First, we present a frame of reference, then introduce the motivation, and the chapter organization is described; see Fig. 1.1. We analyze the demand (need) for designing complex systems by introducing their characteristics and examples in Sect. 1.1, analyze the supply (possibility) of designing complex systems by introducing two typical modeling strategies in Sect. 1.2, discuss the challenges in realizing complex systems regarding the gaps in demand and supply in Sect. 1.3, identify the research gaps in Sect. 1.4, pose the research question in Sect. 1.5, outline our plan of answering the research question in Sects. 1.6 and 1.7, and summarize the role of Chapter 1 in this monograph in Sect. 1.8. We interpret this chapter with respect to the "what, how, and why" questions in Table 1.1.

In Table 1.2, we summarize the logic flow in this monograph and the role of each chapter. In Fig. 1.2, we illustrate the verification and validation plan. In this chapter we provide the foundation for this monograph by identifying key research gaps (RG)

Table 1.1 Summary of architecture for design evolution cycle

Architecture Constituents	Summary interpretation for a computational framework in the design of complex systems
Elements	*What?*
Components	Stages of modeling complex systems (Fig. 1.2)
Connectors	The algorithms that connect and integrate the components
Form	*How?*
Component roles	Different roles that the algorithms play in the loop
Properties	Attributes or features of the modeling stages and connecting algorithms
Relationship	The relationships among different components in the framework
Rationale	*Why?*
Motivation	The underlying reason for using this framework, driven by requirements
Assumptions	Generalizations or simplifications used to guide the selection of elements
Interpretation	Understanding the big picture of the design evolution framework

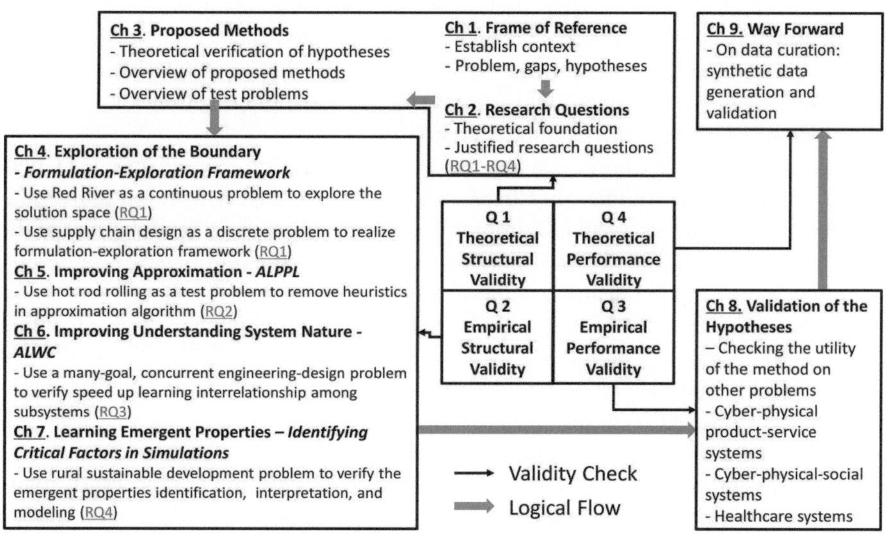

Fig. 1.2 Monograph layout and plan of verification and validation

Table 1.2 The logic flow and the role of each chapter

Chapters	Chap. 1	Ch2	Ch3	Ch4 –7	Ch8	Ch9
Actions	RG H	RD RQ SH	TVe M	EVe SQT AQ	CQ EVa	TE
Nomenclature	AQ—Answer research question CQ—closure of the answers to research questions EVa—empirically validate hypotheses EVe—empirically verify hypotheses H—give hypotheses M—introduce methods RD—tie to robust design RG—give research gaps RQ—pose research questions SH—state hypotheses SQT—state research questions in the context of test problems TE—theoretically extend the research TVe—theoretically verify hypotheses					

and outlining the hypotheses (H) that will drive subsequent investigations. In Chap. 2 we build upon this foundation by explicitly posing research questions (RQ) that guide the development of a satisficing strategy in model-based design. Here, we transition from identifying the need for design evolution to examining how we can realize model evolution—a shift that is crucial for implementing satisficing in engineering design.

In Chap. 2, we explore the theoretical underpinnings that justify a satisficing strategy over conventional optimization approaches. This includes an analysis of the Karush–Kuhn–Tucker (KKT) conditions and their implications for engineering design, illustrating why satisficing provides a more robust and adaptable alternative. We also introduce the compromise Decision Support Problem (cDSP) construct, which serves as the computational framework for satisficing, setting the stage for methodological advancements in later chapters.

Moving forward, in Chap. 3 we theoretically verify the hypotheses posed in Chap. 2, ensuring structural validity. In Chaps. 4 through 7 we deal with empirical validation of these hypotheses using test problems, demonstrating the practical utility of the satisficing approach. Finally, in Chap. 8 we consolidate the findings by empirically validating research conclusions, while in Chap. 9 we extend the theoretical implications beyond the scope of this study, exploring broader applications of satisficing strategies in complex system design.

Table 1.2 is expanded by adding a summary of each action in the later chapters.

What are Complex Systems?

A system is a set of entities which form a unified whole through their interactions, relationships, or dependencies. A complex system is a system composed of many components which may interact with each other,[1] such as a manufacturing system,

[1] This definition is from Wikipedia: https://en.wikipedia.org/wiki/Complex_system.

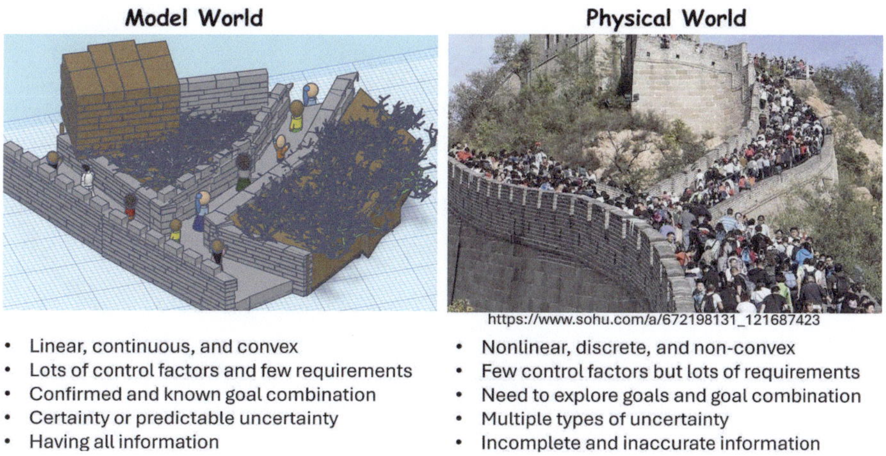

Fig. 1.3 The model world and its corresponding physical world[2]

a village, or an economic entity. The relationships between the parts of a complex system are hard to predict and control (Ladyman et al., 2013).

The most prominent feature of the realization of complex systems is the real-time information sharing and system evolution based on a highly integrated human-cyber-physical system. As “All models are wrong, but some are useful”—Attributed to Box (1979). Box and Draper (1987); models are often incomplete, inaccurate, and embody different levels of fidelity, a solution may be optimal to the model but may not be optimal to the real problem (the real complex system) which is way more complicated than the model.

If one categorizes everything in two worlds—model and physical worlds, see Fig. 1.3, he or she can observe that the model world is simplified, whereas the physical world is much more complicated. In the model world, designers desire problems to be linear, continuous, and convex. Many factors can be controlled, uncertainties can be predicted accurately and managed well, and all the necessary information can be collected as needed. However, the physical world is quite the opposite. Although designers try to model the physical world by incorporating as many complexities as possible, there is always a conceptual disconnect between the two worlds.

Although there is an intellectual disconnect between physical complex systems and their corresponding decision models, designers need their models to be useful for giving decision support. The principal motivation in this monograph is to make decision models of complex systems more useful by enabling model evolutions to make those models adapt to the rapidly changing requirements in the physical world.

The four stages of the evolution cycle of complex systems realization are formulation, approximation, exploration, and evaluation; see Fig. 1.4. By acting forward from formulation to approximation, exploration, and evaluation, deductive methods are applied, and heuristics are used to proceed with system realization. With the

[2] The picture on the right-hand side is from: https://www.sohu.com/a/672198131_121687423.

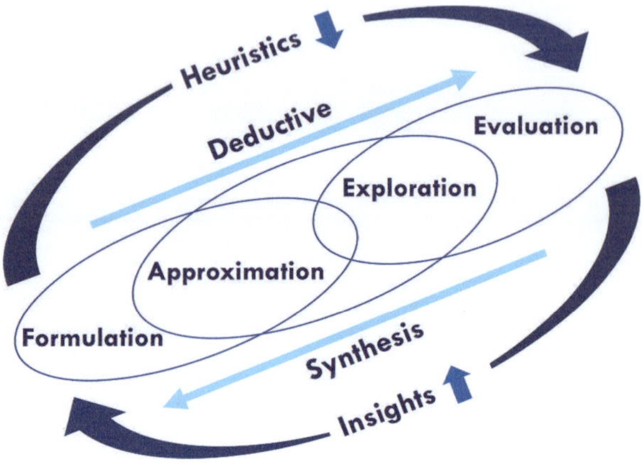

Fig. 1.4 The evolution cycle of complex systems realization

results and observations obtained, calculations and analyses can be done, insight is acquired, and feedback can be given to make improvements or adjustments in the previous stages. By acting backward, synthesis methods are applied, and insight can be used to replace heuristics. By managing the interactions between different stages of the evolution cycle, knowledge of the connections between stages and general knowledge of the evolution of decision models can be acquired and synthesized.

1.2 Characteristics of Complex Systems

1.2.1 The Behaviors of a Complex System

The behaviors of a complex system are difficult to infer from the properties of the system. Ignoring these difficulties can lead to models that are neither accurate nor useful. Complex systems are studied across many disciplines, but they share some common characteristics. The major domain-independent characteristics of complex systems include:

1. Multiple, conflicting goals.
2. Evolving preferences among goals during design or operation.
3. A large number of goals that are hard to visualize using traditional methods.
4. Nonlinear functional relationships among elements.
5. Coupled decisions, such as selecting a material and its corresponding geometry.
6. Emergent properties.

These characteristics are addressed in this monograph. Further descriptions of these characteristics are as follows.

Multiple conflicting goals. As there can be multiple users or stakeholders of a complex system, they may compete for the same limited system resources, and their expectations of the system may conflict. A designer needs to consider users' different goals. For example, the users of a dam network include the residents in the neighborhood, the farmers or industrial workers near the dam network, and the fish and other wild animals that live along the river basin. They all use the water in the river. Meeting each user's water demand represents the multiple conflicting goals of the dam-network system.

Evolving preferences among the goals. As a complex system needs to adapt to the changing environment, and the users' urgency of demand may be different over time, the priority of the goals may evolve. For example, in the introduction stage of a new product, the stability of its supply chain should be prioritized. In contrast, profit should be the most important goal in its maturity.

There are a large number of goals that are hard to visualize. As many elements and components can be incorporated into a complex system, the number of goals to be achieved by designing and operating the system can be large. Some goals may be highly orthogonal and thus cannot be linearly combined. Visualizing the tradeoffs among the goals and presenting solutions that satisfy multiple goals is challenging.

Nonlinearity. Complex systems often have nonlinear behavior. Nonlinearity is a term which is used to describe a situation with no straight-line or direct relationship between an independent and dependent variable.[3] Nonlinearity also indicates that a change in a system's input size does not produce a proportional change in the output size. For example, a step cost does not change steadily with activity changes but changes at discrete points.

Coupled decision making. Multiple decision types are usually used in designing and operating a complex system. The two typical decision types in systems realization are selection decisions and compromise decisions (Mistree et al., 1993). The decision variables reflect such decision types—integer variables indicate selection decisions (such as selecting a material), and continuous variables indicate compromise decisions (such as designing the dimensions). When there are mixed variables, selection, and compromise decisions should be made simultaneously, and each decision affects the other.

Emergent properties. Emergent properties are properties of several elements that cannot be identified in any individual element. For example, in a village where 40% of the population refuses to change their lifestyle, if the remaining 60% adopt new technology and improve their social and economic status, almost everyone will quickly change their lifestyle because of social pressure. However, each individual can still claim that he or she may not accept such change.

[3] This definition is from Investopedia: https://www.investopedia.com/terms/n/nonlinearity.asp.

1.2.2 Examples of Complex Systems

The examples in this section are used in Chaps. 4, 5, 6, and 7 to test the proposed methods. Each example may not contain all the characteristics of complex systems identified in Sect. 1.2.1 but each contains at least one of them.

A dam-network system. The dams along a river basin form a network. Managing fresh water supply and sensible distribution to support human activity while sustaining vigorous, effective ecosystems is a major ecological challenge. The reservoir behind each dam stores water. The water in each reservoir is released downstream by controlling the dam. The basin has three user groups—people, fish in the reservoirs, and fish in the streams between reservoirs. To meet water demands, there are three goals (Guo et al., 2019):

(1) To reach the target for water storage in reservoirs.
(2) To meet people's demand for water—including agricultural and municipal demand.
(3) To meet the water requirements for the fish.

All the users compete for the water resources in the river basin. There are uncertainties in the water inflow, such as precipitation and tributary inflow, as well as the water outflow, such as the demand of each user group. Besides this complexity, the demand of each user group varies with the season, and the priority of each goal evolves. Therefore, in a dam network system, the complexity includes managing multiple conflicting goals and the evolving preferences of the goals (Guo et al., 2019). In Chap. 4, the dam-network planning problem is used as a test problem to testify to the utility of the Three-Step Exploration Method in Chap. 4.

A multistage manufacturing system. Hot rod rolling is a multistage manufacturing process in which a reheated billet, slab, or bloom that is produced after the casting process is further thermo-mechanically processed by passing through a series of rollers (Nellippallil et al., 2017, 2018). In this problem, the requirement is to produce steel rods with improved mechanical properties like yield strength, tensile strength, and hardness. These mechanical properties are defined by the microstructure after cooling, which includes the phase fractions (ferrite and pearlite phases are only considered in this problem), pearlite interlamellar spacing, ferrite grain size, and chemical compositions (Nellippallil et al., 2018). The microstructural requirements are to achieve a high ferrite fraction value, low pearlite interlamellar spacing, and low ferrite grain size value within the defined ranges. The requirement is to carry out the integrated design of the material and the process by managing the cooling rate (cooling process variable), final austenite grain size after rolling (rolling microstructure variable), and the chemical composition of the material. In a multi-step manufacturing system, there are relatively complicated relationships among the independent and dependent variables, and many such functional relationships are nonlinear. The goals for different steps and sub-processes conflict, and their priorities change along with the circumstances. Therefore, in a multistage manufacturing system, the complexity includes multiple conflicting goals, evolving preferences

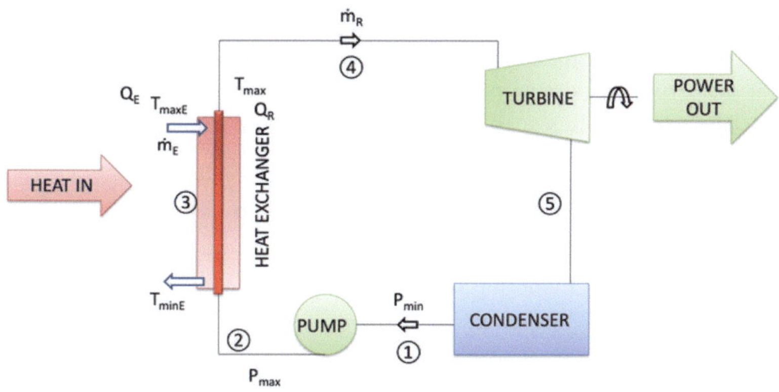

Fig. 1.5 The thermal system around the rankine cycle

among the goals, and nonlinearity among variables. In Chap. 5, the hot rod rolling problem is used to demonstrate the utility of the Adaptive Linear Programming (ALP) Algorithm with Parameter Learning (ALPPL) (Guo et al., 2024a, 2024b).

A concurrent engineering system. The thermal system around the Rankine cycle is a concurrent engineering system. The Rankine cycle is a mathematical representation of a heat engine that converts heat into mechanical work while undergoing phase change (Macquorn Rankine, 1853; Wikipedia, 2019). The system's major components are a power-producing turbine, a pump to pressurize the flow to the turbine, two heat exchangers, a condenser, and a heater, Fig. 1.4. The system designer should deal with heat source issues (left side of Fig. 1.5), power use issues (right side of Fig. 1.5), and the choice of working fluids. The common working fluid in a Rankine cycle is water. Uses of other fluids (often organic chemicals) have led to the development of "organic Rankine cycles." Of course, geometric specifications and design analysis of physical elements in the system also represent opportunities for model and design space exploration. A designer needs to achieve six goals in this Rankine cycle (Smith et al., 2015):

(1) Achieve zero moisture in steam leaving the turbine.
(2) Maximize Rankine cycle efficiency.
(3) Maximize temperature exchanger efficiency.
(4) Maximize system efficiency indicator 1.
(5) Maximize system efficiency indicator 2.
(6) Maximize heat transfer effectiveness in the exchanger.

Therefore, the complexity of a concurrent engineering system, such as a thermal system around the Rankine cycle, includes multiple conflicting goals, many goals, and coupling decision-making. In Chap. 6, the thermal system design problem is used to demonstrate the utility of the Adaptive Leveling Weighting (ALW) Algorithm (Guo et al., 2023).

A cyber-socio-technical system. A small human society, such as a relatively isolated, underdeveloped village is a complex system. When social entrepreneurs want to improve the villagers' social and economic status by promoting new technologies, the village is to become a cyber-socio-technical system. Such a system may have emergent properties–collective properties that cannot be predicted from individual behavior. The complexities social entrepreneurs should manage are the cognition and acceptance of new technologies and lifestyles resulting from new technologies. Simulations and predictions of the emergent properties and social behaviors should be addressed to manage such complexities. In Chap. 7, the cyber-socio-technical system problem is used to demonstrate the utility of the social development framework using agent-based modeling and systems dynamics, (SDF_ABM_SD) (Guo et al., 2020).

1.3 Modeling Strategies and Their Foci

Design methods and solution algorithms for dealing with complex problems fall into two categories; see Fig. 1.6 and Table 1.3: (i) formulate a complex problem exactly and solve it approximately, or (ii) approximate a complex problem and solve it exactly. We call the first strategy the "optimizing" strategy and the second the "*satisficing*" strategy. Examples of an optimizing strategy are gradient-based methods (Williams and Zipser, 1995), pattern search methods (Rios and Sahinidis, 2013), penalty function methods (Viswanathan and Grossmann, 1990), etc.; According to Herbert A. Simon, the decision maker has a choice between an optimal decision from an imaginary world or decisions that are "good enough," that *satisfice* for a world approximating the complex real one more closely (Simon and Kadane, 1975).

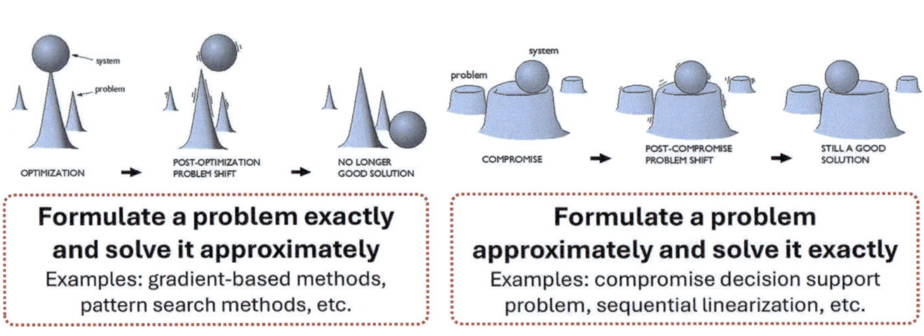

Fig. 1.6 The optimization strategy and the *satisficing* strategy

Table 1.3 Advantages and disadvantages of the two categories of solution algorithms

#	Category	Example methods	Advantages	Disadvantages
Optimizing Strategy	Formulate a problem exactly and solve it approximately	Gradient-based methods, pattern search methods, penalty function methods, etc.	– Maintaining a relatively accurate model along the solution search (given the information that the designer has on hand)	– The solution is still an approximate, inaccurate one – Cannot get the information of the dual and use it to facilitate problem-solving or post-solution analysis – Heuristics are used in solution algorithms, which may result in premature convergence or unnecessarily high computational complexity
Satisficing Strategy	Approximate a problem and solve it exactly	ALP, SLP, SQL, etc.	– Solutions are on the vertices of the approximated problem, so the dual of the approximated problem can be explored – Solutions may be away from the boundary of the original problem, so they are relatively insensitive to variations – Accumulating the linearized constraints during iteration can improve the problem's approximation, making it possible to obtain an approximate problem with acceptable accuracy	– Introducing information loss while doing approximation, making the solution inaccurate – Heuristics are used in approximation algorithms, which may result in premature convergence or unnecessarily high computational complexity

1.3.1 Optimizing Strategy and Satisficing Strategy

Satisficing strategies in complex system design enable decision-makers to navigate uncertainty while balancing multiple competing objectives. Unlike optimization, which often leads to computationally expensive searches for an elusive global optimum, satisficing prioritizes adaptability and robustness. When dealing with problems characterized by nonlinearity, nonconvexity, and multi-objective trade-offs, satisficing provides solutions that are 'good enough' within feasible design spaces without the strict requirements of meeting both the necessary and sufficient Kuhn-Tucker conditions. By embracing satisficing, designers can achieve resilient solutions that respond effectively to real-world constraints and evolving system requirement (Guo and Chen, 2024; Guo et al., 2024a, 2024b). These conditions make the solution set small, hard to find, and sensitive to uncertainties captured or not captured in the decision model.

In contrast, the *satisficing* strategy methods, such as Sequential Linear Programming, allow designers to obtain vertex solutions of an approximated linear problem (Guo et al., 2024a, 2024b). Such solutions meet the necessary Kuhn-Tucker conditions but may not meet the sufficient Kuhn-Tucker conditions, so we call these solutions "good enough" solutions or *satisficing* solutions. The *satisficing* solutions set contains the optimal solutions set, see Fig. 1.7. Therefore, the *satisficing* set is easier to find, especially when the problem is complex. If solutions lie on the vertices, this enables designers to use the duality embodied in linear programming to explore the solution space without having to perform numerical differentiation which is required if the solution is obtained using methods identified in Category 1 (Hajihashemi, 2023; Mistree et al., 1981). Further, *satisficing* methods facilitate the rapid identification of robust solutions. Solutions that are relatively insensitive to approximations make the problem's solution tractable. However, like most metaheuristic methods, satisficing methods rely on fixed parameter heuristics to set parameters.

Satisficing strategy. *Satisficing* is a decision-making strategy of cognitive heuristics that entails searching through the available alternatives until an acceptability

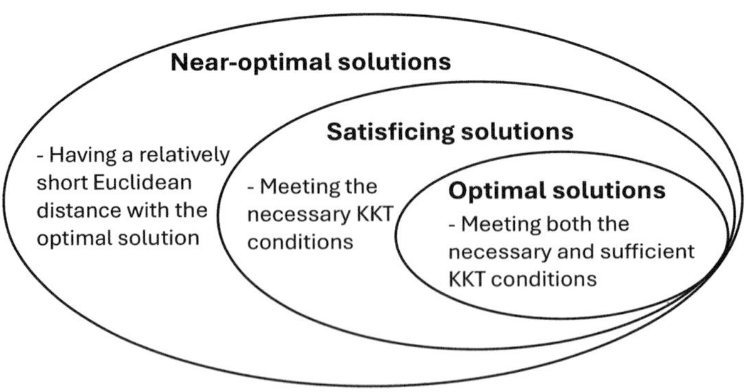

Fig. 1.7 The optimal solutions are included in the satisficing solution space

threshold is met. The term *satisficing*, a portmanteau of satisfy and suffice, was introduced by Herbert A. Simon in 1956 (Simon, 1956). The *satisficing* solutions contain optimal solutions.

How can designers identify *satisficing* solutions and accomplish design evolution using the s*atisficing* strategy? In this monograph, compromise Decision Support Problems (cDSP) are used as the construct, and the Adaptive Linear Programming (ALP) algorithm is used as the approximation and solution algorithm to identify *satisficing* solutions. The Formulation-Exploration framework, the Adaptive Linear Algorithm with Parameter Learning (APPLL), and the Adaptive-Leveling-Weighting-Clustering (ALWC) algorithms are used to improve decision models for complex systems under *the satisficing* strategy (Guo et al., 2023).

In a subsequent paper, the authors (Guo et al., 2024a, 2024b) explore the dilemma of whether to optimize or satisfice in engineering design. The authors address the challenges of dealing with nonlinear, nonconvex objectives and constraints, and the complexities of computational models that are abstractions of reality. They propose a method to identify satisficing solutions that satisfy the necessary Karush–Kuhn–Tucker (KKT) conditions but not the sufficient conditions. These solutions are relatively robust, easy to acquire, and often good enough. The authors demonstrate the combined use of the compromise Decision Support Problems and the adaptive linear programming algorithm; see (Hajihashemi, 2023; Karandikar and Mistree, 1992; Lyon et al., 1982). This method is appropriate for formulating design problems and obtaining solutions that meet the necessary KKT conditions, circumventing the complications associated with gradient-based optimization algorithms1. The authors illustrate the efficacy of their proposed method using four test problems, showing how the satisficing strategy can outperform the optimizing strategy in model-based engineering design.

1.3.2 Why is Satisficing Desired in Engineering Design?

Why is *a satisficing* strategy desired in realizing complex systems and design evolution? Because a complex system may encounter multiple types of complexity and uncertainty, making it extremely difficult or impossible to "make things work" using an optimizing strategy, a *satisficing* strategy is often effective (Guo et al., 2024a, 2024b). In this monograph that means making the system run efficiently, healthily, and sustainably. Their definitions of these terms in the context of complex systems and the interpretation in model language are given as follows.

Efficient—Stakeholders, users, operators, and designers of complex systems can achieve maximum system performance with minimum expense. In decision models, this means achieving the objective with minimum computational complexity and no or minimum violation of constraints and bounds.

Healthy—A system can reach a status with no potential danger or failure and low-security risk. In decision models, this means no sacrifice or loss of performance that temporarily affects system health to meet certain requirements.

Sustainable—A system can be used constantly. For decision models, this means a model can run in multiple iterations with a relatively good and stable performance.

In Table 1.4, we list three typical methods and their features. Reviewing the literature reveals that four categories of features are being well studied: (1) managing complexity, (2) convergence or features of the solution, (3) solution algorithms, and (4) decision support.

The compromise Decision Support Problem (cDSP) is one of the *satisficing* constructs. Mistree and the coauthors (Mistree et al., 1993) implement a solution algorithm, the Adaptive Linear Programming (ALP) algorithm in DSIDES. Using DSIDES, designers can explore the solution space of nonlinear, nonconvex, multi-goal engineering-design problems and manage four types of uncertainties (Choi et al., 2005; Nellippallil et al., 2020). However, they may not find optimal solutions that satisfy the sufficient Kuhn-Tucker condition. The benefit of meeting only the necessary Kuhn-Tucker conditions but not the sufficient conditions is that the solution can be relatively insensitive to the errors in the model or the changes that affect the sufficient Kuhn-Tucker conditions. The details are given in Sect. 2.1. The research gaps using *a satisficing* strategy are presented in Sect. 1.4.2.

Goal Programming is another *satisficing* method but using it does not guarantee *satisficing* solutions. The performance of Goal Programming dependent on the solution algorithm or the solver that a designer uses. For example, using NSGA-II or NSGA-III, the nonlinear and nonconvex problems can be well managed if the extent of the search is sufficient, whereas using a Conjugate Gradient may not guarantee a feasible solution.

Mathematical programming is a construct using the optimizing strategy. Like Goal Programming, the performance of an optimization model regarding the features that a designer can manage depends on the solution algorithm and the solver. Since, at the optimal solution point, both the necessary Kuhn-Tucker condition and the sufficient Kuhn-Tucker condition must be met. The optimal solution is relatively hard to identify and easy to lose in an infeasible area. The details are given in Sect. 2.1. The research gaps using the optimizing strategy are stated in Sect. 1.4.1. The differences among the methods of seeking optimal solutions in cDSP, Goal Programming, and mathematical programming are discussed in Sect. 2.3.1.

1.4 The Challenges in the Model-Based Realization of Complex Systems

The realization of systems is a process of implementing a given input–output behavior. In other words, given a system's input–output relationship, the realization of the system is a combination of four (time-varying) matrices. In this monograph, we focus on the model-based realization of complex systems, using mathematical models to realize the combination of the time varying matrices. Models are approximations of the physical world. However, there is a conceptual disconnect between

Table 1.4 Several representative methods and their features

Construct	Method	Managing complexity				
		Managing nonlinearity	Managing nonconvexity	Managing uncertainty	Managing multiple/ many objectives (goals)	Managing hard requirements (constraints)
cDSP	cDSP using ALP in DSIDES		Nonconvexity < 0.15		Multi-goal	
	cDSP using EMI			Uncertainty in parameters, variables		
	cDSP using DCI			Uncertainty in design capacity		
	cDSP using ALPPL			Uncertainty in approximation		
	cDSP using ALWC			Uncertainty in design preferences	Many-goal	
Goal programming	Goal programming using regular solver		It depends on the solution algorithm or the solver	It depends on the solution algorithm or the solver	Many	
	Goal programming using a nonlinear solver					
Mathematical Programming Seeking Optimal Solutions	Mathematical programming using nonlinear solver		It depends on the solution algorithm or the solver	It depends on the solution algorithm or the solver	Multi-goal	
	Mathematical programming using NSGA					

(continued)

Table 1.4 (continued)

Construct	Method	Managing complexity				
		Managing nonlinearity	Managing nonconvexity	Managing uncertainty	Managing multiple/many objectives (goals)	Managing hard requirements (constraints)
	Mathematical programming using NSGA-II				Many-goal	
	Mathematical programming using NSGA-III					
	Stochastic programming			Uncertainty in parameters and variables		

Converge condition/solution feature		Solution algorithms			Decision support		
Necessary Kuhn-Tucker conditions	Sufficient Kuhn-Tucker conditions	Interior-point searching	(Linearizing and) obtaining vertex solutions	Comprehensively using interior-point searching and searching for vertex solutions	Can apply visualization tools to facilitate decision-making	Managing knowledge discovery and reuse	Supporting evolving design preferences
		Target feasible solution space	Return vertex solutions		Tradeoffs among multiple goals; *satisficing* design space		Manual Support different types of goal combination

(continued)

Table 1.4 (continued)

Converge condition/solution feature		Solution algorithms			Decision support		
Necessary Kuhn-Tucker conditions	Sufficient Kuhn-Tucker conditions	Interior-point searching	(Linearizing and) obtaining vertex solutions	Comprehensively using interior-point searching and searching for vertex solutions	Can apply visualization tools to facilitate decision-making	Managing knowledge discovery and reuse	Supporting evolving design preferences
					Tradeoffs among multiple goals; *satisficing* design space; EMI distribution	Knowledge of margin	
					Tradeoffs among multiple goal; *satisficing* design space; DCI distribution	Knowledge of capacity	
					Tradeoffs among multiple goals; *satisficing* design space	Knowledge of reduced move coefficient	
					Clusters of the many goals	Knowledge of interrelationships among many goals	Automatically Support different types of goal combination
				It depends on the solution algorithm or the solver	Tradeoffs among multiple goals	Knowledge of goal achievement	Manually Support changing weight vectors

(continued)

Table 1.4 (continued)

Converge condition/solution feature		Solution algorithms			Decision support		
Necessary Kuhn-Tucker conditions	Sufficient Kuhn-Tucker conditions	Interior-point searching	(Linearizing and) obtaining vertex solutions	Comprehensively using interior-point searching and searching for vertex solutions	Can apply visualization tools to facilitate decision-making	Managing knowledge discovery and reuse	Supporting evolving design preferences
				It depends on the solution algorithm or the solver	Tradeoffs among multiple goals; near Pareto front	Knowledge of search strategy	Manually Support changing weight vectors
						Knowledge of stochasticity	

the physical and model worlds. That is one of the limitations of model-based realization of complex systems. In this monograph, it is hypothesized that designers can use design evolution to reduce the impact of the conceptual disconnect between the physical world and the model world.

1.4.1 Models Are Approximations of the Real World

"All models are wrong, but some are useful" (Box, 1979; Box and Draper, 1987). All models are approximations of the real world. An example of the mapping and correspondence between the physical world and the model world is illustrated in Fig. 1.8. No matter how accurate a model is, there is a conceptual disconnect between the physical and model worlds. The assumptions, simplifications, and heuristics are used to model, solve, and approximate the physical world. There is not a single model that can capture all the information in the physical world. Some solutions are sensitive to the incompleteness and inaccuracy of a model, whereas other solutions are relatively insensitive to the incompleteness and inaccuracy of the model. Given that, the evolution of the model regarding the improvement in model formulation, approximation, exploration, and evaluation is important in obtaining the solution space that is relatively insensitive to model inaccuracy.

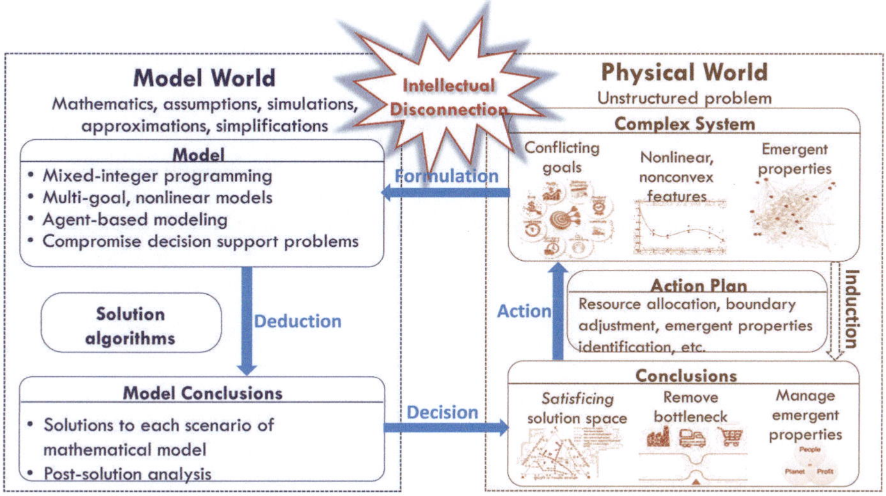

Fig. 1.8 The correspondence between physical world and model world

1.4.2 The Purpose of Design Evolution

The concept of an "evolutionary model" comes from biological evolution, based on the evolution of DNA. Inspired by this conception, we expand "design evolution" to all complex systems in this monograph. In design evolution, two capabilities of a model are improved. (1) The capability of a model to capture and incorporate more useful information of the physical world, or in other words, model accuracy. (2) The capability of a model to deliver solutions that are relatively insensitive to uncertainties, or in other words, model robustness. These uncertainties include variation in parameters variations (Type I), variation in decision variables (Type II), uncertainty in model structure (Type III), the uncertainty due to simultaneously managing the previous three types of uncertainty (Type IV) (Choi et al., 2005), due to model errors, model inaccuracies, changes in the environment, etc.

In this monograph, the purpose of design evolution is to improve the solution space's insensitivity to the uncertainties encountered in the model.

1.5 Problem Statement—Problems in Both Strategies

To design complex systems with limited information, multiple interactions among the four stages of the evolution cycle (Fig. 1.4) are expected to be established; see Fig. 1.9. Information can be passed through different design stages by establishing connections, and corresponding actions to improve the design can be taken iteratively. Therefore, the conceptual disconnect between the model and physical worlds can be managed.

As introduced in Sect. 1.2, there are two strategies for designing a complex system: optimizing and *satisficing*. However, there are limitations to both strategies. We summarize them into the research gaps that we identify in this monograph, which are briefly introduced regarding each strategy—optimizing (Sect. 1.4.1) and *satisficing* (Sect. 1.4.2). The mathematical explanations of the research gaps are introduced in detail in Chap. 2.

1.5.1 Problems in Optimizing Strategy

Albert Einstein said, "So far as the theories of mathematics are about reality, they are not certain; so far as they are certain, they are not about reality."

Using an optimizing strategy, designers assume that their abstractions of reality, which are their mathematical models, are certain. Even when they incorporate uncertainties in their mathematical models, they assume they are mathematically representable. The general representation of a mathematical model using an optimizing strategy is given as follows: Eqs. 1.1–1.3. f is the objective function that

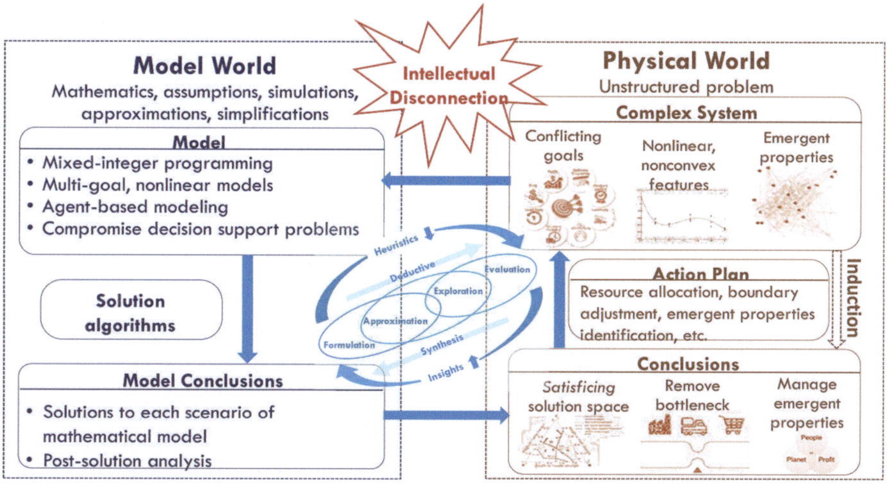

Fig. 1.9 Establishing connections among multiple stages of the engineering-design evolution cycle, formulation, approximation, exploration, and evaluation

maps the variable vector x from the n-dimensional real number set $\mathbb{R}^n$ to the one-dimensional real number set $\mathbb{R}$. Ω is the n-dimensional feasible solution space that bounded by all constraints and bounds, including m number of inequality constraints $g_i(x), i = 1, \ldots, m$, k number of equality constraints $h_j(x) = 0, j = 1, \ldots, k$, and the upper and lower bounds of x. x^* is the optimal solution to the optimization problem. x^* makes the objective function value $f(x^*)$ superior to any $f(x)$ resulted from other feasible solution x.

Given

$$\boldsymbol{f} : \mathbb{R}^n \rightarrow \mathbb{R}, \boldsymbol{\Omega} \subseteq \mathbb{R}^n \tag{1.1}$$

$$\boldsymbol{\Omega} = \left\{\boldsymbol{x} \in \mathbb{R}^n | \boldsymbol{g}_i(\boldsymbol{x}) \geq \boldsymbol{0}, \boldsymbol{i} = \boldsymbol{1}, \ldots, \boldsymbol{m}, \boldsymbol{h}_j(\boldsymbol{x}) = \boldsymbol{0}, \boldsymbol{j} = \boldsymbol{1}, \ldots, \boldsymbol{k}\right\} \tag{1.2}$$

Find

$$\boldsymbol{x}^* : \boldsymbol{f}\left(\boldsymbol{x}^*\right) \succeq \boldsymbol{f}(\boldsymbol{x}), \forall \boldsymbol{x} \in \boldsymbol{\Omega} \tag{1.3}$$

The objective of an optimization problem is to maximize a function consisting of decision variables (Eq. 1.1) while satisfying the constraints (Eq. 1.2). By critically reviewing the literature using optimizing methods in Sect. 1.2 and identifying the limitations of decision models in Sect. 1.3, we summarize the problems of using optimizing strategies in Table 1.5.

As the boundary of the physical system changes, design preferences evolve, unpredictable uncertainties, unknown interactions among subsystems or between

Table 1.5 Problems in methods of optimizing strategy

Uncontrollable factors	List of requirements	Problems (phenomenon)	Mathematical explanation location
Boundary changes	Accurate and exact boundary	The optimal solution to the mathematical model may not work for its corresponding physical engineering-design problem	Sections 2.1–2.3
Design preferences evolve	Evolving design preferences can be captured, quantified, and represented in the model		
Unpredictable, unparameterizable uncertainties	Uncertainties can be parameterized and predicted		
Unknown interactions	Interactions can be represented in the model		
Emergent properties	Emergent properties can be captured and incorporated into the model		

the system and its environment cannot be represented in the model, emergent properties add complexity to the modeling and solution analysis, etc. We cannot control or understand these factors while abstracting physical systems into models and interpreting model solutions into physical phenomena. A design may fail because we have limited knowledge of the associations among those factors. The designers may lose the optimal solution. In other words, the optimal solution to the mathematical model may not work for its corresponding physical engineering-design problem. Why? Because the designers seek optimal solutions for the wrong models. Therefore, using an optimizing strategy to manage engineering-design problems is like trying to maximize the optimality of the solution to an incorrect model without knowing that the model is inaccurate and without considering the robustness of the solutions to the inaccuracy. However, this is the basis of the research gap. The mathematical explanation of the research gap is presented in Sects. 2.1 to 2.3.

1.5.2 Problems in Satisficing Strategy

Using *a satisficing* strategy, designers know that their abstractions of reality in their decision models may be incomplete or inaccurate. So, they seek solutions that are "good enough" and relatively insensitive to the model's errors instead of optimal. One of the representations of a decision model under *the satisficing* strategy is given as follows—Eqs. 1.4–1.6. f is the objective function that maps the variable vector x from the n-dimensional real number set $\mathbb{R}^n$ to the one-dimensional real number set $\mathbb{R}$. Ω is the n-dimensional feasible solution space that bounded by all constraints and bounds, including m number of inequality constraints $g_i(x), i = 1, \ldots, m$, k number of equality constraints $h_j(x) = 0, j = 1, \ldots, k$, and the upper and lower bounds of x. By assigning a target value to the objective, we make it as a goal, $f(x) = Target$. Suppose that the goal is projected onto the feasible solution space Ω, for any feasible solution x, if x makes the Euclidean distance between the goal and the projection no greater than any other solutions, we define that x is a satisficing solution x^s, represented as $x^s : \mathcal{P}_{x\in\Omega}(f(x) = Target)$.

Given

$$\boldsymbol{f} : \mathbb{R}^n \rightarrow \mathbb{R}, \boldsymbol{\Omega} \subseteq \mathbb{R}^n \tag{1.4}$$

$$\Omega = \{x \in \mathbb{R}^n | g_i(x) \geq 0, i = 1, \ldots, m, h_j(x) = 0, j = 1, \ldots, k,\} \tag{1.5}$$

Find

$$\boldsymbol{x^s} : \mathcal{P}_{\boldsymbol{x}\in\boldsymbol{\Omega}}(\boldsymbol{f}(\boldsymbol{x}) = \boldsymbol{Target}) \tag{1.6}$$

The merit function of a decision model using *a satisficing* strategy is used to identify the nearest projection of the objective function $f(x)$ onto the feasible space bounded by constraints. In other words, the aim is minimizing the deviation between the target and the actual achieved value of a goal, as Eqs. 1.7–1.9.

Given

$$\boldsymbol{f} : \mathbb{R}^n \rightarrow \mathbb{R}, \boldsymbol{\Omega} \subseteq \mathbb{R}^n \tag{1.7}$$

$$\begin{aligned}\Omega = \{&x \in \mathbb{R}^n | g_i(x) \geq 0, i = 1, \ldots, m, h_j(x) = 0, j = 1, \ldots, k,\\ &f(x) + d^- - d^+ = Target\}\end{aligned} \tag{1.8}$$

Find

$$\boldsymbol{d^*} : \boldsymbol{d^*}(\boldsymbol{x}) \leq \boldsymbol{d}(\boldsymbol{x}) \tag{1.9}$$

Although using *a satisficing* strategy, designers may identify solutions that are relatively insensitive to model errors and uncertainties, there are limitations in methods under *the satisficing* strategy. By reviewing the literature using *satisficing* methods in Sect. 1.2 and identifying the limitations of decision models in Sect. 1.3, we describe the problems using *satisficing* strategy in Table 1.6.

Given the problems in both strategies, we summarize the research gaps and pose the hypotheses we will study in Sect. 1.5.

Table 1.6 Problems in methods of *satisficing* strategy

Uncontrollable factors	List of requirements	Problems (phenomenon)	Mathematical explanation location
Boundary changes	Solutions are relatively insensitive to boundary variations	Adding too much buffer to ensure feasibility; relying on heuristics, metaheuristics, and domain knowledge to make decisions; no information is passed through different stages of the design; no mechanism to evaluate and update the heuristics and metaheuristics	Sections 2.1–2.3
Design preferences evolve	Solutions are relatively insensitive to design preferences evolving		
Unknown interactions	Solutions are relatively insensitive to unknown interactions		
Emergent properties	Solutions are relatively insensitive to the emergent properties even if they cannot be incorporated into the model		
Relying on heuristics and metaheuristics to make rules	Boundary variations, evolving design preferences, unknown interactions, and emergent properties do not break the equilibrium of the sufficient Karush–Kuhn–Tucker (KKT) condition—see Chap. 2 for more details		
Relying on domain knowledge to make rules	Evolving design preferences can be captured, quantified, and represented in the model		
No information passing through different stages of engineering designs	Uncertainties can be parameterized and predicted		

1.6 Research Gaps and Hypotheses

1.6.1 Research Gaps

Based on the problems in the two design strategies, optimizing and *satisfying*, we summarize the research gaps to be filled in this monograph. That is, how can designers manage the problems in both strategies while finishing the tasks of engineering design? The research gaps and the potential contributions of filling these research gaps are illustrated in Fig. 1.10.

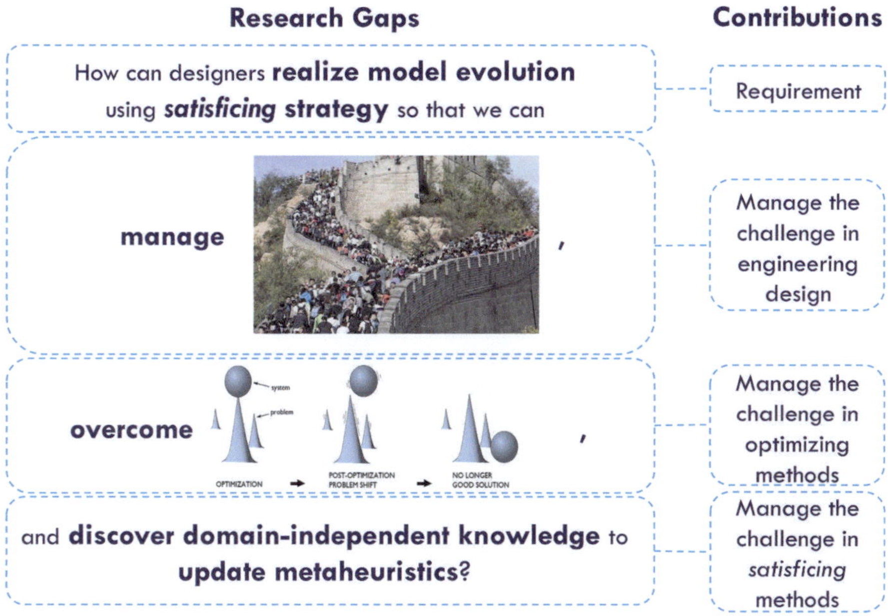

Fig. 1.10 The research gaps and the potential contributions by filling the research gaps[4]

- ***Research Gaps—How can designers realize design evolution using a satisfying strategy so that they can manage chaos in the physical world, reduce the risk of losing an optimal solution, and discover domain-independent knowledge to update metaheuristics?***

1.6.2 Hypotheses to Bridge the Research Gaps

In this monograph, it is hypothesized that ***by connecting the multiple stages of design and passing information through them, designers can improve their decision models iteratively***, defined as “design evolution.” During the design evolution, designers can incorporate more chaos in the physical systems into their decision models and manage their impact on the solution space, identifying solutions that are relatively insensitive to errors and uncertainties that the decision models may encounter, and discover domain-independent knowledge to update metaheuristics. The research gap and hypothesis are filled into the Table (Table 1.2) as Table 1.7.

[4] The picture in the “Manage” block is from: https://www.sohu.com/a/672198131_121687423.

Table 1.7 The Research Gaps (RG) and Hypotheses (H)

Chapters	Ch1	Ch2	Ch3	Ch4–7	Ch8	Ch9
Actions	*RG: How can designers realize design evolution using a satisficing strategy to manage chaos in the physical world, reduce the risk of losing an optimal solution, and discover domain-independent knowledge to update metaheuristics?* *H: By connecting the multiple design stages and passing information through them, designers can improve their decision models in iterations*	RD RQ SH	TVe M	EVe SQT AQ	CQ EVa	TE
Nomenclature	AQ—answer research questions CQ—closure of the answers to research questions EVa—empirically validate hypotheses EVe—empirically verify hypotheses H—give hypotheses M—introduce methods RD—tie to robust design RG—give research gaps RQ—pose research questions SH—state hypotheses SQT—specify research questions in the context of test problems TE—theoretically extend the research TVe—theoretically verify hypotheses					

1.6.3 Expected Contribution by Testifying the Hypotheses

By exploring the hypotheses, we expect to contribute a design method that allows designers to manage the following attributes by exploring the solution space and identifying a set of solutions that weighs all attributes, which is defined by a *satisficing* solution space; see Fig. 1.11.

Accurate—The mapping from the physical system to a mathematical model is relatively accurate, and the accuracy can be improved during the exploration of the solution space.

Robust—the *satisficing* solutions are relatively insensitive to the errors and uncertainties embodied in the model. There are four types of uncertainty (Choi et al., 2005): noise factors, control factors, uncertainties in the model structure, and uncertainties in the process chain when managing the first three types of uncertainty. A thorough introduction of the four types of uncertainty and the corresponding robust design methods is given in Sect. 2.4.1.

Diverse—the solutions identified as *satisficing* solutions are relatively different, allowing designers to have more alternatives in various situations.

Simple—the computational complexity of the method is relatively low.

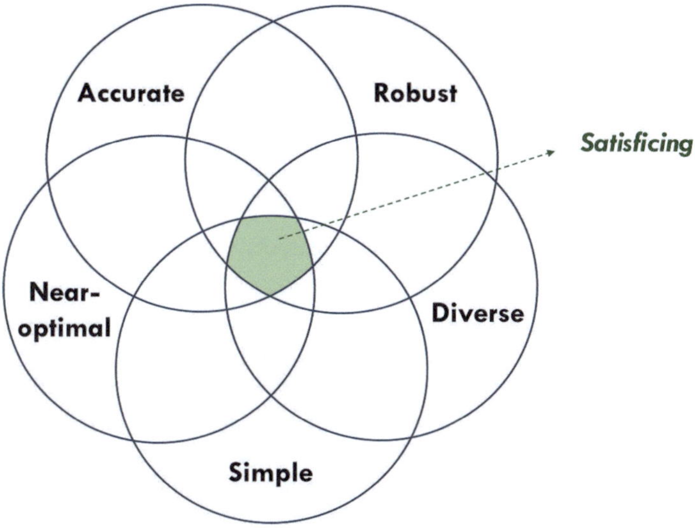

Fig. 1.11 Expected contributions

Optimal—the achieved value of the goals is close enough to the target of the goals.

Designers may use interior-point searching algorithms to identify optimal and diverse solutions. Stochastic optimization methods allow designers to take into account the solutions' accuracy, optimality, and diversity. However, the accuracy and exhaustiveness of the randomness of a stochastic model may rely on assumptions and heuristics which may not be correct or accurate. Using the methods belonging to the *satisficing* strategy methods, such as cDSP-ALP-DSIDES (in Table 1.3), designers have mechanisms to obtain robust solutions and maintain an acceptable level of computational complexity. However, they do not improve the accuracy of their decision models.

In this monograph, the contribution is to identify a satisficing solution space in the decision model of complex engineering-design problems and provide decision support based on knowledge of the tradeoffs among the five attributes.

1.7 Plan for Verification and Validation

Validation of engineering research has traditionally been anchored in the context of scientific inquiry that demands formal, rigorous, and quantitative validation. Logical induction and deduction play key roles in this formalism, making it particularly useful for validating internal consistency within the framework of the scientific method. Since much engineering research is based on mathematical modeling, this kind of validation has worked—and still works—very well. However, other areas of engineering research rely on subjective statements (qualitative information) as well as mathematical modeling, making formal, rigorous, and quantitative validation problematic. One such area is design methods within the field of engineering design. The Validation Square is a comprehensive framework developed to address the challenges of validating design methods in engineering research; see (Pedersen et al., 2000; Seepersad et al., 2006).

The Validation Square consists of four quadrants, each used to address a different aspect of validation:

- **Theoretical Structural Validity (TSV)**: To ensure the internal consistency of the theoretical foundations that underpin the method.
- **Theoretical Performance Validity (TPV)**: To assess the efficacy, using numerous test problems, of the method to achieve the intended theoretical outcomes.
- **Empirical Structural Validity (ESV)**: To verify the structural soundness of the method through empirical testing using one or more comprehensive problems.
- **Empirical Performance Validity (EPV)**: To assess the generalizability of the method for other applications.

By addressing both theoretical and empirical aspects, the Validation Square provides a for validating design methods. It bridges the gap between internal logic and external applicability, ensuring that design methods are both scientifically sound and practically relevant. The Validation Square framework has been adapted in this monograph to ensure clarity of the message; see Fig. 1.2

Quadrant 1: Theoretical Structural Validity: In this chapter, the context of the monograph is given—models are approximations of the physical world. Thus, design evolution is required to improve the model realization's robustness. In Chap. 2, the theoretical foundation and justified research questions are documented. In Chap. 3, an overview of the methods is provided.

Quadrant 2: Empirical Structural Validity: In Chaps. 4 to 7, test problems are used to verify the proposed methods and algorithms that constitute new knowledge and contributions. The research questions are answered one by one in Chaps. 4 to 7.

Quadrant 3: Empirical Performance Validity: In Chap. 8 the answers to the research questions are summarized and the utility of the proposed methods to other problems highlighted.

Quadrant 4: Theoretical Performance Validity: In Chap. 9, we summarize the contributions, limitations, and the generalizability of what is documented in this monograph.

1.8 Organization of the Monograph

There are nine chapters in this monograph. An overview of this monograph is presented as a roadmap in Fig. 1.12. The figure is intended to help readers navigate through the monograph and develop an overall picture of what is discussed in

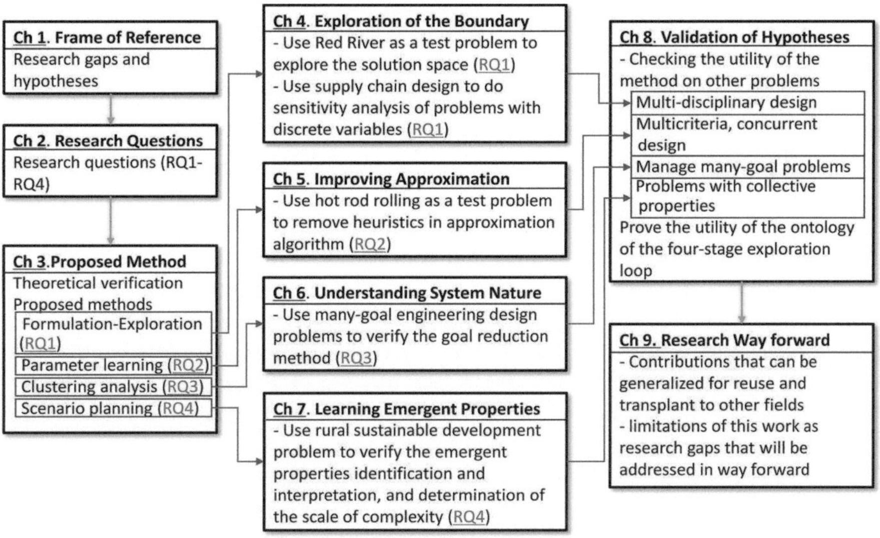

Fig. 1.12 Organization of the chapters

each chapter, thereby providing context and a way to rapidly glean what is of interest to the individual.

The monograph is structured as follows.

Chapter 1: Why?—Why do we need design evolution?

Chapter 2: What?—In this chapter, we delve into the concept of the satisficing strategy, explaining its mathematical foundation and differentiating it from the optimizing strategy.

Chapter 3: How?—We propose the methods and the validation plan, setting the stage for the detailed methods and applications that follow.

Chapters 4–7: Methods and applications—We present four methods incorporated into the design evolution cycle. In each chapter, we demonstrate the application of a method to a distinct field, showcasing its effectiveness in designing complex systems.

Chapter 8: Validation—In this chapter, we provide a holistic validation of the design evolution cycle, summarizing and generalizing the validations conducted in Chaps. 4–7.

Chapter 9: Way Forward—In this final chapter, we explore the future research possibilities, potential application fields, and broader impact of the design evolution cycle.

Through this monograph, we aim to equip STEM graduate students and researchers with the knowledge and tools to implement satisficing strategies in their work, ultimately advancing the field of complex systems design.

1.9 Role of This Chapter in This Monograph

In Chapter 1, we establish the context of this monograph. We answer the "why question"—"why we need design evolution. give the motivation of design evolution and describe why we manage complex systems in a certain way—using satisficing methods.

In this chapter, we provide the context of the realization of complex systems—an introduction to the characteristics and challenges of model-based realization of complex systems. By analyzing the methods of an optimizing strategy and a *satisficing* strategy, we identify research gaps in both methods and pose the primary research question to be answered here. We give the verification and validation plan and the organization of the monograph. We use Fig. 1.13 to illustrate the knowledge foundation of the "Design Evolution Cycle." We will revisit this figure in later chapters as we introduce each element of the figure.

Some sections in this chapter are referred in later chapters, especially when we identify the research gaps in complex systems design and provide the foundation of our work (Chap. 2), propose our methods (Chap. 3), and use test problems to demonstrate our proposed methods (Chaps. 4–7). Those internal references also demonstrate the soundness of the structure of this monograph. The literature review, design approach, developed method, and validation of hypotheses are discussed in the following chapters.

Glossary

Compromise Decision Support Problem: A hybrid modeling approach in which designers combine mathematical programming and goal programming to handle multiple goals that may conflict. In this approach, we seek design solutions that balance those conflicting objectives to achieve an acceptable compromise.

Design evolution: We use Design evolution to describe how designers refine a solution over successive iterations to improve accuracy, robustness, and computational efficiency as well as other performance measures.

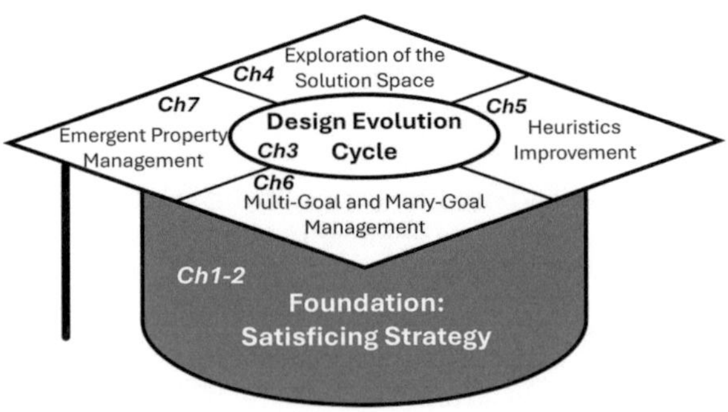

Fig. 1.13 The knowledge foundation of this monograph

Diverse (of solutions): We define Diverse solutions as sets of solutions that differ substantially from one another so that designers can choose among multiple alternatives for different scenarios. When we achieve high solution diversity, we expand the range and richness of possible designs and thus can accommodate evolving preferences.

Satisficing: We define *Satisficing* as a decision-making strategy or a cognitive heuristic that entails searching through available alternatives until they met an acceptability threshold rather than maximizing any specific objective (Byron, 1998).

References

Box, G. E. P., & Draper, N. R. (1987). *Empirical model-building and response surfaces*. Wiley.

Box, G. E. P. (1979). Robustness in the strategy of scientific model building. In R. L. Launer & G. N. Wilkinson (Eds.), *Robustness in statistics* (pp. 201–236). Elsevier.

Byron, M. (1998). Satisficing and optimality. *Ethics, 109*(1), 67–93.

Choi, H.-J., Austin, R., Allen, J. K., McDowell, D. L., Mistree, F., & Benson, D. J. (2005). An approach for robust design of reactive powder metal mixtures based on non-deterministic micro-scale shock simulation. *Journal of Computer-Aided Materials Design, 12*(1), 57–85.

Guo, L., & Chen, S. (2024). Satisficing strategy in engineering design. *Journal of Mechanical Design, 146*(5), Article 050801.

Guo, L., Zamanisabzi, H., Neeson, T. M., Allen, J. K., & Mistree, F. (2019). Managing conflicting water resource goals and uncertainties in a dam network by exploring the solution space. *Journal of Mechanical Design, 141*(3), Article 031702.

Guo, L., Mohebbi, S., Das, A., Allen, J. K., & Mistree, F. (2020). A framework for the exploration of critical factors on promoting two-season cultivation in India. *Journal of Mechanical Design, 142*(12), Article 124503.

Guo, L., Milisavljevic-Syed, J., Wang, R., Huang, Y., Allen, J. K., & Mistree, F. (2023). Managing multi-goal design problems using adaptive leveling-weighting-clustering algorithm. *Research in Engineering Design, 34*(1), 39–60.

Guo, L., Allen, J. K., & Mistree, F. (2024a). Optimize or satisfice in engineering design? *Research in Engineering Design, 35*(3), 239–267.

Guo, L., Nellippallil, A. B., Smith, W. F., Allen, J. K., & Mistree, F. (2024b). An adaptive linear programming algorithm with parameter learning. *Algorithms, 17*(2), 88.

Hajihashemi, S. (2023). A user-friendly wrapper for DSIDES (Decision Support in the Design of Engineering Systems). Industrial and Systems Engineering, University of Oklahoma. MS Thesis.

Karandikar, H., & Mistree, F. (1992). Designing a composite material pressure vessel for manufacture: A case study in concurrent engineering. *Engineering Optimization, 18*(4), 235–262.

Ladyman, J., Lambert, J., & Wiesner, K. (2013). What is a complex system? *European Journal for Philosophy of Science, 3*(1), 33–67.

Lyon, T., Mistree, F., & Shupe, J. (1982). Computer-based design synthesis: An approach to problem solving. In *Frontiers in education conference, ASEE frontiers in education conference proceedings* (pp. 102–108). University of South Carolina.

Macquorn Rankine, W. J. (1853). XVIII. On the general law of the transformation of energy. *The London, Edinburgh, and Dublin Philosophical Magazine and Journal of Science, 5*(30), 106–117.

Mistree, F., Hughes, O., & Phuoc, H. (1981). An optimization method for the design of large, highly constrained complex systems. *Engineering Optimization, 5*(3), 179–197.

Mistree, F., Hughes, O. F., & Bras, B. (1993). The compromise decision support problem and the adaptive linear programming algorithm. In M. P. Kamat (Ed.), *Structural optimization: Status and promise* (pp. 247–286). AIAA.

Nellippallil, A. B., Song, K. N., Goh, C.-H., Zagade, P., Gautham, B., Allen, J. K., & Mistree, F. (2017). A goal-oriented, sequential, inverse design method for the horizontal integration of a multistage hot rod rolling system. *Journal of Mechanical Design, 139*(3), Article 031403.

Nellippallil, A. B., Rangaraj, V., Gautham, B., Singh, A. K., Allen, J. K., & Mistree, F. (2018). An inverse, decision-based design method for integrated design exploration of materials, products, and manufacturing processes. *Journal of Mechanical Design, 140*(11), Article 111403.

Nellippallil, A. B., Allen, J. K., Gautham, B., Singh, A. K., & Mistree, F. (2020). *Architecting robust co-design of materials, products, and manufacturing processes.* Springer.

Pedersen, K., Emblemsvag, J., Bailey, R., Allen, J., & Mistree, F. (2000). The 'Validation Square'-validating design methods. In *ASME design theory and methodology conference.* ASME DETC2000/DTM-14579.

Rios, L. M., & Sahinidis, N. V. (2013). Derivative-free optimization: a review of algorithms and comparison of software implementations. *Journal of Global Optimization, 56*(3), 1247–1293.

Seepersad, C. C., Pedersen, K., Emblemsvåg, J., Bailey, R., Allen, J. K., & Mistree, F. (2006). *The validation square: how does one verify and validate a design method? Decision making in engineering design* (pp. 303–314).

Seppersad, C. C. et al. (2006). *The Validation Square: How Does One Validate Design Methods?* Decision-Based Design: Making Effective Decisions in Product and Systems Design. In: W. Chen, K. Lewis, and L. Schmidt (Eds.) ASME Press, New York Chapter 25, pp. 305–326.

Simon, H. A. (1956). Rational choice and the structure of the environment. *Psychological Review, 63*(2), 129.

Simon, H. A., & Kadane, J. B. (1975). Optimal Problem-solving search: All-or-none Solutions. *Artificial Intelligence, 6*(3), 235–247.

Smith, W. F., Milisavljevic, J., Sabeghi, M., Allen, J. K., & Mistree, F. (2015). The realization of engineered systems with considerations of complexity. In *ASME design automation conference.* Boston, MA, Paper Number DETC2015–46211.

Viswanathan, J., & Grossmann, I. E. (1990). A combined penalty function and outer-approximation method for MINLP optimization. *Computers and Chemical Engineering, 14*(7), 769–782.

Wikipedia. (2019). Rankine Cycle. from https://en.wikipedia.org/wiki/Rankine_cycle. Retrieved June 1, 2025.

Williams, R. J., & Zipser, D. (1995). Gradient-based learning algorithms for recurrent networks and their computational complexity. *Backpropagation: Theory, Architectures, and Applications, 1*, 433–486.

Chapter 2
Research Questions: How Can We Realize Model Evolution

Theoretical Foundation and Justified Research Questions

In designing complex systems, the quest for optimality often leads to frustration and inefficiency. The use of the optimization construct using the gradient-based solution algorithms result in single-point optimal solution that meet rigorous Karush–Kuhn–Tucker (KKT) conditions. However, these algorithms need to be revised when faced with the complexities of engineering design problems: nonlinear, nonconvex objectives, constraints and objectives with different units, and mathematical models that are abstractions of reality. In this monograph, we propose architecting a satisficing strategy using the compromise Decision Support Problem (cDSP) construct (Mistree et al., 1981, 1993) that is exercised using DSIDES (Hajihashemi, 2023) to identify design solutions that meet the necessary but not the sufficient KKT condition.

Architecting a Satisficing Strategy to Design Complex Systems

Imagine you are designing a dam network to manage water resources effectively. The challenges are multifaceted: preventing floods, ensuring water supply during droughts, and balancing ecological needs. Optimization methods, such as gradient-based algorithms, may provide an optimal solution to the mathematical model. However, they often fall short in exploring alternative designs that could be more robust and practical. This is where satisficing comes into play.

Satisficing, derived from the words "satisfy" and "suffice," is a decision-making strategy introduced by Herbert A. Simon in 1947. It aims for solutions that are good enough to meet the desired criteria without necessarily being the best. Satisficing is a decision-making strategy or a cognitive heuristic that entails searching through available alternatives until they met an acceptability threshold rather than maximizing any specific objective (Byron, 1998). By applying this concept, designers recognize the inherent limitations and uncertainties of complex systems, allowing for more flexible and adaptive solutions.

L. Guo et al., *Architecting a Satisficing Strategy for Designing Complex Systems*,
https://doi.org/10.1007/978-3-032-04577-5_2

In the context of designing complex systems, satisficing involves:

- **Balancing Multiple Objectives**: Instead of seeking a single optimal solution, satisficing allows for the consideration of multiple, often competing objectives, ensuring that the solution meets a broad range of criteria.
- **Flexibility and Adaptability**: Satisficing strategies are inherently more adaptable to changing conditions and uncertainties, making them suitable for dynamic and complex environments.
- **Practicality and Robustness**: By focusing on solutions that are good enough, satisficing encourages the exploration of practical and robust designs that can withstand real-world challenges.

Through this monograph, we aim to equip readers with the knowledge and tools to implement satisficing strategies in their work, ultimately advancing the field of designing complex systems. We hope this comprehensive guide will serve as a valuable resource for those seeking to navigate the complexities of designing robust and adaptive systems in an ever-changing world.

Satisficing in Model-Based Engineering Design

In this chapter, we delve into the work of Mistree and coauthors, who propose a framework for satisficing in engineering design. Our approach incorporates the compromise Decision Support Problem (cDSP) and the Adaptive Linear Programming (ALP) algorithm, as well as the software, Decision Support in the Design of Engineering Systems (DSIDES). These methods are particularly adept at handling nonlinear, multi-goal problems where optimization falters. We use five "toy problems" to illustrate the power of the satisficing strategy—the problems are simple, without any context, but adequate to test the validity of the approach; see Fig. 1.2, Quadrant 1 of the Validation Square.

2.1 Organization of This Chapter

In Fig. 2.1, we illustrate what is covered in this chapter.

- Section 2.2: We provide the mathematical explanations of the differences between optimizing and satisficing strategies.
- Section 2.3: We use five toy problems to illustrate the advantages of using the satisficing strategy in engineering design.
- Section 2.4: We summarize the mathematical and practical differences between the strategies.
- Section. 2.5: We justify the primary research question in the context of "what should we do" (robust design) and "what would we do" (model evolution) based on the discussions.
- Section 2.6: We review the role of this chapter, interpreting it in terms of the "what, how, and why" questions as summarized in Table 2.1.

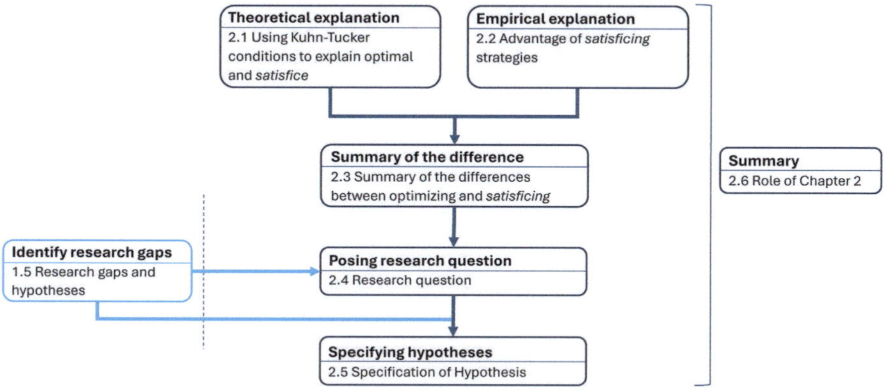

Fig. 2.1 Organization of this chapter

Table 2.1 Summary interpretation of the problem investigated in this chapter

Elements	What?
Components	Foundational concepts, methods, and technologies to realize satisficing strategy
Connectors	The elements and representations of satisficing strategy along the modeling stages
Form	How?
Component roles	Provides the foundational concepts for the satisficing strategy
Properties relationship	Theoretical foundation
	Methods for modeling, design evolution, and knowledge management
Rationale	Why?
Motivation	The need and necessity of applying the satisficing strategy in the evolution of complex systems and the methods that realize the satisficing strategy
Assumptions	Applying the satisficing strategy is one way to make models insensitive to errors and uncertainties, thereby improving the robustness of complex systems
Interpretation	Understanding the methods upon which the satisficing strategy is realized

The main issues in designing complex systems include:

1. Dealing with high complexity, uncertainty, and rapidly changing requirements.
2. Proposing a method to capture system features and analyze system behavior.
3. Automating the method to minimize the need for human input and outputting reusable knowledge.

In this chapter, we review the literature regarding these issues and identify research opportunities.

2.2 The Karush–Kuhn–Tucker (KKT) Conditions—Optimize and Satisfice

2.2.1 *The History of the KKT Conditions*

The Kuhn-Tucker conditions, also known as the Karush–Kuhn–Tucker (KKT) conditions, result from generalizing the method of Lagrange multipliers. A nonlinear problem with constraints can be represented as a Lagrange function. Harold W. Kuhn and Albert W. Tucker proposed the Kuhn-Tucker approach in 1951 (Kuhn and Tucker, 1951). The optimal point of the Lagrange function is a saddle point, so the Kuhn-Tucker approach is also known as the saddle-point theorem. Later, it was discovered that William Karush had summarized the necessary conditions in his master's thesis in 1939 (Karush, 1939). Therefore, Kuhn-Tucker conditions are also referred to as the Karush–Kuhn–Tucker (KKT) conditions. Hereafter, we will refer to them as KKT conditions.

2.2.2 *Necessary and Sufficient KKT Conditions*

The KKT conditions include first-order necessary conditions (Eqs. 2.1–2.5) and second-order sufficient conditions (Eqs. 2.6–2.8).

First-order necessary conditions.

Stationary:

$$\nabla f\left(x^{*}\right)+\sum_{i=1}^{m}\mu_i\nabla g_i\left(x^{*}\right)-\sum_{j=1}^{\ell}\lambda_j\nabla h_j\left(x^{*}\right)=0 \tag{2.1}$$

Primal feasibility:

$$g_i\left(x^{*}\right)\geq 0, \forall i=1,\ldots,m \tag{2.2}$$

$$h_j\left(x^{*}\right)=0, \forall j=1,\ldots,\ell \tag{2.3}$$

Dual feasibility:

$$\mu_i\geq 0, \forall i=1,\ldots,m \tag{2.4}$$

Complementary slackness:

$$\mu_i g_i(x^*) = 0, \forall i = 1, \ldots, m \tag{2.5}$$

Second-order sufficient conditions:

For the Lagrangian:

$$L(x, \lambda, \mu) = f(x) + \sum_{i=1}^{m} \mu_i g_i(x) - \sum_{j=1}^{\ell} \lambda_j h_j(x) \tag{2.6}$$

$$=> s^T \nabla^2_{xx} \mathrm{L}(x^*, \lambda^*, \mu^*) \mathrm{s} \geq 0 \ \ where\ s \neq 0 \tag{2.7}$$

and

$$[\nabla_x g_i(x^*), \nabla_x h_j(x^*)]^T s = 0. \tag{2.8}$$

2.2.3 The Physical Meaning of the KKT Conditions

The essence of the necessary conditions is that at the solution point, x^*, where both the primal and the dual are feasible, the gradient vector of the objective $\nabla f(x^*)$ can be represented as the nonzero linear combination of the gradient matrix of all equality constraints $\nabla h_j(x^*)$ and the active inequality constraints[1] $\nabla g_i(x^*)$ for i iff $g_i(x^*) = 0$, as shown in Fig. 2.2. There are two situations—whether the necessary condition can be met: Situation 1 is illustrated in Fig. 2.2a, as the necessary (first-order) KKT condition is met, meaning the optimal solution is the tangent point of (i) the linear combination (using the Lagrange multipliers) of all equality constraints and the surface of all inequality constraints (hereafter, referred to "the surface"), and (ii) a contour of the objective. Situation 2 is illustrated in Fig. 2.2b, as the necessary KKT condition cannot be met, meaning one cannot find a tangent point of the feasible surface and (ii) a contour of the objective anywhere in the problem's feasible space.

The essence of the sufficient conditions is that at the solution point, x^*, there exists a nonzero vector, s, that is orthogonal to the gradient matrix of all active inequality and equality constraints, that the second-order matrix of the Lagrange equation concerning decision variables x^* and Lagrange multipliers λ^* and μ^* is conditionally positive semidefinite: $s^T \nabla^2_{xx} \mathrm{L}(x^*, \lambda^*, \mu^*)\mathrm{s} \geq 0, \forall s \in \mathcal{S}$.

[1] In this monograph, we define an active constraint (or an active inequality constraint) as the constraint with zero slack or surplus when inserting in the solution point. Alternatively, if an inequality constraint's left-hand side value equals its right-hand side value when being inserted in the solution point, then such an inequality constraint is an active constraint; it is also known as a binding constraint; see "DEFINITION OF TERMS."

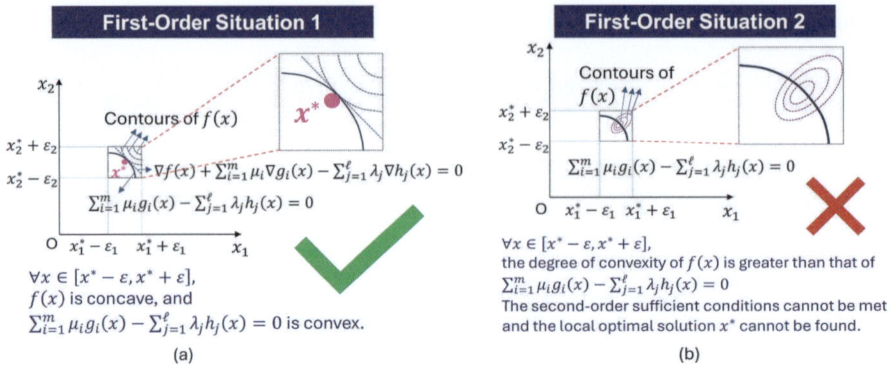

Fig. 2.2 The first-order necessary KKT conditions are satisfied at x^*

Why is it conditionally positive semidefinite? What is the condition?

The second-order sufficient condition requires that the Lagrange equation be convex at x^*; that is, in a local range that contains the solution point, x^*, the convexity degree of the objective should not exceed the convexity degree of the constraints combined by Lagrange multipliers; there can be three situations—see Fig. 2.3. The first situation, shown in Fig. 2.3a, is when the feasible surface is convex, and the objective contours are concave. The second situation, shown in Fig. 2.3b, is when both the feasible surface and the objective contours are convex, but the convexity of the feasible surface is higher than the convexity of the objective contours. The third situation, shown in Fig. 2.3c, is when the objective contours are more convex than the feasible surface within the feasible space but the objective contours are less convex than the feasible surface outside of the feasible space. In any of these three situations, can the sufficient KKT condition be met.

However, if at least one of the objectives and the feasible surface is nonconvex and nonconcave, as the one in Fig. 2.4 (which is an example of nonconvex-nonconcave objective), then the sufficient condition cannot be met. Therefore, the requirement of meeting the sufficient condition significantly hinders the access to a solution for the

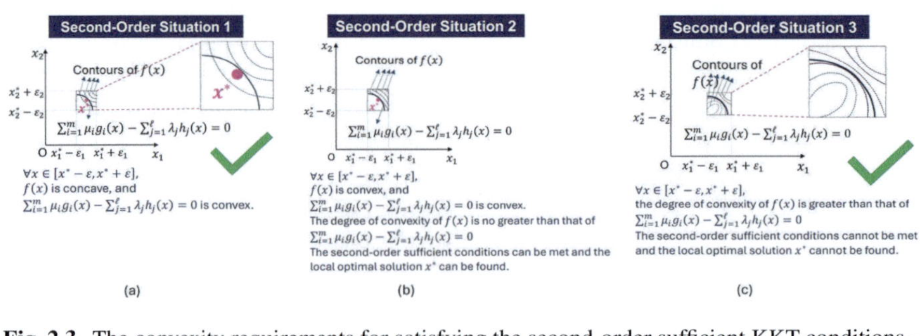

Fig. 2.3 The convexity requirements for satisfying the second-order sufficient KKT conditions—three situations: **(a)**–**(c)**

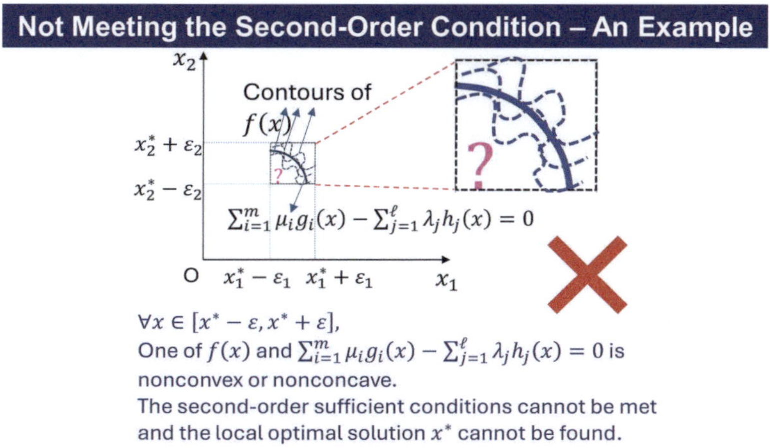

Fig. 2.4 Not meeting the sufficient (second-order) KKT condition—an example: at least one of (i) the feasible surface and (ii) the objective contours is neither convex nor concave

problems with (1) an objective with a different convexity in the feasible region and in the infeasible region, or (2) a nonconvex and nonconcave objective or constraints.

2.2.4 Assumptions Behind the KKT Conditions

There are assumptions behind applying the KKT conditions to solve a nonlinear programming problem, especially when designers seek optimal solutions which meet both necessary and sufficient KKT Conditions. These assumptions are also mathematical requirements for the problem that can be applied with KKT conditions to obtain an optimal solution. We automatically accept these assumptions using the KKT approach or Lagrange function to solve a nonlinear problem.

Assumption/Requirement 1—mathematical models are 100% complete and accurate representations of physical problems.

To use Lagrange functions to solve nonlinear problems, one naturally assumes that the mathematical models are complete and accurate and capture all the necessary details of the physical problems. All segments of a model have the same level of fidelity. Thus, an optimal solution to to a mathematical problem is also optimal to its corresponding physical problem or at least feasible for the physical problem.

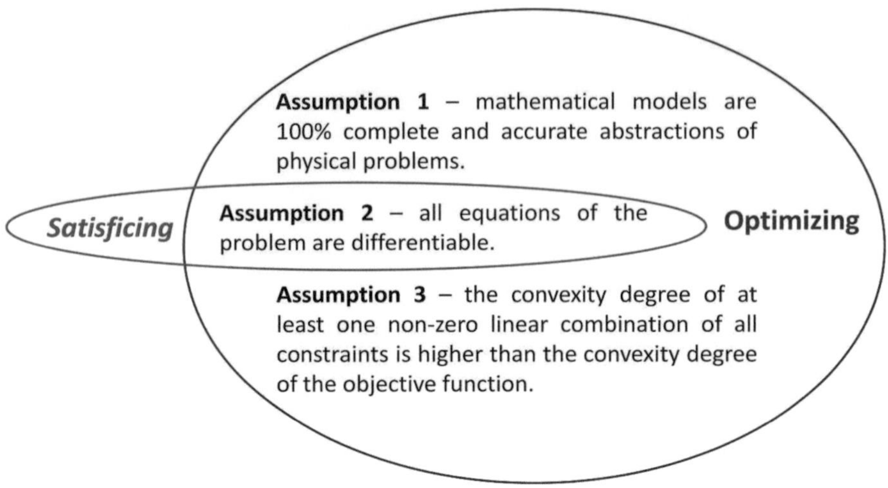

Fig. 2.5 The assumptions when using the optimizing strategy and the *satisficing* strategy

Assumption/Requirement 2—all equations of the problem are differentiable.

However, according to KKT conditions, the objective function, active constraints, and equality constraints are only required to be differentiable within a small finite area around the optimal solution. As we need to solve the Lagrange function globally to obtain the optimal solution, all problem equations must must be differentiable.

Assumption/Requirement 3—the convexity degree of at least one nonzero linear combination of all constraints is higher than the convexity degree of the objective function.

This assumption is especially true for KKT sufficiency. The sufficient conditions only guarantee optimal solutions to problems like those in Fig. 2.3 but cannot find an optimal solution to problems like those in Fig. 2.4.

In summary, designers make all three assumptions when using an optimizing strategy to seek an optimal solution to a decision model. In contrast, they make only one assumption when using a satisfying *strategy*. The assumptions that underlie the two strategies are illustrated in Fig. 2.5.

2.3 Advantages of *Satisficing* Strategy

– ***Advantages in Each Stage of Model Evolution***

In using the optimizing strategy, designers use the Lagrange function to find solutions that satisfy both necessary and sufficient KKT conditions. In contrast, when designers adopt a *satisficing* strategy, they only seek to satisfy the necessary KKT condition. Therefore, when seeking optimal solutions, designers have three assumptions; in other words, the mathematical model needs to meet the three often-mentioned requirements. In other words, when seeking optimal solutions, designers accept the three assumptions behind KKT Conditions: models are 100% complete and accurate, all equations are differentiable, and the problem is convex.

When seeking *satisficing* solutions, the assumptions are different. Designers are aware that their models can be incomplete and inaccurate, so they aim at good enough solutions that are relatively insensitive to the error and incompletion of the models. When their decision models are discrete or non-convex, we describe the difference between optimizing and satisficing strategy in Sect. 2.3.

First, we use five representative "toy problem" to illustrate the differences between the two strategies regarding the returned solutions when managing engineering design problems, especially complex systems design and improvement[2] to illustrate the differences in appearance between optimizing and *satisficing*. Then, we analyze, in principle, why there are such differences.

Four advantages of using the *satisficing* strategy, which in this monograph is realized by using cDSP, ALP, and DSIDES, include formulation, approximation, solution, and exploration. In Table 2.2, we summarize the advantages of using *satisficing* in the four stages of engineering design. The differences are illustrated through examples from Sect. 2.3.1 to 2.3.10.

2.3.1 Possible Features of Engineering-Design Problems

In this monograph, we define the *satisficing* strategy as the strategy in the formulation, construction, and solution algorithm (Guo et al., 2024) that allows designers to manage those typical features of engineering-design problems. In this monograph, we identify six features, listed in the first column of Table 2.3, as typical features that designers often encounter in engineering-design problems. We choose the five toy problems and add an additional degree of complexity to each toy problem. The toy problems are simple but represent some basic, primary complexity of complex engineering-design problems. So, we use these toy problems to do experiments and comparisons. There can be many more features, challenges, and difficulties of complex engineering-design problems. In this section, we only tackle five basic features because many complexities are combinations and evolution of the six features in different times and spaces. The *satisficing* methods can be foundations or cornerstones of more complex methods or technologies.

---

[2] The problems in Sect. 2.3 are "toy problems" to illustrate the difference in the results that the methods in the two strategies can return. They are separated from the test problems in Chaps. 4, 5, 6, 7, and 9. The test problems in the later chapters verify the proposed method and the new knowledge in this monograph.

Table 2.2 The advantages of realizing *satisficing* strategy using cDSP and ALP in each stage of engineering design

Stage	Feature	Advantage	Introduction and discussion
Formulation	Using Goals and Minimizing Deviation Variables Instead of Objectives	At a solution point, only the necessary KKT conditions are met, whereas the sufficient KKT conditions do not have to be met Therefore, designers have a higher chance of finding a solution and a lower chance of losing a solution due to parameterizable and unparameterizable uncertainties	Section 2.3.7
Approximation	Using second-order sequential linearization	Designers can have a balance between linearization accuracy and computational complexity	Sections 2.3.5 and 5.2.1
	Using accumulated linearization	Designers can manage nonconvex problems in a way that is relatively more accurate than highly convex, nonlinear problems	
Exploration	Combining interior-point searching and vertex searching	Designers can avoid being stuck into local optima to some extent and identify *satisficing* solutions relatively insensitive to changing starting points	Section 2.3.3
Evaluation	Allowing some violations of soft requirements, such as the bounds of deviation variables	Designers can manage rigid and soft requirements in different ways to ensure feasibility As a result, goals and constraints with different scales can be managed	Section 2.3.9

Table 2.3 The features of the Toy Problems (TP)

Feature	Toy Problem (TP)				
	I	II	III	IV	V
1. Two objectives	*	*	*	*	*
2. Nonlinear	*	*	*	*	*
3. Non-convex		*	*	*	*
4. Objectives with various units (scale)			*	*	*
5. Target of goals with various levels of achievability				*	*
6. More than two objectives					*

The optimization methods we use in each strategy are listed in Table 2.4. In the "scipy.optimize" package, besides the three algorithms (COBYLA, Trust-constr, and SLSQP), there are seven other algorithms, Nelder-Mead, Powell, CG (Conjugate Gradient), BFGS, Newton-CG, L-BFGS-B, and TNC. However, Nelder-Mead, Powell, CG, and BFGS cannot manage problems well with constraints. Newton-CG, L-BFGS-B, and TNC require the Jacobian, which is not user-friendly for engineering designers; as for nonlinear engineering-design problems, usually, designers do not want to identify the Jacobian matrix of the nonlinear constraints and formulate the problems by presenting the coefficient matrix. Therefore, we use COBYLA, Trust-constr, and SLSQP to solve toy problems with one or more features, as shown in Table 2.3.

We use NSGA II/III as a benchmark or verification method. NSGA II/III is a well-known solution algorithm in the optimizing category. It is a fast sorting and elite multi-objective genetic algorithm. Unlike the single objective optimization technique, NSGA II/III simultaneously optimizes each objective without being dominated by any other solution. As NSGA II/III can manage all six features listed in Table 2.3, we use it as a benchmark tool to verify the solution quality of the *satisficing* algorithm.

Since NSGA II//III can manage the five typical features of engineering-design problems, why do we study satisficing algorithms? Because NSGA II/III has drawbacks that prevent designers from further exploring the solutions space and improving the decision models (Guo et al., 2024).

1. **Lack of Insight**: First, NSGA II/III cannot give designers insight into the nature of the decision model or the possible ways to improve the model. NSGA II/III is an interior search algorithm that uses metaheuristics to search for solutions

Table 2.4 Methods for comparison of the two strategies

Item	Strategy	
	Optimizing	Satisficing
Model formulation construct	Mathematical programming or Goal programming	Compromise decision support problem
Solution algorithm	Constrained Optimization by Linear Approximation (COBYLA) algorithm	Adaptive Linear Programming (ALP) algorithm
	Trust-region constrained (trust-constr) algorithm	
	Sequential Least Squares Programming (SLSQP) algorithm	
	Nondominated Sorting Generation Algorithm II/III (NSGA II/III)	
Solver	Python Scipy.optimize	DSIDES

that generationally improve the optimality and diversity of the solutions, but for information, such as the bottleneck of the model or the sensitivity of each part of the model or anything else that may indicate model improvement, cannot be provided along with the searching.

2. **Sensitivity to Hyperparameters:** The performance of NSGA II/III (including convergence speed, optimality of solutions, and diversity of solutions) is sensitive to hyperparameter settings. Typical hyperparameters, such as population size and generation number, should be predefined by designers. However, designers only believe that a larger population or a larger generation number can return better solutions. However, they may not know how large "good enough" is. In Toy Problems II and III, we can show how different the solutions can be when setting the population to 20 and 50 for the same problem.
3. **Computational Power**: NSGA II/III requires much more computational power than satisficing algorithms such as the Adaptive Linear Programming (ALP) algorithm. In this section, we will discuss it in detail in the "summary of observations" for each toy problem.

The toy problems and solutions using different algorithms, comparisons, and conclusions are as follows.

2.3.2 *Toy Problem I (TP-I)*

– ***Problem with multiple, nonlinear objectives (goals)***

Formulation: The two formulations for Toy Problem I (TP-1) are given in Table 2.5: the optimization formulation and its corresponding compromise DSP (or cDSP). x_i are decision variables (design variables). d_i^- and d_i^+ are deviation variables, which measure the percentage of under-achievement and over-achievement of Goal i versus its target. The two objectives are visualized in the solution space (decision-variable space, X-Plane) in Fig. 2.6. The solutions to the problem using the corresponding algorithms are given in Table 2.6.

Target of the goals: For compromise DSP, a right-hand side value is needed for each objective. Designers must assign a target value for each goal and minimize the deviation between the achieved value and the goal's target. In this problem, the targets of Goal 1 and Goal 2 are 14 and 8, respectively.

Weight to combine the goals or objectives: The problem has multiple goals (objectives), but the priority of the goals still needs to be determined. So, for the toy problems, we use the Archimedean strategy (scalarization) to combine the multiple goals. By exploring different weight scenarios, the tradeoffs among the goals can be better studied. Therefore, designers can be provided with further support for weight scenario selection.

Table 2.5 The optimization model and compromise DSP of TP-I

TP	Strategy	
	Optimizing	Satisficing
TP-I	<u>Objective Functions</u> $f_1(x) = (x_1 - 1)^2 + (x_2 - 1)^2 + 3 \cdot x_1 \cdot x_2$ $f_2(x) = \frac{1}{2} \cdot (x_1 - 2)^2 + (x_2 - 2)^2 + x_1 \cdot x_2^2$ <u>Constraints and Bounds</u> $s.t. \begin{cases} x_1 \cdot x_2 \le 1 \\ 0 \le x_1 \le 2 \\ 0 \le x_2 \le 2 \end{cases}$ <u>Combination of Objective Functions</u> $Max \sum_{i=1}^{2} w_i \cdot f_i(x)$	<u>Given</u> $x_1, x_2, d_1^{\pm}, d_2^{\pm}$ $f_1(x) = (x_1 - 1)^2 + (x_2 - 1)^2 + 3 \cdot x_1 \cdot x_2$ $f_2(x) = \frac{1}{2} \cdot (x_1 - 2)^2 + (x_2 - 2)^2 + x_1 \cdot x_2^2$ <u>Find</u> $x_1, x_2, d_1^{\mp}, d_2^{\mp}$ <u>Satisfy</u> <u>Goals:</u> $\frac{f_1(x)}{14} + d_1^- - d_1^+ = 1$ $\frac{f_2(x)}{8} + d_2^- - d_2^+ = 1$ <u>Constraints:</u> $x_1 \cdot x_2 \le 1$ $d_i^- \cdot d_i^+ = 0, i = 1, 2$ <u>Bounds:</u> $0 \le x_1, x_2 \le 2$ $0 \le d_i^{\pm} \le 1$ <u>Minimize</u> $Z = \sum_{i=1}^{2} w_i \cdot (d_i^- + d_i^+)$

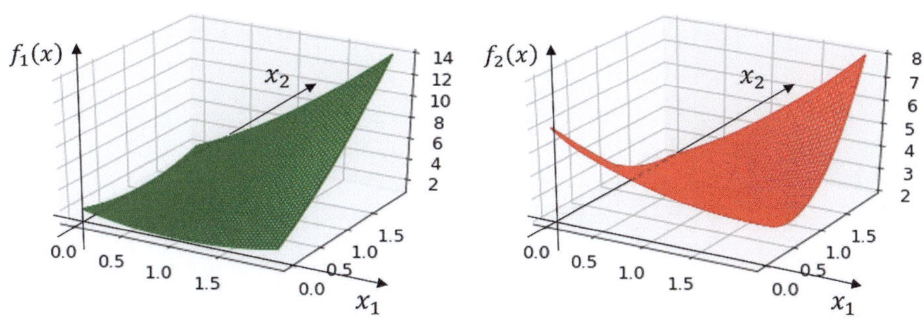

Fig. 2.6 The two objective functions in the X-f(X) space of TP-I—**a** Objective 1, and **b** Objective 2

<u>Results</u>: By using five weight scenarios (1, 0), (0, 1), (0.5, 0.5), (0.7, 0.3), and (0.3, 0.7), no matter which solution algorithm is used, three solutions are always obtained, as visualized on objective space in Fig. 2.7 and on *X–f(X)* Space in Fig. 2.8. As the objective of TP-I is to maximize the two objective functions $f_1(x)$ and $f_2(x)$, in Fig. 2.7, the ideal solutions should be close to the upper right hand corner of the objective space. In Fig. 2.8, the value of the objective should be as high as possible.

Table 2.6 Solutions to TP-I (*the best solution in each scenario is in italics*)

Weight	Starting point	COBYLA		Trust-constr		SLSQP		ALP		NSGA II	
		Solution	$\sum_{i=1}^{2} w_i \cdot f_i(x)$	Solution	$\sum_{i=1}^{2} w_i \cdot f_i(x)$	Solution	$\sum_{i=1}^{2} w_i \cdot f_i(x)$	Solution	$\sum_{i=1}^{2} w_i \cdot f_i(x)$	Solution	$\sum_{i=1}^{2} w_i \cdot f_i(x)$
(1, 0)	(0.5, 1)	(2, 0.5)	4.25	(2, 0.5)	4.25	(2, 0.5)	4.25	(2, 0.5)	4.25	(2, 0.5)	4.25
	(0, 0)	*(0, 0)*	*2*	*(0, 0)*	*2*	*(0, 0)*	*2*				
	(2, 0.5)	(2, 0.5)	4.25	(2, 0.5)	4.25	(2, 0.5)	4.25				
(0, 1)	(0.5, 1)	*(0.5, 2)*	*3.13*	(0, 0)	6	(0, 0)	6	(0, 0)	6	(0, 0)	6
	(0, 0)	(0, 0)	6								
	(2, 0.5)					*(2, 0)*	*4*				
(0.5, 0.5)	(0.5, 1)	*(0.5, 2)*	*3.69*	*(0.5 2)*	*3.69*	*(0.5,2)*	*3.69*	*(0.5, 2)*	*3.69*	(0, 0)	4
	(0, 0)	(0, 0)	4	(0, 0)	4	(0, 0)	4				
	(2, 0.5)	*(2, 0.5)*	*3.5*	*(2, 0.5)*	*3.5*	*(2,0.5)*	*3.5*				
(0.7,0.3)	(0.5, 1)	(0.5, 2)	3.91	(0.5, 2)	3.91	(0.5, 2)	3.91	(0.5, 2)	3.91	(0.5, 2)	3.91
	(0, 0)	*(0, 0)*	*3.2*	*(0, 0)*	*3.2*	*(0, 0)*	*3.2*				
	(2, 0.5)	*(2, 0.5)*	*3.8*	*(2, 0.5)*	*3.8*	*(2,0.5)*	*3.8*				
(0.3, 0.7)	(0.5, 1)	*(2, 0)*	*3.4*	(0, 0)	4.8	(0, 0)	4.8	(0, 0)	4.8	(0, 0)	4.8
	(0, 0)	(0, 0)	4.8								
	(2, 0.5)	*(2, 0.5)*	*3.2*	*(2, 0.5)*	*3.2*	*(2, 0.5)*	*3.2*				

The two axes, x_1 and x_2, are marked with arrows. $f_1(x)$ and $f_2(x)$ are the vertical axis of Fig. 2.8a and b, respectively.

Observation: Thesame solutions can be obtained using an optimizing and satisficing strategy for a test problem with nonlinear functions and multiple objectives without any non-convex function or objective functions with various units and scales. The results of using optimizing solution algorithms such as COBYLA, Trust-constr, and SLSQP are relatively sensitive to the starting point.

So, in Sect. 2.3.3, we address "Why are the solutions insensitive to the starting point when using the ALP?"

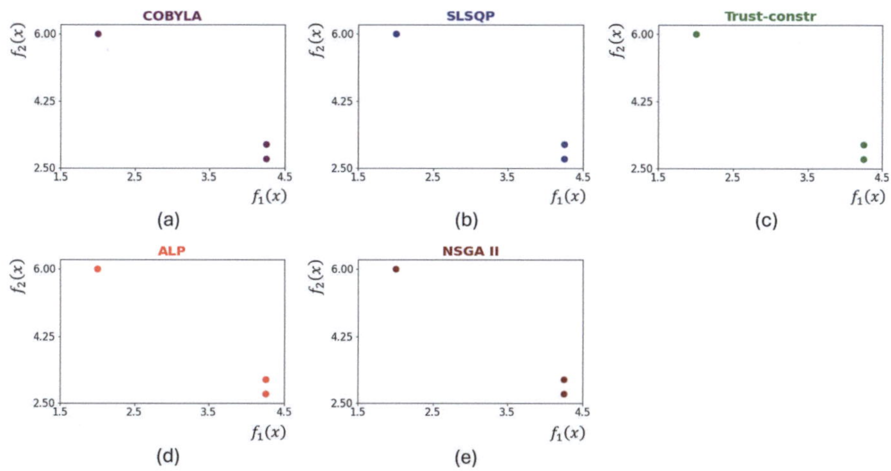

Fig. 2.7 The solution points to TP-I on the objective space using five algorithms—the same

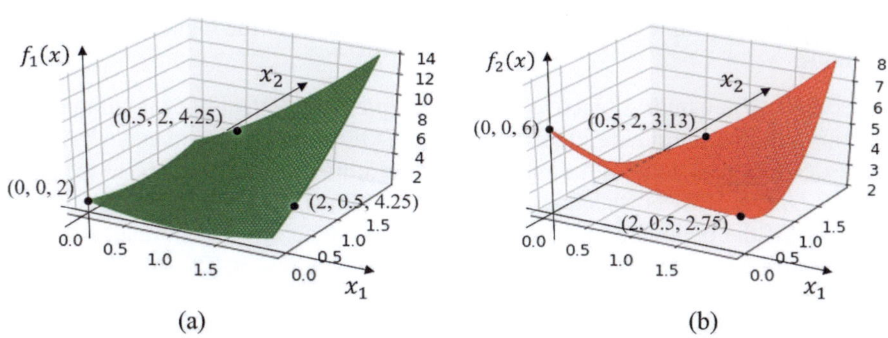

Fig. 2.8 The solution points to TP-I on the x–$f(x)$ space—**a** is the 3D illustration of Objective 1, $f_1(x)$. **b** is the 3D illustration of Objective 2, $f_2(x)$. Since the solutions are the same when using different formulations and algorithms, all three points are the same for all five methods (in Table 2.4)

2.3.3 Method Requirement 1: Combination of Interior-Point Searching and Vertex Searching

- *Why are the solutions insensitive to the starting point when using the ALP?*
- The combination of interior-point search and Simplex—In the ALP, there is a module named "XPLORE." When using XPLORE, "*m*" starting points within the bounds of each variable are randomly selected and tested for the goal-achieved value and feasibility. Then, feasible solutions are ranked based on their goal-achieved value. The top "*n*" solutions are used one by one as the starting point, around which the problem is linearized to a linear problem and then solved by the Dual Simplex. Usually, "m" is set as a number in [800, 1200], and "*n*" is set as a number in [5, 20]. Therefore, no matter what starting point is assigned by the user, using "XPLORE" allows the feasible points with relatively good goal-achievement values to be selected as the starting points, hence, to some extent, avoid it being stuck into the local optima.

2.3.4 Toy Problem II (TP-II)

– ***Problem with multiple, nonlinear, and non-convex objectives (goals)***

In TP-II, we use a non-convex objective (goal) and a cubic polynomial objective (goal). The formulation, visualization, solutions, and visualization of solutions on objective space and on *X–f(X)* Space are given in Table 2.7, Fig. 2.9, Table 2.8, Figs. 2.10, and 2.11, respectively. For NSGA II/III, when we set the "population[3] as 20, the solutions are often dominated solutions, and when we set the "population" as 50, the solutions are much better and can converge to nondominated solutions under each scenario. We only give the results by using NSGA II with population 50 and generation 100 in Table 2.8, Figs. 2.10, and 2.11 because of the higher quality of the solutions.

Observation: For a multi-objective (multi-goal) problem with nonlinear, non-convex functions, some optimizing algorithms (e.g., COBYLA) cannot manage the non-convexity, whereas some other optimizing algorithms converge easily at the local optima. NSGA II/III is sensitive to parameter settings but can return high-quality solutions (if the population size is large enough) which are nondominated but not quite diverse and require relatively high computational power. The ALP can return "good enough" and relatively diverse solutions.

Why should the diversity of solutions be associated with *robustness?* "Considering the goodness of a solution set regarding its distribution quality is key to multi-objective problems" (Farhang-Mehr and Azam, 2002). It is important to balance the convergence of the search algorithm and the diversity of solutions for multi-objective problems (Xu et al., 2022). Engineering designers often need different solutions under

[3] The "population" in NSGA II is the population size, a parameter predetermined by the user.

Table 2.7 The Optimization Model and Compromise DSP of the TP-II

TP	Strategy	
	Optimizing	Satisficing
II	<u>Objective Functions</u> $f_1(x) = cos\left(x_1{}^2 + x_2{}^3\right)$ $f_2(x) = \frac{1}{2} \cdot (x_1 - 2)^3 + (x_2 - 2)^3 + x_1 \cdot x_2{}^2$ <u>Constraints and Bounds</u> $s.t. \begin{cases} x_1 \cdot x_2 \leq 1 \\ 0 \leq x_1 \leq 2 \\ 0 \leq x_2 \leq 2 \end{cases}$ <u>Combination of Objective Functions</u> $Max \sum_{i=1}^{2} w_i \cdot f_i(x)$	<u>Given</u> $x_1, x_2, d_1{}^{\pm}, d_2{}^{\pm}$ $f_1(x) = \cos\left(x_1{}^2 + x_2{}^3\right)$ $f_2(x) = \frac{1}{2} \cdot (x_1 - 2)^3 + (x_2 - 2)^3 + x_1 \cdot x_2$ <u>Find</u> $x_1, x_2, d_1{}^{\mp}, d_2{}^{\mp}$ <u>Satisfy</u> <u>Goals:</u> $\frac{f_1(x)}{1.2} + d_1{}^- - d_1{}^+ = 1$ $\frac{f_2(x)}{8} + d_2{}^- - d_2{}^+ = 1$ <u>Constraints:</u> $x_1 \cdot x_2 \leq 1$ $d_i{}^- \cdot d_i{}^+ = 0, i = 1, 2$ <u>Bounds:</u> $0 \leq x_1, x_2 \leq 2$ $0 \leq d_1{}^{\pm}, d_2{}^{\pm} \leq 1$ <u>Minimize</u> $Z = \sum_{i=1}^{2} w_i \cdot (d_i{}^- + d_i{}^+)$

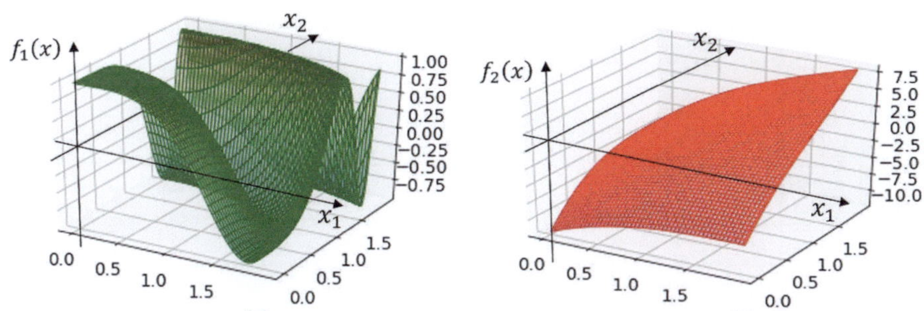

(a) (b)

Fig. 2.9 The two objective functions on the *X–f(X)* plane of TP-II—**a** the first objective $f_1(x)$ is non-convex, and **b** the second objective $f_2(x)$ is convex

multiple design scenarios that represent their various design preferences among the multiple goals, meanwhile the diversity of solutions brought about by different solution processes, such as different search starting points, should be minimized. When design preferences evolve, solutions can be adapted. High solution diversity means that range and richness of solutions has been expanded therefore accommodating evolving design preferences.

Table 2.8 Solutions to TP-II—the best solution for each weight is in italics; for each method, the best solution of each design scenario is marked using a capital letter (A–H and J)

Weight	Starting point	COBYLA		Trust-constr		SLSQP		ALP		NSGA II	
		Solution	$\sum_{i=1}^{2} w_i \cdot f_i(x)$	Solution	$\sum_{i=1}^{2} w_i \cdot f_i(x)$	Solution	$\sum_{i=1}^{2} w_i \cdot f_i(x)$	Solution	$\sum_{i=1}^{2} w_i \cdot f_i(x)$	Solution	$\sum_{i=1}^{2} w_i \cdot f_i(x)$
(1, 0)	(0.5, 1)	A (0, 0)	1	A (0, 0)	1	A (0, 0)	1	*F (0.5, 1.84)*	*0.97*	A (0, 0)	1
	(0, 0)										
	(2, 0.5)			*(2, 0.5)*	*−0.55*	*(2, 0.5)*	*−0.55*				
(0, 1)	(0.5, 1)	B (0.5, 2)	0.31	B (0.5, 2)	0.31	B (0.5, 2)	0.31	G (0.51, 1.96)	0.31	B (0.5, 2)	0.31
	(0, 0)										
	(2, 0.5)	*(2.0, 0.5)*	*−2.88*								
(0.5, 0.5)	(0.5, 1)	*(0.9, 1.12)*	*−0.42*	*D (0.9, 1.12)*	*−0.42*	*(0.9, 1.12)*	*−0.42*	*H (0.52, 1.87)*	*0.54*	C (0.55, 1.82)	0.64
	(0, 0)					C (0.55, 1.82)	0.64				
	(2, 0.5)	C (0.55, 1.82)	0.64			*(2, 0.5)*	*−1.71*				
(0.7, 0.3)	(0.5, 1)	C (0.55, 1.82)	0.79	*E (0.73, 0.88)*	*−0.31*	*(0.73, 0.88)*	*−0.31*	*F (0.5, 1.84)*	*0.68*	J (0.56, 1.8)	0.79
	(0, 0)					C (0.55, 1.82)	0.79				
	(2, 0.5)	*(0.73, 0.88)*	*−0.31*	*(2, 0.5)*	*−1.25*	*(2, 0.5)*	*−1.25*				
(0.3, 0.7)	(0.5, 1)	C (0.55, 1.82)	0.5	C (0.55, 1.82)	0.5	C (0.55, 1.82)	0.5	C (0.55, 1.82)	0.5	C (0.55, 1.82)	0.5
	(0, 0)										
	(2, 0.5)	*(2, 0.5)*	*−2.18*								

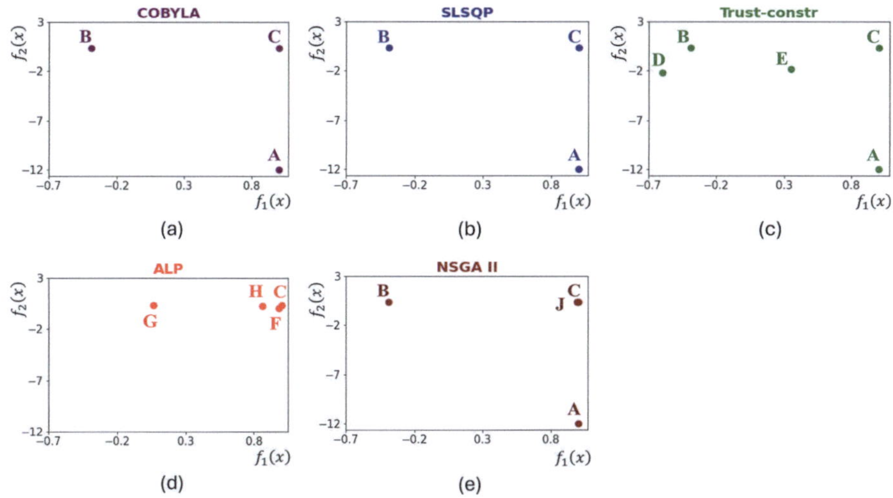

Fig. 2.10 The solution points to TP-II on the objective space using four algorithms—solutions returned by trust-constr and SLSQP are not "good enough," solutions returned by ALP are "good enough" and diverse, and solutions returned by NSGA II contain nondominated solutions but are not diverse

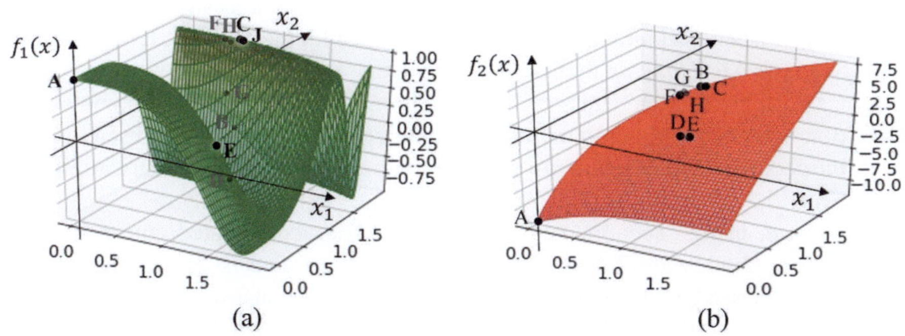

Fig. 2.11 The solution points to TP-II on the x–$f(x)$ space—using Trust-constraint and SLSQP, it is easy to fall into local optima. Green, blue, red, and dark red dots are the solutions using trust-constr, SLSQP, ALP, and NSGA II, respectively

In Sect. 2.3.5, we address the question, "*Why does the ALP manage non-convex problems and return solutions close to the nondominated solutions (the solutions returned by NSGA II/III)?*"

2.3.5 Method Requirements 2 and 3: Second-Order Sequential Linearization and Accumulated Linearization

- *Why does the ALP manage non-convex problems and return solutions close to the nondominated solutions (those returned by NSGA II/III)?*
  - Two mechanisms of the ALP allow it to linearize the non-convex function relatively accurately and converge with good enough solutions.
  - ***First***, using "second-order sequential linearization." The second-order derivatives function (when the paraboloid is used to approximate the nonlinear function has two real roots) and the first-order derivatives (when the paraboloid has no real root) of the nonlinear functions make the linear problem relatively robust. The nonlinear equation is first approximated into a paraboloid and then approximated into a linear equation. For example, as shown in Fig. 2.12. First, the surface of the nonlinear constraint is approximated into a paraboloid at point x^0, represented as Paraboloid AB^*C^* in Fig. 2.12c, using the second-order derivative (diagonal items of the Hessian matrix of the nonlinear constraint). Then, if the paraboloid has real roots, B^* and C^*, which means the paraboloid intersects with the X-plane, the paraboloid is linearized as Plane AB^*C^*. On the contrary, if the paraboloid does not have a real root, as shown in Fig. 2.13, that is, when the paraboloid does not intersect with the X-plane, it is linearized as Plane AB^0C^0 using first-order derivatives.
  - Using second-order sequential linearization, designers can balance linearization accuracy and computational complexity. A more detailed discussion of the accuracy and flexibility of using second-order sequential linearization is in Sect. 5.2.1.
  - ***Second***, using "accumulated constraints." The accumulated constraints are only applied when the local convexity of an equation is ≥ -0.15. Such an equation can be a goal or an equality or inequality constraint. We define local convexity as the gradient of a function in a small neighborhood around a linearization point. We choose -0.15 based on experiments and heuristics. When the local convexity at a linearization point is greater than or equal to -0.15, the function is convex or only slightly non convex in that neighborhood. In Fig. 2.14a and b, we observe convex or slightly non convex behavior. Therefore, we linearize the function around a new linearization point and then replace the nonlinear constraint with linear constraints obtained from multiple such linearization iterations. We call these combined linear constraints accumulated constraints. If the local convexity at a point is less than -0.15 (as in Fig. 2.14c), the function is significantly non convex there. In that case, we form only a single linearized constraint per iteration.

    We use accumulated constraints as a tool to manage non convex problems and to address situations where functions are only slightly non convex more accurately. Section 5.2.1 provides further details on this mechanism.

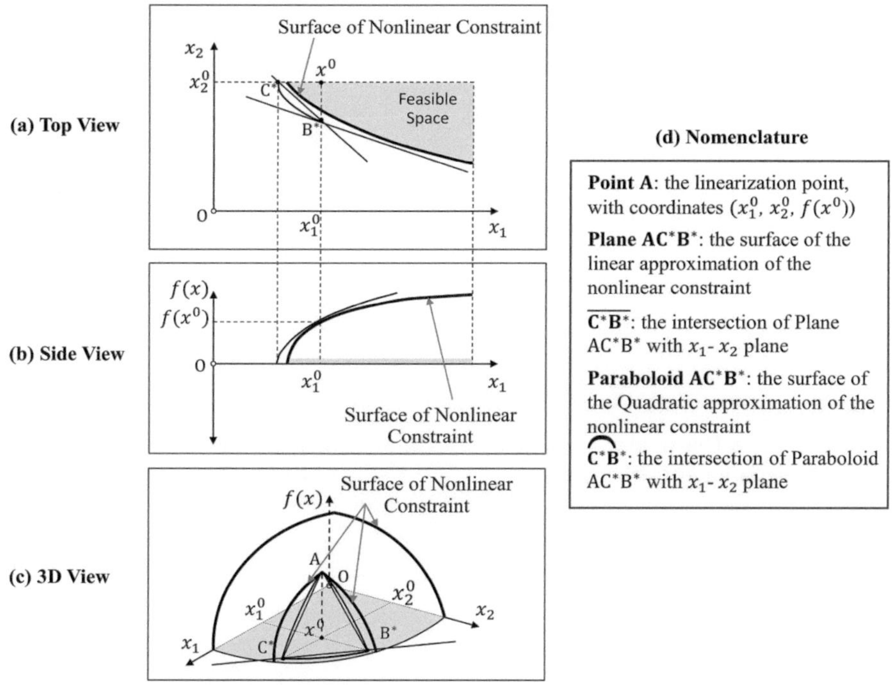

Fig. 2.12 Illustration of the sequential linearization using the ALP with different views when the quadratic approximated paraboloid has real roots

Fig. 2.13 Linearization using the ALP When the quadratic approximated paraboloid has no real roots

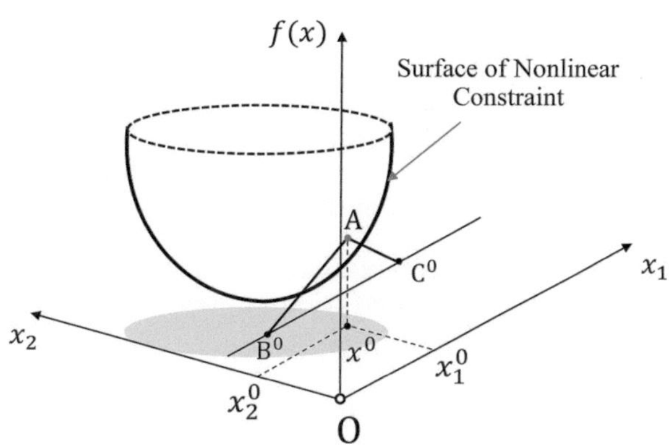

– Designers can use accumulated constraints to manage nonconvex problems and deal with highly convex, nonlinear problems more accurately. The details of this mechanism are further discussed in Sect. 5.2.1.

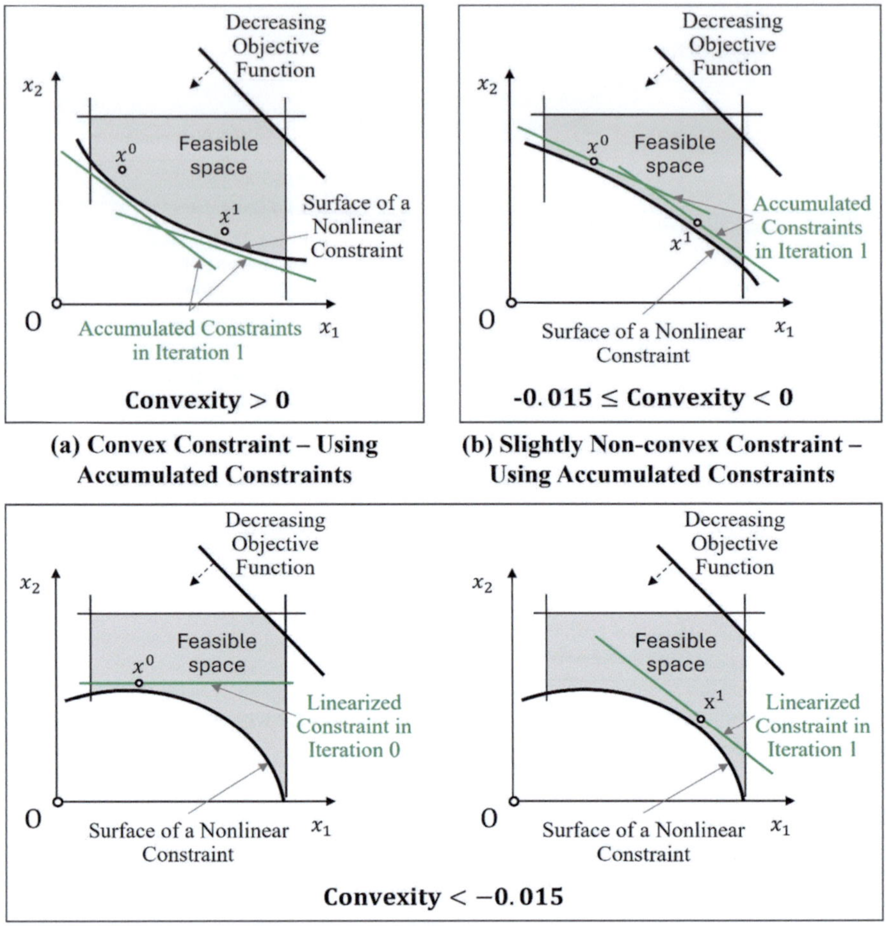

(a) Convex Constraint – Using Accumulated Constraints

(b) Slightly Non-convex Constraint – Using Accumulated Constraints

(c) Significantly Non-convex Constraint – Using Single Constraint

Fig. 2.14 Using the accumulated constraints from multiple linearization iterations for convex or slightly non-convex equations and using single linearized constraint for significantly non-convex constraint

2.3.6 *Toy Problem III (TP-III)*

– ***Problem with multiple, nonlinear, and non-convex objectives (goals) and the objectives (goals) are with various scales***

When the scale of one objective of TP-II is adjusted, it becomes TP-III. Accordingly, the target of the goal of the cDSP is enlarged at the same time, as it is shown in Table 2.9. We add a lower bound of the two objectives to avoid the negative achieved value of objectives. The formulation, visualization, solutions, and visualization of solutions on objective space and the X-Plane of TP-III are given in Table 2.9, Fig. 2.15, Table 2.10, Figs. 2.16, and 2.17, respectively. Because the scale of the objectives varies dramatically, combining the objectives using weights makes the problem "lose sense," which includes several situations: (1) the objective with a large scale "dominates" the other objective, (2) the linearized function of the weighted combined objective at a local area is singular, (3) the scale of the second objective is so large that it dominates the penalty added as the constraint, which makes the constraints always be violated.

Observation: For a multi-objective (multi-goal) problem with nonlinear, non-convex functions, and the scale of the objectives varies largely, *why some optimizing algorithms (e.g., COBYLA) cannot work it out? The answer is given as follows.*

Table 2.9 The optimization model and compromise DSP of the TP-III

TP	Strategy	
	Optimizing	Satisficing
III	Objective Functions $f_1(x) = cos(x_1{}^2 + x_2{}^3)$ $f_2(x) =$ $25 \cdot (x_1 - 2)^3 + 50 \cdot (x_2 - 2)^3 + 50 \cdot x_1 \cdot x_2{}^2$ Constraints and Bounds $s.t. \begin{cases} x_1 \cdot x_2 \leq 1 \\ f_1(x) \geq 0 \\ f_2(x) \geq 0 \\ 0 \leq x_1 \leq 2 \\ 0 \leq x_2 \leq 2 \end{cases}$ Combination of Objective Functions $Max \sum_{i=1}^{2} w_i \cdot f_i(x)$	Given $x_1, x_2, d_1{}^{\pm}, d_2{}^{\pm}$ $f_1(x) = \cos(x_1{}^2 + x_2{}^3)$ $f_2(x) =$ $25 \cdot (x_1 - 2)^3 + 50 \cdot (x_2 - 2)^3 + 50 \cdot x_1 \cdot x_2{}^2$ Find $x_1, x_2, d_1{}^{\mp}, d_2{}^{\mp}$ Satisfy Goals: $\frac{f_1(x)}{1.2} + d_1{}^- - d_1{}^+ = 1$ $\frac{f_2(x)}{400} + d_2{}^- - d_2{}^+ = 1$ Constraints: $x_1 \cdot x_2 \leq 1$ $f_1(x) \geq 0$ $f_2(x) \geq 0$ $d_i{}^- \cdot d_i{}^+ = 0, i = 1, 2$ Bounds: $0 \leq x_1, x_2 \leq 2$ $0 \leq d_1{}^{\pm}, d_2{}^{\pm} \leq 1$ Minimize $Z = \sum_{i=1}^{2} w_i \cdot (d_i{}^- + d_i{}^+)$

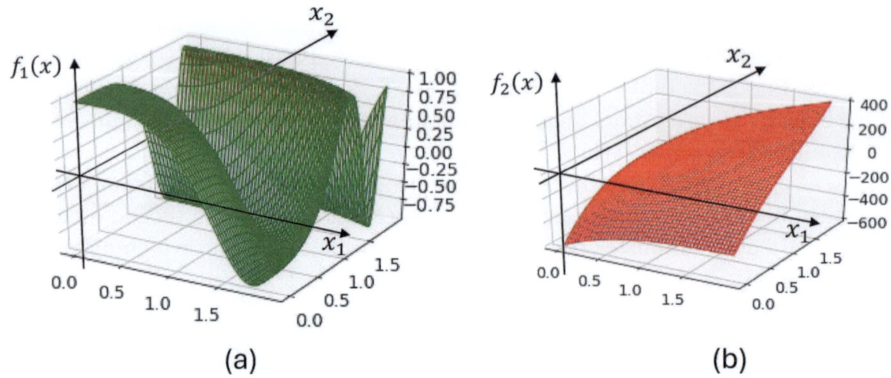

Fig. 2.15 The two objective functions of TP-III on the *x*–*f(x)* space—**a** the first objective $f_1(x)$ is the same as TP-II, whereas **b** the second objective $f_2(x)$ is enlarged by 50 times versus that of TP-II

2.3.7 Method Requirement 4: Using Goals and Minimizing Deviation Variables Instead of Objectives

- Using cDSP, designers minimize the deviation variables that measure the distance between the real achieved value of a goal and the target of the goal.

$$\min Z = \sum_{i=1}^{2} w_i \cdot \left(di^- + di^+\right) \tag{2.9}$$

- So, the Lagrange equation does not depend on decision variables X:

$$\frac{\partial L(X, \mu, \lambda)}{\partial X} \equiv 0 \tag{2.10}$$

- So, the variation of the decision variables does not affect the feasibility of a solution.
- The goals do not dominate one another in completion (or achievement, or fulfillment) due to the difference in the actual achieved value:

$$\frac{f_i(x)}{T_i} \cong \frac{f_j(x)}{T_j}, \quad \text{given that} \quad T_i^{>>} T_j \tag{2.11}$$

The first order and the second-order Lagrange equations are functions of parameters $\mathcal{P}$ and decision variables x of the decision model, and Lagrange multipliers μ, λ.

$$\boldsymbol{\nabla}_{\boldsymbol{x}} \boldsymbol{L}(\boldsymbol{x}, \boldsymbol{\mu}, \boldsymbol{\lambda}) = \boldsymbol{\mathcal{Y}}(\boldsymbol{\mathcal{P}}, \boldsymbol{x}, \boldsymbol{\mu}, \boldsymbol{\lambda}) \tag{2.12}$$

Table 2.10 Solutions to TP-III—dominated solutions of each weight are in *italics*; for each method, the best solution of each design scenario is marked using a capital letter (A′–H′, and J′)

Weight	Starting point	COBYLA		Trust-constr		SLSQP		ALP		NSGA II/III—Population (P) = 20/50	
		Solution	$\sum_{i=1}^{2} w_i \cdot f_i(x)$	Solution	$\sum_{i=1}^{2} w_i \cdot f_i(x)$	Solution	$\sum_{i=1}^{2} w_i \cdot f_i(x)$	Solution	$\sum_{i=1}^{2} w_i \cdot f_i(x)$	Solution	$\sum_{i=1}^{2} w_i \cdot f_i(x)$
(1, 0)	(0.5, 1) (0, 0) (2, 0.5)	Cannot manage nonconvex equations with bounds		All solutions violate one or more constraints		All solutions violate one or more constraints		A′ (0.51, 1.82)	1	*P = 20:* *G′ (0.55, 1.85)* P = 50: A’ (0.51, 1.82)	*P = 20:* *0.99* P = 50: 1
(0, 1)	(0.5, 1) (0, 0) (2, 0.5)							B′ (0.51, 1.96)	15.27	*P = 20:* *H′ (0.52, 1.92)* P = 50: B′ (0.51, 1.96)	*P = 20:* *14.86* P = 50: 15.36
(0.5, 0.5)	(0.5, 1) (0, 0) (2, 0.5)							*C′ (0.55, 1.82)*	*7.5*	*P = 20:* *H′ (0.52, 1.92)* P = 50: D′ (0.53, 1.87)	*P = 20:* *7.69* P = 50: 7.78
(0.7, 0.3)	(0.5, 1) (0, 0) (2, 0.5)							*C′ (0.55, 1.82)*	*4.9*	*P = 20:* *C′ (0.55, 1.82)* P = 50: E′ (0.53, 1.88)	*P = 20:* *4.9* P = 50: 5.01
(0.3, 0.7)	(0.5, 1) (0, 0) (2, 0.5)							*C′ (0.55, 1.82)*	*10.01*	*P = 20:* *J′ (0.54, 1.85)* P = 50: F′ (0.53, 1.89)	*P = 20:* *10.49* P = 50: 10.6

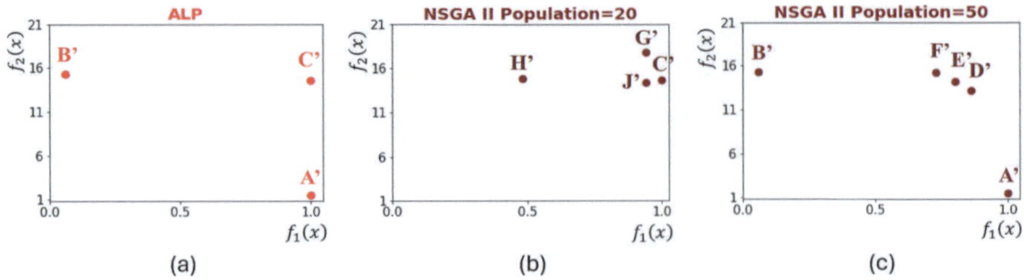

Fig. 2.16 The solution points to TP-III on the objective space using two algorithms—solutions returned by NSGA II are closer to the nondominated solution and more diverse but sensitive to parameter setting and require higher computational power

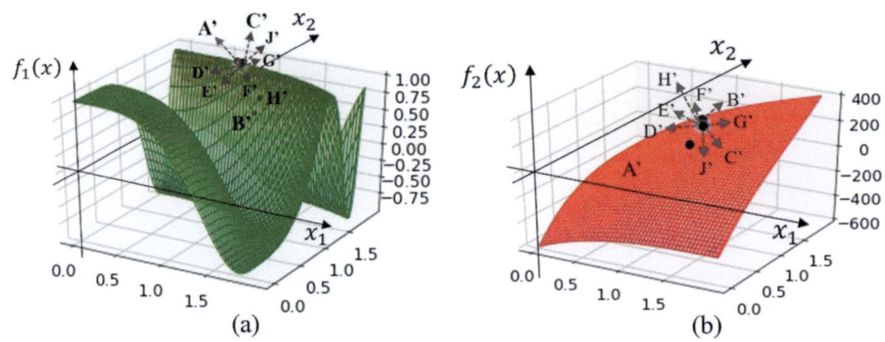

Fig. 2.17 The solution points to TP-III on the X–$f(X)$ plane—little performance differences between ALP and NSGA II

$$\boldsymbol{\nabla}^2_{xx}\boldsymbol{L}(\boldsymbol{x},\boldsymbol{\mu},\boldsymbol{\lambda}) = \boldsymbol{\nabla}_x\boldsymbol{\mathcal{Y}}(\boldsymbol{\mathcal{P}},\boldsymbol{x},\boldsymbol{\mu},\boldsymbol{\lambda}) = \boldsymbol{\mathcal{Y}}'(\boldsymbol{\mathcal{P}}',\boldsymbol{x}) \tag{2.13}$$

Hence, whether a locally optimal solution, x^*, can be maintained as an optimal solution (given that the model may be incomplete, and something may change but may not be captured in the decision model) depends on the balance of the two equations and such a balance depends on the value and stability of the variables and coefficients $\{x, \mathcal{P}, \mu, \lambda\}$. Suppose that mathematically, we identify a locally optimal solution, x^*, to an engineering-design problem; however, if any uncertainty occurs in any parameter or any unparameterized factors that break the balance of Eqs. (2.12) and (2.13), as Eqs. (2.14) and (2.15), then we lose, x^*, as the solution. This often happens in engineering problems because models are always incomplete and inaccurate and have different fidelity levels. Suppose that for an N-dimension, Q-parameter problem, the probability of uncertainty for a parameter or an equation is P. The probability of such uncertainty violating the KKT conditions is p. The probability of a locally optimal solution x^* to stay optimal under uncertainty is $\mathbb{P}(x^*|P)$, as Eq. (2.16) shows, assuming the superposition and mutual influence do not cause any greater uncertainty. However, Eq. (2.16) only represents the parameterizable uncertainties, so we use "almost equal to" instead of "equal to." The actual $\mathbb{P}(x^*|P)$, considering the mutual influence among parameterizable uncertainties and unparameterizable uncertainties, can be even larger.

$$\nabla_x L(x^*, \tilde{\mu}, \tilde{\lambda}) \neq \mathcal{Y}(\tilde{\mathcal{P}}, \tilde{x}^*, \tilde{\mu}, \tilde{\lambda}) \tag{2.14}$$

$$\nabla^2_{xx} L(x^*, \tilde{\mu}, \tilde{\lambda}) \neq \nabla_x \mathcal{Y}(\tilde{x}^*, \tilde{\mu}, \tilde{\lambda}) \tag{2.15}$$

$$\mathbb{P}(x^*|P) \approx \prod_{q=1}^{Q}[1 - p(\tilde{\mathcal{P}}_q|P)] \prod_{n=1}^{N}[1 - P(\tilde{x}_n|P)] \prod_{i=1}^{m}[1 - P(\tilde{\mu}_i|P)] \prod_{j=1}^{\ell}[1 - P(\tilde{\lambda}_j|P)] \tag{2.16}$$

To avoid "no solution" caused by the strong convexity of the objective or to avoid losing a solution due to uncertainties that break the KKT conditions, we turn to *satisficing* solutions, x^s, instead of optimal solutions, x^*. We obtain *satisficing* solutions by using goals instead of objectives. A target value, t, is assigned to the right-hand side of the objective. The objective, $f(x)$, becomes a goal, $G(x)$, whose position is fixed in the feasible solution space. In $\mathcal{F}$,

$$\mathcal{F} = \left\{x \in \mathbb{R}^n | g_i(x) \geq 0, i = 1, \ldots m, h_j(x) = 0, j = 1, \ldots \ell\right\} \tag{2.17}$$

$\mathcal{F}$ is the feasible space bounded by all active constraints and bounds, in which, a point, or several points, or an area, that is/are on the goal set $G(x)$ or closest to it (using Euclidean distance in this chapter) are the *satisficing* solution(s), x^s, see Eqs. (2.18) and (2.19) and Fig. 2.18. Using deviation variables, $d = (d^-, d^+)$, to measure the under-achievement and over-achievement of a goal versus its target and minimizing the deviation variables, the goal $G(x)$ becomes an equality constraint to be satisfied, the deviation variables, d, become decision variables that form the new objective function $z(d)$. The original decision variables x become auxiliary variables that do not appear in the objective.

$$G(x) : f(x) + d^- - d^+ = t, \quad \text{where } 0 \leq d^-, d^+ \leq 1, d^- \cdot d^+ = 0 \tag{2.18}$$

$$\text{Minimize: } z(d) \tag{2.19}$$

Such a construct is the compromise Decision Support Problem (cDSP) (Mistree et al., 1993).

Given: $\mathcal{P}, t$

Find: x, d

Satisfy:

$$x \in \mathcal{F} : \{g(x) \geq 0, h(x) = 0, \textit{ lower bound} \leq x \leq \textit{ upper bound}\},$$

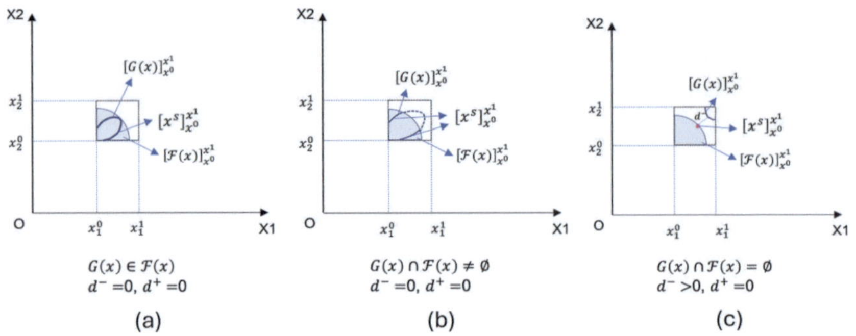

Fig. 2.18 Satisficing solutions in different cases—**a** the goal is fully incorporated in the feasible region, **b** the goal is partially incorporated in the feasible region, and **c** the goal has no intersection with the feasible region

$$G(x), 0 \leq d^-, d^+ \leq 1, d^- \cdot d^+ = 0$$

Minimize: $z(d)$.

Using the cDSP construct, the nonlinear problem is linearized using the Adaptive Linear Programming (ALP) algorithm (Hajihashemi, 2023; Mistree et al., 1993) so that the linearized problem can be solved using the Simplex algorithm and a vertex solution is obtained. The benefits of using a vertex solution versus an interior point solution are (1) a vertex solution is a good enough solution that guarantees the closest distance to the target, and (2) it does not require computing power to search for better interior point solutions.

There are other benefits of using cDSP to manage nonlinear, multi-goal problems, such as the capacity to identify *satisficing* solutions that are relatively insensitive to uncertainties. Suppose there are parameters $\boldsymbol{p}$ and deviation variables d to formulate the merit function $\boldsymbol{z(p, d)}$. Therefore, we identify *satisficing* solution x^s that obtain the minimal $\boldsymbol{z(p, d)}$. x^s is relatively insensitive to uncertainties because the probability of any uncertainty breaking the KKT conditions of a cDSP is much less (than that of an optimization problem without a right-hand side of the objective), as in Eqs. (2.20)–(2.23). Usually, the combination form of d is linear because, for typical engineering problems, we do not expect a merit function that is too complex unless any domain knowledge on goal(s) requires the designer to do so. Hence, the first-order Lagrange equation only consists of $\boldsymbol{p}$, and the second-order Lagrange equation constantly equals zero. So, this formulation significantly increases $\mathbb{P}(x^s|P)$, the probability of maintaining x^s as a *satisficing* solution under uncertainties.

$$\begin{aligned}\boldsymbol{\nabla_d L(x^s, d, \mu, \lambda) = \nabla_d z(d) + \textstyle\sum_{i=1}^{m} \mu_i \nabla_d g_i(x^s) - \sum_{j=1}^{\ell} \lambda_j \nabla_d h_j(x^s) - \lambda_{\ell+1} \nabla_d G(x^s, d) =}\\ \boldsymbol{\nabla_d z(d) + 0 - 0 \pm 1 = \nabla_d z(d) \pm 1 = y(p) \pm 1}\end{aligned} \tag{2.20}$$

$$\nabla_{dd}^2 L\left(x^s, d, \mu, \lambda\right) = \nabla_{dd}^2 z(d) \equiv 0 \tag{2.21}$$

$$\mathbb{P}(\boldsymbol{x^s}|\boldsymbol{P}) \approx \prod_{r=1}^{R}[\mathbf{1} - \boldsymbol{p}(\widetilde{\mathcal{P}}_r|\boldsymbol{P})] \tag{2.22}$$

$$\mathbb{P}(x^*|P) \ll \mathbb{P}(x^s|P) \tag{2.23}$$

For an n-dimension, K-goal problem, by adding deviation variables, we increase the dimensionality of a design problem, from $[x_1, x_2, \ldots x_n]^T$ to $[x_1, x_2, \ldots x_n, d_1^-, d_1^+, d_2^-, d_2^+, \ldots d_K^-, d_K^+]^T$, thus making it possible to absorb the risk of uncertainty that breaks the second-order sufficient conditions. This results in a robust solution that is relatively insensitive to uncertainties. We decrease the dimensionality by returning solutions consisting only of the original decision variables, $x = [x_1, x_2, \ldots x_n]^T$. Such "dimension expansion and reduction" ensures a solution that is relatively insensitive to the uncertainties embodied in the modeling of an optimization problem. Therefore, we use compromise Decision Support Problems (cDSP) to formulate many-goal problems. See Mistree et al. (1994) for additional information.

For a many-goal problem, the formulation of the merit function reflects how the goals are managed, especially their priorities. The ways of combining the goals impact the design performance. Hereafter, we focus on exploring ways to combine the goals in detail in Chap. 6.

2.3.8 Toy Problem IV (TP-IV)

- ***Problem with multiple, nonlinear, and non-convex objectives (goals), the objectives (goals) are with various scales, and the target of the goals with various levels of achievability (the aspiration of one goal is not in feasible space)***

To determine whether the formulation construct, the solution algorithm, or both make the solutions different, we formulate the problem into a cDSP and use the optimizing and the satisficing algorithms to solve the cDSP and compare the results. If the results are the same, then the formulation format (objective versus goal) leads to the difference; otherwise, the solution algorithm leads to the difference.

In TP-IV, one more complexity is managed—the goals' targets have various levels of achievability, which often occurs in engineering-design problems. Hence, the aspiration of one goal is not feasible. This means the upper limit of some deviation variables (which is 1) must be violated.

The cDSP formulation, visualization, solutions, and visualization of solutions on the objective space and X-Plane of TP-IV are given in Table 2.11, Fig. 2.19, Table 2.12, Figs. 2.20, and 2.21, respectively.

Observation: cDSP and ALP are designed to formulate and explore engineering-design problems with various abilities to complete the goals:

Table 2.11 The compromise DSP of TP-IV

TP	The compromise DSP
IV	Given
	$x_1, x_2, d_1{}^{\mp}, d_2{}^{\mp}$
	$f_1(x) = \cos\left(x_1{}^2 + x_2{}^3\right)$
	$f_2(x) = 25 \cdot (x_1 - 2)^3 + 50 \cdot (x_2 - 2)^3 + 50 \cdot x_1 \cdot x_2{}^2$
	Find
	$x_1, x_2, d_1{}^{\mp}, d_2{}^{\mp}$
	Satisfy
	Goals:
	$\frac{f_1(x)}{1.2} + d_1{}^- - d_1{}^+ = 1$
	$\frac{f_2(x)}{400} + d_2{}^- - d_2{}^+ = 5$
	Constraints:
	$x_1 \cdot x_2 \le 1$
	$d_i{}^- \cdot d_i{}^+ = 0, i = 1, 2$
	Bounds:
	$0 \le x_1, x_2 \le 2$
	$0 \le d_1{}^{\mp}, d_2{}^{\mp} \le 1$
	Minimize
	$Z = \sum_{i=1}^{2} w_i \cdot (d_i{}^- + d_i{}^+)$

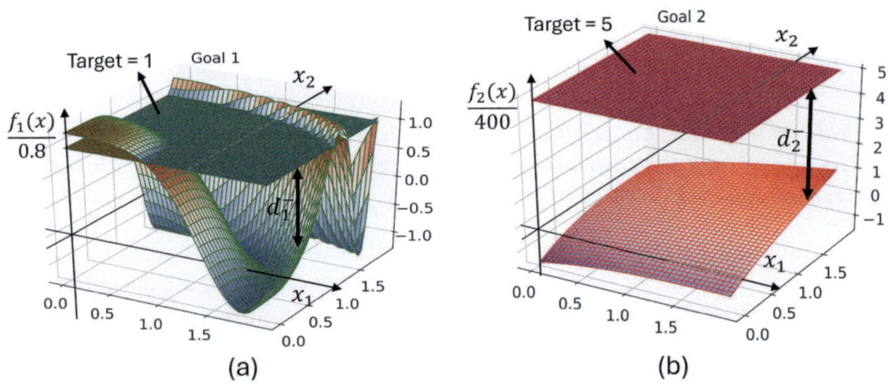

Fig. 2.19 The left-hand side (objective function) and the right-hand side (target) of the two goals of TP-IV on the $X\text{–}f(X)$ space

$$\frac{f_i(x)}{T_i} \gg \frac{f_j(x)}{T_j} \tag{2.24}$$

Using the ALP, good enough solutions (comparing with the solutions from NSGA II) can be obtained, whereas using optimizing algorithms, no feasible solutions are returned.

Table 2.12 Solutions to TP-IV (close-to-nondominated solutions or good-enough solutions)

Weight	Starting point	COBYLA		Trust-constr		SLSQP		ALP		NSGA II—population (P) = 50	
		Solution	$\sum_{i=1}^{2} w_i \cdot f_i(x)$	Solution	$\sum_{i=1}^{2} w_i \cdot f_i(x)$	Solution	$\sum_{i=1}^{2} w_i \cdot f_i(x)$	Solution	$\sum_{i=1}^{2} w_i \cdot f_i(x)$	Solution	$\sum_{i=1}^{2} w_i \cdot f_i(x)$
(1, 0)	(0.5, 1)	Cannot manage nonconvex equations with bounds		All solutions violate one or more constraints		All solutions violate one or more constraints		*A″ (0.53, 1.75)*	*0.8*	D″ (0.51, 1.88)	0.81
	(0, 0)										
	(2, 0.5)										
(0, 1)	(0.5, 1)							B″ (0.51, 1.97)	16.26	*E″ (0.51, 1.96)*	*15.26*
	(0, 0)										
	(2, 0.5)										
(0.5, 0.5)	(0.5, 1)							C″ (0.57, 1.74)	7.29	*F″ (0.53, 1.88)*	*7.21*
	(0, 0)										
	(2, 0.5)										
(0.7, 0.3)	(0.5, 1)							C″ (0.57, 1.74)	4.69	*G″ (0.57, 1.75)*	*4.6*
	(0, 0)										
	(2, 0.5)										
(0.3, 0.7)	(0.5, 1)							*C″ (0.57, 1.74)*	*9.88*	F″ (0.53, 1.88)	10.54
	(0, 0)										
	(2, 0.5)										

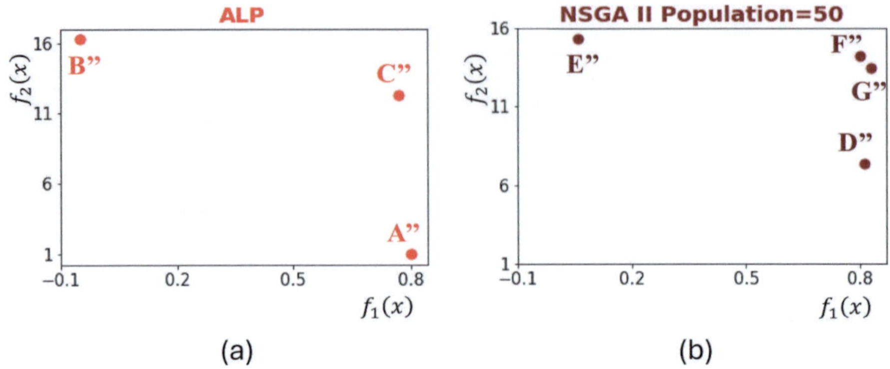

Fig. 2.20 The solution points to TP-IV on the objective space using two algorithms—NSGA II finds more nondominated solutions, whereas ALP finds solutions close to nondominated solutions but with better weighted combined goal-achieved value

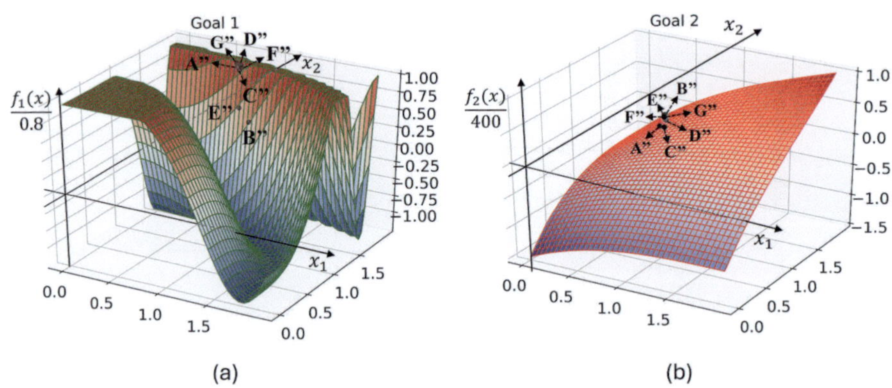

Fig. 2.21 The solution points to TP-IV on the $X{-}f(X)$ space—little performance differences between ALP and NSGA II

In Sect. 2.3.9, we address “*Why does using the ALP allow designers to get good enough solutions to a cDSP with various abilities to complete the goals, but using optimizing algorithms (other than NSGA II) does not?*”

2.3.9 Method Requirement 5: Allowing Some Violations of Soft Requirements, Such as the Bounds of Deviation Variables

- *Why does using ALP allow designers to get good enough solutions to a cDSP with various completions of the goals, but using optimizing algorithms (other than NSGA II) does not?*
  - When $\frac{f_i(x)}{T_i} \frac{"f_j(x)}{T_j}$, especially when d_j^- cannot meet its upper limit, that is, this bound must be violated:

$$d_j^- \leq 1 \tag{2.25}$$

  - Since optimization algorithms (such as Trust-constr and SLSQP) treat all constraints and bounds priorities equally, if at least one of the constraints and bounds must be violated, the problem is considered infeasible. No feasible solutions can be returned using optimizing algorithms.
  - In contrast to optimizing algorithms, we prioritize satisfying the constraints and the bounds on system variables when we use the ALP. Next, we give secondary priority to satisfying the bounds on deviation variables. Suppose the violation of any deviation bound allows a point X^S in the feasible area bounded by the constraints and the bounds of system variables to be found. In that case, X^S is returned as a solution. So, for an *n*-dimension, *m*-constraint (with *p* inequality constraints and *m-p* equality constraints), *k*-goal cDSP:

X^S is a *satisficing* solution,

if and only if

$$\frac{f_i(X^S)}{T_i} + d_i^- - d_i^+ = 1, \ \ \forall i = 1, \ldots k \ \ \text{//Goal functions hold at } X^S$$

and

$$g_i(X^S) \geq 0, \ \forall i = 1, \ldots p \ \text{//Inequality constraints are satisfied at } X^S$$

and

$$g_i(X^S) \geq 0, \ \forall i = p+1, \ldots m \ \text{//Equality constraints are satisfied at } X^S$$

and

$$\textit{lower bound} \leq X^S \leq \textit{upper bound}$$

and

$min(lower\ bound - d_i^{\mp}),\ \forall i = 1, \ldots k$ //Minimize the violation of deviation bunds

and

$min\left(d_i^{\mp} - upper\ bound\right),\ \forall i = 1, \ldots, k$ // Minimize the violation of deviation bunds.

2.3.10 Toy Problem V (TP-V)

- ***Problem with three nonlinear objectives (goals), non-convex functions, the objectives (goals) are with various scales, and the target of the goals with various levels of achievability***.

Problem statement of TP-V: In the late 1800s, Ringling Bros and Barnum and Baily Circus sought to establish dimensions of a new pedestal for their circus elephant Jumbo; see Fig. 2.22. They would play a trick that involved a support pedestal where Jumbo would perform a one-legged handstand. The cost of manufacturing must be minimized, depending on its thickness, width, and the amount of material it will consume. Moreover, it must be as tall as possible for a wow factor, and the pedestal must be wide enough to ensure Jumbo has enough room to stand safely on one foot. This means the goals of the design are to minimize the manufacturing cost, maximize the height, and maximize the outer radius. A material of 2024 Aluminum with a modulus of 10,600 ksi and a density of 0.1 $lb/(in)^3$ has been selected. Jumbo's foot is approximately 25″ in diameter, so the pedestal must also be >25″. Jumbo weighs 13,560 lb and stands 13.5ft tall. Use a factor of safety of 1.5.

Given a certain type of material, design a cylinder (the "elephant stand") as shown in Fig. 2.23. The cylinder has two parts joined together. The upper half is a tube. The designer's interest is determining the thickness, radius, and height that best satisfies the identified goals. The lower half is a 4-inch-height solid base. The goals identified by the designer include minimizing the manufacturing cost, maximizing the height, and maximizing the outer diameter. Requirements include: the upper and lower limits of the parameters the cylinder can physically reach.

The word and mathematical formulations of the cDSP of TP-V are in Table 2.13. There are intermediate variables in TP-V that make the problem a high-dimensional problem hard to visualize using a three-dimensional coordinate system, so we skip the visualization and show the solutions in Table 2.14. Because the optimizing algorithms other than NSGA II cannot deal with TP-V due to its features (three nonlinear goals, non-convex functions, the goals with various scales, and the target of the goals with various levels of achievability), we only show the results by using ALP and NSGA II. Since both ALP and NSGA II are insensitive to the starting points, we do not try

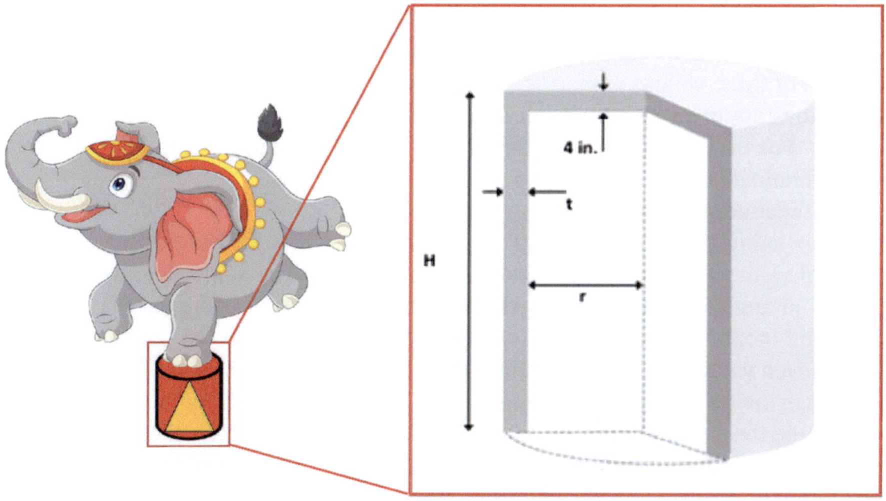

Fig. 2.22 The functionality of an elephant stand (TP-V)

Fig. 2.23 The dimension of an elephant stand (TP-V)

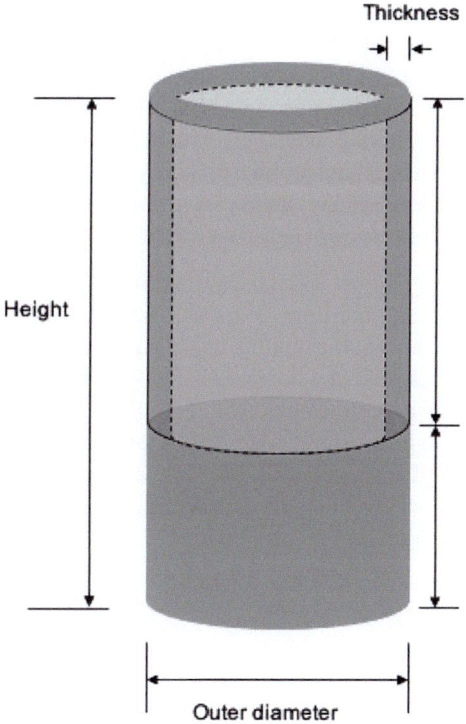

different starting points for TP-V. We use 13 weight scenarios to explore the goals' different priorities. We only show the goal achieved value $f_i(x)$ to simplify the table format. For three weight scenarios, (1, 0, 0), (0, 1, 0), and (0.2, 0.6, 0.2), the NSGA II solutions dominate the ALP solutions; that is, for every goal, the achieved value is better. For the other ten weight scenarios, NSGA II solutions and ALP solutions do not dominate each other.

We separately visualize each goal's achieved values under thirteen weight scenarios using box charts (Fig. 2.24). In that case, the distribution of the goal-achieved values from using ALP and NSGA II varies. The achieved value of Goal 1 (Minimizing Cost) and Goal 2 (Maximizing Height) by using ALP is slightly better and more stable than that of using NSGA II; see Fig. 2.24a and b. However, the achieved value of Goal 3 (Maximizing Outer Radius) by using NSGA II is much better than that of using ALP; see Fig. 2.24c. Therefore, even using the same weights to combine the goals, solutions, and goal-achievement values are different.

We visualize the solution points on the "deviation space" from different angles in Fig. 2.25. The darker the color of the dot, the closer it is to the viewer. The three axes are d_1^+, d_2^-, and d_3^-. The ideal solution point (or the Utopian point) not in the feasible space is the origin O (0, 0, 0) of this 3D coordinate system because all three goals reach the target at this point. So, the solution points close to O are more desirable because they have better goal-achieved values. The round dots are solutions using the ALP. The triangle dots are solutions using NSGA II. From Fig. 2.25, we cannot conclude that one algorithm performs better than the other, but for Goal 3, NSGA II solutions are much closer to O. However, the solutions with low d_3^- have much higher d_1^+, which means an extremely high cost is required to obtain a better outer diameter of the elephant stand. So, using NSGA II allows designers to identify solutions to better achieve Goal 3 by sacrificing Goal 1.

Observation: For an engineering-design problem with three nonlinear goals, non-convex functions, goals with various scales, and the target of the goals with various levels of achievability, using ALP allows designers to get good enough solutions that are close to the nondominated solutions obtained by using NSGA II. However, as we give the drawbacks of using NSGA II earlier in this section, designers cannot obtain insight or knowledge on the problem formulation and improvement, we conclude that the formulation of cDSP and the use of ALP can give *satisficing* solutions and knowledge on problem formulation and system behavior within a relatively low computational complexity.

Table 2.13 The word form and math form of the compromise DSP of TP-V

TP-V	The word form	The math form
	Given System parameters Material—2014 Aluminum The elastic modulus for the material Safety factor Yield stress for the material Density of the material Load (elephant's weight) Moment of inertia for the cylindrical section Maximum normal stress Maximum buckling stress for a fixed free column Cost target Height Target OR target (the sum of the radius and the thickness) **Find** System variables Radius Thickness Height Deviation variables **Satisfy** System constraints Reaching the minimum outer diameter Not exceeding a certain height-to-width ratio Reaching the stress requirement Not exceeding the maximum weight Not exceeding the maximum load in the stand System goals Goal 1: Reaching minimum cost target Goal 2: Reaching the maximum height target Goal 3: Reaching the maximum outer radius target Bounds The upper and lower limits of Radius Thickness Height **Minimize** The deviation function	**Given** $E = 10{,}600{,}000$ $OR = R + T$ $SF = 1.5$ $SIGY = 11{,}000$ $P = 12{,}000$ PI = 2*ACOS(0.0) $RHO = 0.1$ $I = \frac{\pi}{4} * \left[(R+T)^4 - R^4\right]$ $W1 = \pi * \left[(R+T)^2 - R^2\right] * (H-4)$ $W2 = \pi * (R+T)^2 * 4$ $W = (W1 + W2) * RHO$ $STR = \frac{P}{\pi * \left[(R+T)^2 - R^2\right]}$ $PCR = \frac{\pi^2 * E * I}{4 * H^2}$ $COST = e^{\frac{2.5}{T}} * e^{\frac{3}{R}} * W$ Cost target = 5000 Height target = 180 OR target = 15 **Find** System variables R: Radius T: Thickness H: Height Deviation variables d_i^+ : over achievement of Goal i, where i = 1,2,3 d_i^- : under-achievement of Goal i, where i = 1,2,3 **Satisfy** System constraints minOD: minimum outer diameter: $2 * R + 2 * T \geq 6$ (CO1) heiwid: height to width ratio: $R + T - 0.03 * H \geq 0$ (CO2) stress: minimum stress in stand: $\frac{SIGY}{SF} - STR \geq 0$ (CO3) weight: maximum weight: $1000 - W \geq 0$ (CO4) buckle: maximum load in stand: $\frac{PCR}{F} - P \geq 0$ (CO5) System goals Goal 1: minimum cost: $\frac{cost}{cost\ target} + d_1^- - d_1^+ = 1$ (G1) Goal 2: maximum height: $\frac{height}{height\ target} + d_2^- - d_2^+ = 1$ (G2) Goal 3: maximum outer radius: $\frac{OR}{OR\ target} + d_3^- - d_3^+ = 0$ (G3) Bounds $3 \leq R \leq 45$ $0.5 \leq T \leq 2.5$ $100 \leq H \leq 120$ **Minimize** The deviation function $Z = \sum_{i=1}^{3} w_i \cdot (di^- + di^+), \sum_{i=1}^{3} w_i = 1$

Table 2.14 Solutions to TP-V—the best solution of each scenario is highlighted, and the solution that gives a better-achieved value of a goal but is not a nondominated solution is underlined

Weight scenario	Weight			ALP			NSGA II/III—population (P) = 50		
	w_1	w_2	w_3	$\min f_1(x)$	$\max f_2(x)$	$\max f_3(x)$	$\min f_1(x)$	$\max f_2(x)$	$\max f_3(x)$
1	1	0	0	5507.8	100	4.38	***5000***	***102.56***	***4.98***
2	0	1	0	53530.4	118.8	3.59	***25,768***	***120***	***4.89***
3	0	0	1	80580.1	100	10.44	94,863	114.156	15
4	0.6	0.2	0.2	5340.06	110.6	5.09	5000	100	7.44
5	0.2	0.6	0.2	6079.2	117.5	6.08	***5892.7***	***120***	***7.05***
6	0.2	0.2	0.6	6547.6	114.7	6.82	51,197	120	15
7	0.5	0.35	0.15	6079.2	117.5	6.08	5000	100	7.44
8	0.15	0.5	0.35	6114.2	119.96	6.12	7848.7	108.15	11.29
9	0.35	0.15	0.5	5564.9	106.1	6.33	5208.9	100	8.02
10	0.7	0	0.3	5269.2	109.2	5.13	5000	100	7.43
11	0.3	0.7	0	5821.1	120	5.06	5691.7	120	5.81
12	0	0.3	0.7	94608.4	110	10.33	149,720	120	15
13	0.33	0.33	0.33	6079.2	117.5	6.08	5060.5	100	7.66

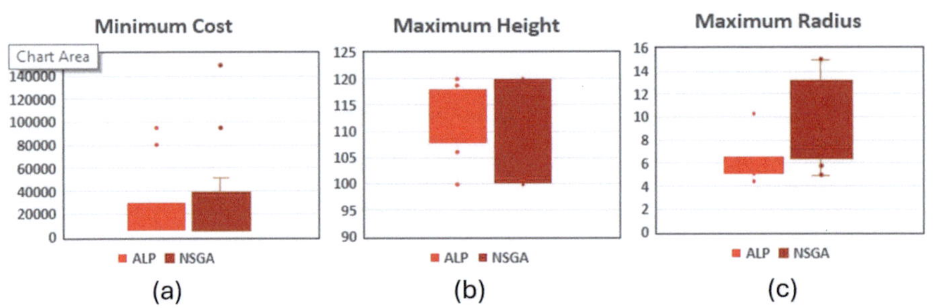

Fig. 2.24 The box chart solution points to TP-V on the objective space using two algorithms

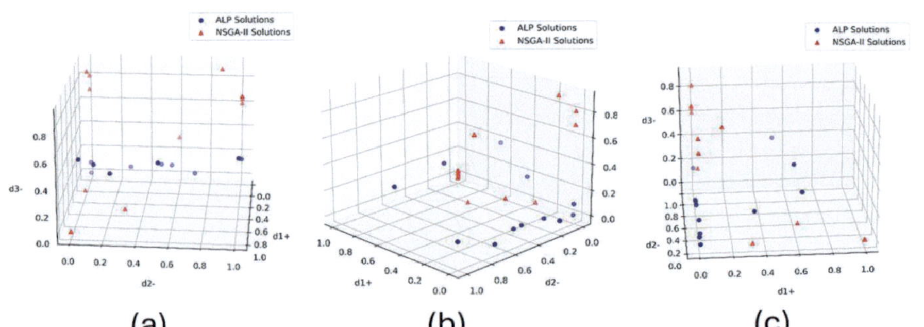

Fig. 2.25 The solution points to TP-V on the deviation space using two algorithms—round dots are solutions using the ALP, and Triangle dots are solutions using NSGA II

2.4 Summary of Differences Between Optimizing and *Satisficing* Strategy

2.4.1 Differences Between Optimizing and Satisficing Strategy

Why do cDSP, ALP, and DSIDES work for nonlinear, non-convex, multi-objective, multi-unit, and evolving-target Problems?

In summary, there are four advantages of using a *satisficing* strategy, which are realized in this monograph by using cDSP, ALP, and DSIDES. The advantages include (this is an extension of Table 2.2).

The advantage of the formulation:

- Using Goals and Minimizing Deviation Variables Instead of Objectives.
  - The benefits are that only the necessary KKT conditions are met at a solution point, whereas the sufficient KKT conditions do not have to be met.
  - Therefore, designers have a higher chance of finding a solution and a lower chance of losing a solution due to parameterizable and unparameterizable uncertainties.

The advantages of approximation:

- Using second-order sequential linearization
  - The benefit is that designers can balance linearization accuracy and computational complexity.
- Using accumulated linearization
  - The benefit is that designers can more accurately manage nonconvex problems and deal with highly convex, nonlinear problems.

The advantage of exploration:

- Combining interior-point searches and vertex searches:
  - The benefit is that designers can avoid being stuck into local optimum to some extent and identify satisficing solutions relatively insensitive to starting points changing.

The advantage in evaluation:

- We are allowing some violations of soft requirements, such as the bounds of deviation variables.
  - The benefits are that designers can manage rigid and soft requirements differently to ensure feasibility.
  - As a result, goals and constraints with different scales can be managed.

2.4.2 Summary of Differences Among cDSP, Goal Programming, and Mathematical Programming

– *Why We Choose the Compromise Decision Support Problem (cDSP) Construct?*

Based on the previous discussion—the features each method can or cannot manage in Sect. 1.22, Table 1.4, the mathematical explanation regarding meeting the KKT conditions in Sect. 2.2, and using different methods to solve five toy problems and get different solutions in Sect. 2.3, we summarize the differences among cDSP, Goal Programming, and Mathematical Programming seeking optimal solutions as follows.

Stage 1: Formulation. First, the cDSP is a hybrid of mathematical programming (seeking optimal solutions) and goal programming. In a cDSP, there are constraints and goals, while in goal programming, there are no constraints.[4] In a cDSP, the constraints are requirements (demands) that cannot be violated, whereas the goals are soft requirements (wishes) whose targets may not be reached, but we want to minimize the distance between the targets and our results. The constraints and goals can be linear or nonlinear (convex or non-convex), equality or inequalities, or both. The benefit of being able to model both demands and wishes in one formulation is attractive in design. Due to the complexity of the supply chain with mass customization, the resulting cDSP usually entails dealing with non-convex and convex constraints and goals. The algorithm for solving the cDSP is documented in Mistree et al. (1993).

Second, in a cDSP, we use deviation variables to assess how much we over-achieve or under-achieve a goal. By adding the deviation variables, we ensure that the solutions to a cDSP satisfies the necessary KKT condition. The solutions meet the sufficiency test to guarantee a "true" or "global" optimum. A *satisficing* solution to a cDSP is mapping the optimal solution to a lower-dimensional space. The reduced dimensions consist of the deviation variables $D = [d_1^-, d_1^+, d_2^-, d_2^+, \ldots d_k^-, d_k^+]^T$. By adding deviation variables, we increase the dimensionality of a design problem, from $[x_1, x_2, \ldots x_n]^T$ to $[x_1, x_2, \ldots x_n, d_1^-, d_1^+, d_2^-, d_2^+, \ldots d_k^-, d_k^+]^T$, thus making it possible to absorb the risk of uncertainty at the constraint boundary. This results in a robust solution that is relatively insensitive to uncertainties. We decrease the dimensionality by returning solutions consisting only of system variables, as in solving an optimization problem, $X = [x_1, x_2, \ldots x_n]^T$. Such "dimensionality reduction" does not result in a solution that is relatively insensitive to the uncertainties embodied in modeling the constraints of an optimization problem. See Mistree et al. (1994) for additional information.

[4] However, in later publications, the formulation of Goal Programming allows for managing constraints. By the time Mistree and the coauthors published their work on cDSP and ALP (Mistree et al., 1993), it was generally accepted that in Goal Programming, there are only "soft requirements" as goals but no "rigid requirements" as constraints.

Third, although a cDSP has similarities with the auxiliary problem of a linear programming problem when using the two-phase method,[5] there are differences. For a linear programming problem (P), when we relax the m equality constraints $A \cdot X = b$ to m inequality constraints $A \cdot X + U = b$ by adding slack variables (or artificial variables) $U = [u_1, u_2, \ldots u_m]^T$, and change the objective function from $\min C^T \cdot X$ to $\min \sum_{i=1}^{m} u_i$, an auxiliary problem (A) of the original linear programming problem (P) is created. If a solution $[x_1, x_2, \ldots x_n, u_1, u_2, \ldots u_m]^T$ is optimal for (A) with $u_i = 0, i = 1, 2, \ldots m$, then the solution $[x_1, x_2, \ldots x_n]^T$ is feasible for (P).

There are similarities between a cDSP and an auxiliary problem (A). The slack variables U in the auxiliary problem are similar to the deviation variables D in a cDSP if we treat the equality constraints $A \cdot X + U = b$ of (A) as the goals of a cDSP $\frac{Goal_i(X)}{Target_i} + d_i^- - d_i^+ = 1$. The objective of (A) is minimizing the sum of the slack variables U. Similarly, the merit function of a cDSP is minimizing the linear combination of the deviation variables D.

However, there are differences between a cDSP and an auxiliary problem (A) of a linear programming problem (P). An auxiliary problem is a linear problem, whereas a cDSP can be nonlinear—both constraints and goals. When solving an auxiliary problem (A), one can only obtain the feasibility of its original problem (P). However, a *satisficing* solution (a good enough solution) is not guaranteed because the original objective function $\min C^T \cdot X$ is not incorporated in (A). On the contrary, in a cDSP, goals are satisfied as equality constraints in a corresponding optimization problem. Thus, a *satisficing* solution close to achieving the goals is identified. In addition, in an auxiliary problem, for any constraint, we only minimize either its under-achievement or over-achievement. In contrast, in a cDSP, we minimize both under-achievement and over-achievement of each goal. Furthermore, in an auxiliary problem, we treat all constraints equally by simply adding the slack variables U. At the same time, in a cDSP, we use weights to linearly combine the deviation variables so that we may assign different priorities to each goal.

In summary, a cDSP differs from goal programming, optimization, or an auxiliary problem in linear programming. Further, a *satisficing* solution to a cDSP is feasible and good enough to achieve the goals.

Stage 2: Exploration of the solution space. In the second stage, the solution space is explored to find *satisficing* solutions associated with each design preference (scenario), in different phases in the product life cycle. Exploring the solution space (ESS) manages Type I and Type II. In this monograph, when we refer to design preferences, we particularly focus on the importance of the different goals.

Weight sensitivity analysis: exploration of the design preferences. We use weight sensitivity analysis to explore how assessing different weights of the goals affects system performance, that is, to identify satisficing solutions that are relatively insensitive to uncertainties.

---

[5] The introduction of the auxiliary problem is given at "http://www.math.uwaterloo.ca/~hwolkowi/henry/teaching/f05/350.f05/L18.pdf".

System capacity analysis—identification and management of the sensitive segment and bottleneck. To overcome the capacity limitation of constraints or bounds, we propose system capacity analysis to identify the sensitive segment(s) and bottleneck(s). If an inequality constraint has zero or tiny surplus or slack compared with its right-hand-side value, we define it as an active constraint. The solution is on or close to the boundary of the active constraint, so the solution is sensitive to the uncertainty of the active constraint. By relaxing this active constraint, suppose the shadow price of an active constraint is lower than other active constraints. In that case, we may not get much improvement in achieving the goals, and we define such a constraint as a "sensitive segment." We then move the solution away from the sensitive segment by restricting the active constraint, that is, by adding a buffer to the constraint to prevent the solution from reaching the boundary defined by the constraint. If the shadow price of an active constraint has the largest value compared to other constraints, relaxing the constraint can result in the greatest improvement in achieving the goals. We define such an active constraint as a "bottleneck." Also, finding ways to relax the physical system bottleneck to boost its potential is important. Once there is no longer the potential of physically relaxing the constraint, we move the solution away from the newly relaxed boundary by restricting the constraint by adding a buffer to the defined boundary. Thus, we balance the need for robustness of the solution with our desire to obtain the best satisficing solution.

2.5 Research Questions (RQ1–RQ4)

In this section, we pose the primary research question based on the design construct we choose as the model evolution construct, the cDSP, and the different types of uncertainty we need to manage through model evolution. First, what are the different types of uncertainty? How can we justify the primary research question concerning managing the four types of uncertainty or realizing the four types of robust design?

2.5.1 Justification of the Primary Research Question Regarding Requirements

– ***Four Types of Robust Design***

There are four types of robust design (Allen et al., 2006; Nellippallil et al., 2018, 2020) for managing the four types of uncertainty in engineering design. The first three types of robust design are illustrated conceptually in Fig. 2.26. Type IV uncertainty combines Types I–III, so it is not represented as an individual area.

In **Type I** robust design, design variable values are identified to satisfy performance requirement targets regardless of noise factors, which are not under a designer's control.

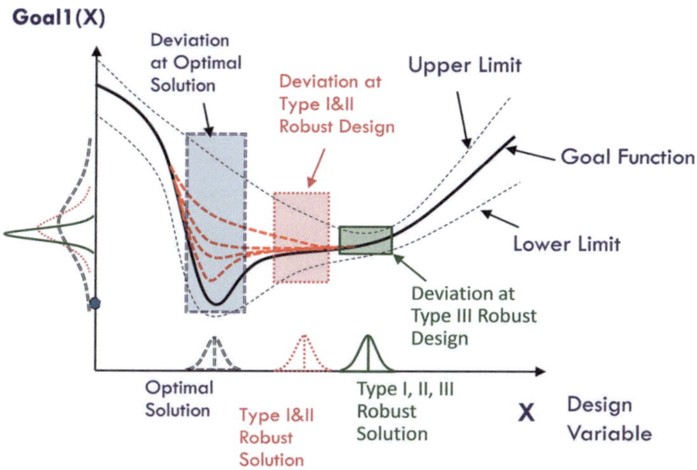

Fig. 2.26 Four types of robust solution

In **Type II** robust design, design variable values are determined that satisfy a set of performance requirement targets regardless of anticipated variations in those *design variables*.

In **Type III** robust design, variable values are determined which satisfy a set of performance requirements regardless of variations in the *mathematical models* used to describe that performance.

In **Type IV (Combination of Types I–III)** robust design variable values are determined, which satisfy a set of performance requirements despite variability introduced by a *hierarchical, multiscale or multidisciplinary formulation of the product*.

Based on the requirement of managing four types of uncertainty in engineering design, we justify the primary research question into three sub-research questions in Table 2.15.

- ***"Primary Research Question—How can designers realize model evolution using a satisficing strategy to manage chaos in the physical world, reduce the risk of losing an optimal solution, and discover domain-independent knowledge to update metaheuristics?"***

The research questions regarding the realization of Type I and Type II Robust Design are marked as RDI-II. The research questions regarding the realization of Type III and Type IV Robust Design are marked as RDIII and RDIV, respectively. Hypotheses are proposed in Chap. 3.

Table 2.15 Justified research questions regarding four types of robust design

Chapter	Ch1	Ch2	Ch3	Ch4	Ch5	Ch6	Ch7	Ch8	Ch9
Actions	RGH	*RDI-II: What mechanisms and procedures enable the exploration of the solutions relatively insensitive to Type I and Type II uncertainty?*	TVe M	EsVe SQT AQ	EVe SQT AQ			CQ EVa	TE
		RDIII: What mathematics in the design method allows exploring solutions that are relatively insensitive to Type III uncertainty?	TVe M		EVe SQT AQ	EVe SQT AQ			
		RDIV: What method allows sensing and managing Type IV uncertainty?	TVe M			EVe SQT AQ	EVe SQT AQ		
Nomenclature	AQ—Answer research questions CQ—Closure of the answers to research questions EVa—Empirically validate hypotheses EVe—Empirically verify hypotheses H—Give hypotheses M—Introduce methods RD—Robust design RG—Give research gaps SQT—Specify research questions in the context of test problems TE—Theoretically extend the research TVe—Theoretically verify hypotheses								

2.5.2 *Justified Research Questions Regarding Tasks*

– ***In the Context of Four Stages of the Design Loop***

Establishing connections among the four stages of the model evolution cycle—formulation, approximation, exploration, and exploration—allows information to pass through different stages, resulting in robust design.

In this monograph, we identify four connections among the stages (see Fig. 2.27): connections between formulation and exploration, connections among approximation, exploration, and evaluation, connections among formulation, exploration, and evaluation, and connections among all four stages. We explain what is expected to happen or the potential contribution of the connections hereafter.

Formulation-exploration: Through exploring the solution space, designers can identify the segments in the model that can be improved to make solutions more insensitive to uncertainties. Based on such findings, designers can improve the model formulation.

Approximation-exploration-evaluation: The solution quality is learned and defined through exploration and evaluation, and its association with the metaheuristics used in the approximation can be identified and quantified. Based on this, designers can update the metaheuristics in approximation and realize a design that is relatively insensitive to approximation uncertainties.

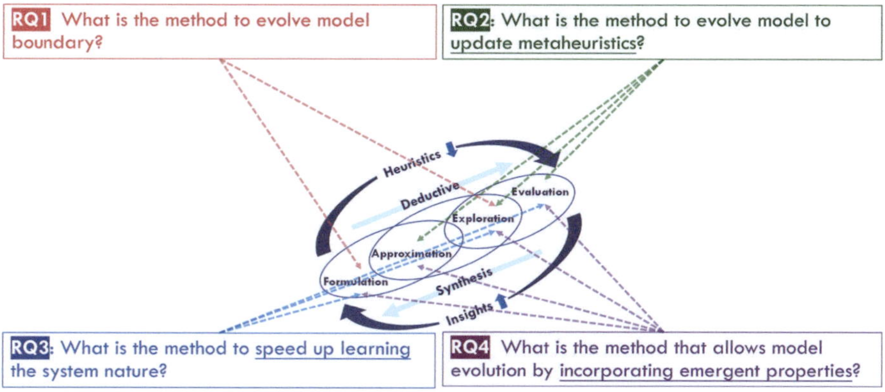

Fig. 2.27 Justified research questions RQ1–RQ4 and their connections with the design loop

Formulation-exploration-evaluation: Through exploration and evaluation, designers learn the quality of the solution associated with different assumptions and rules in the formulation. Based on this, designers can update the formulation rules and make the design relatively insensitive to the formulation's assumptions, errors, uncertainties, and inaccuracies. As the process continues, designers can speed up learning by selecting the most representative scenarios in the formulation.

Formulation-approximation-exploration-evaluation: By connecting all four stages, designers can identify the system's emergent properties and incorporate them into the model by adjusting any stage.

Model evolution includes but is not limited to establishing the connections above, based on which we justify the primary research question into four research questions regarding establishing connections among stages in engineering design in Table 2.16.

The four research questions regarding establishing the four multi-stage connections are marked as RQ1–RQ4, respectively. Hypotheses are proposed in Chap. 3.

2.6 Specification of Hypotheses (H1–H4)

In Sect. 1.1, to fill the research gaps, we hypothesize that ***designers can improve their decision models iteratively by connecting the multiple stages of design and passing information through them, designers can improve their decision models in iterations***.

Given the four tasks and four research questions posed in Sect. 2.5, the hypotheses are specified into four hypotheses, H1–H4, shown in Table 2.17, by testifying which, the research questions can be answered.

In Fig. 2.28, the research questions RQ1–RQ4 and the specified hypotheses H1–H4 are visualized as the connections of different stages of the design cycle.

Table 2.16 Connection between Research Questions (RQs) and Chapters (Ch)

Chapter	Ch1	Ch2	Ch3	Ch4	Ch5	Ch6	Ch7	Ch8	Ch9
Actions	RGH	*RQ1: What is a method to evolve model boundary?*	TVe M	EVe SQT AQ				CQ EVa	TE
		RQ2: What is the method to evolve the model to update metaheuristics?	TVe M		EVe SQT AQ				
		RQ3: What is the method to speed up learning the system's nature?	TVe M			EVe SQT AQ			
		RQ4: What method allows information to be passed through multiple system scales?	TVe M				EVe SQT AQ		
Nomenclature	AQ—Answer research questions CQ—Closure of the answers to research questions EVa—Empirically validate hypotheses EVe—Empirically verify hypotheses H—Give hypotheses M—Introduce methods RG—Give research gaps RQ—Research question SQT—Specify research questions in the context of test problems TE—Theoretically extend the research TVe—Theoretically verify hypotheses								

2.7 Role of This Chapter in This Monograph

In this chapter, we refine the primary research question to make it answerable. We address the how question: How can we realize model evolution using a satisficing strategy? We explain why we select a satisficing strategy, namely cDSP, ALP, and DSIDES, as our foundational approach for model evolution tasks. Next, we translate the primary research question into four research questions in two contexts: robust design (RDI II, RD III, RD IV) and model evolution (RQ1–RQ4). We also state the four hypotheses (H1–H4) that will guide our investigation (see Table 2.18).

Table 2.17 Specification of hypotheses for answering the research questions

Chapter	Ch1	Ch2		Ch3	Ch4	Ch5	Ch6	Ch7	Ch8	Ch9
Actions	RGH	RQ1: What is the method used to evolve the model boundary?	*H1: Explore the sensitivity of the segments of the model boundary and improve accordingly*	TVe M	EVe SQT AQ				CQ EVa	TE
		RQ2: What is the method to evolve the model to update metaheuristics?	*H2: Learn, evaluate, and update metaheuristics to improve model performance*	TVe M		EVe SQT AQ				
		RQ3: What is the method to speed up learning the system's nature?	*H3: Learn the system's nature, such as the interrelationship among subsystems, and reorganize them based on it*	TVe M			EVe SQT AQ			
		RQ4: What method allows information to be passed through multiple system scales?	*H4: Capture and quantify emergent properties through scenario planning in simulation*	TVe M				EVe SQT AQ		
Nomenclature	AQ—Answer research questions CQ—Closure of the answers to research questions EVa—Empirically validate hypotheses EVe—Empirically verify hypotheses H—Give hypotheses M—Introduce methods RG—Give research gaps RQ—Research question SH—Specify hypotheses SQT—Specify research questions in the context of test problems TE—Theoretically extend the research TVe—Theoretically verify hypotheses									

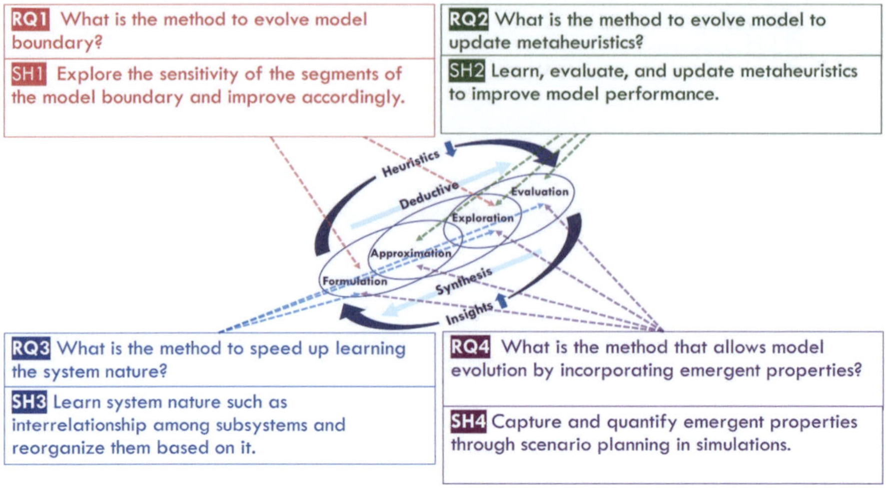

Fig. 2.28 Research Questions RQ1–RQ4 and Hypotheses H1–H4

Finally, we outline how subsequent chapters will proceed. We will present hypotheses to answer the research questions, specify research questions in the context of test problems, use those test problems to answer the research questions, and then extend the research questions in the Way Forward chapter. This completes the foundation pillar of the Design Evolution Cycle (see Fig. 2.29).

In Chap. 3, we verify the feasibility of the hypotheses through theoretical analysis, and we present an overview of the proposed methods and algorithms. In Chaps. 4, 5, 6, and 7, we specify the research questions for each domain, and we use test problems to prove the hypotheses and verify the proposed methods and algorithms. In Chap. 8, we summarize the answers to the research questions, and we provide the closure. In Chap. 9, we explain our research way forward, and we extend the research questions into the new domain.

Table 2.18 Plan for addressing the research questions in each chapter

Chapter	Ch1	Ch2			Ch3	Ch4	Ch5	Ch6	Ch7	Ch8	Ch9
Actions	RGH	RD I–II: What mechanisms and procedures enable the exploration of the solutions relatively insensitive to Type I and Type II uncertainty?	RQ1: What is the method used to evolve the model boundary?	H1: Explore the sensitivity of the segments of the model boundary and improve accordingly	TVe M	EVe SQT AQ				CQ EVa	TE
			RQ2: What is the method to evolve the model to update metaheuristics?	H2: Explore the sensitivity of the segments of the model boundary and improve accordingly	TVe M		EVe SQT AQ				
		RD III: What mathematics in the design method allows exploring solutions that are relatively insensitive to Type III uncertainty?	RQ3: What is the method to speed up learning the system nature?	H3: Learn the system's nature, such as the interrelationship among subsystems, and reorganize them based on it	TVe M			EVe SQT AQ			
		RD IV: What method allows sensing and managing Type IV uncertainty?	RQ4: What method allows information to be passed through multiple system scales?	H4: Capture and quantify emergent properties through scenario planning in simulation	TVe M				EVe SQT AQ		

(continued)

Table 2.18 (continued)

Chapter	Ch1	Ch2	Ch3	Ch4	Ch5	Ch6	Ch7	Ch8	Ch9
Nomenclature	AQ—Answer research questions CQ—Closure of the answers to research questions EVa—Empirically validate hypotheses EVe—Empirically verify hypotheses H—Give hypotheses M—Introduce methods RD—Robust design RG—Give research gaps RQ—Research question SH—Specify hypotheses TVe—Theoretically verify hypotheses SQT—Specify research questions in the context of test problems TE—Theoretically extend the research SQT—specify research questions in the context of test problems TE—Theoretically extend the research								

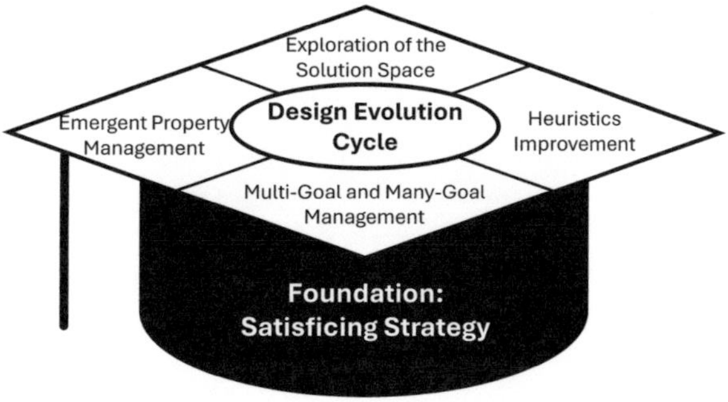

Fig. 2.29 The role of this chapter in the knowledge foundation

References

Allen, J. K., Seepersad, C., Choi, H., & Mistree, F. (2006). Robust design for multiscale and multidisciplinary applications. *Journal of Mechanical Design., 128*(4), 832–843.

Byron, M. (1998). Satisficing and optimality. *Ethics, 109*(1), 67–93.

Farhang-Mehr, A., & Azarm, S. (2002). Diversity assessment of pareto optimal solution sets: An entropy approach. In *Proceedings of the 2002 Congress on Evolutionary Computation.* CEC'02 (Cat. No. 02TH8600). IEEE.

Guo, L., Allen, J. K., & Mistree, F. (2024). Optimize or satisfice in engineering design? *Research in Engineering Design., 35*(3), 239–267.

Hajihashemi, S. (2023). *A user-friendly wrapper for DSIDES (Decision support in the design of engineering systems).* MS Thesis. Industrial and Systems Engineering, University of Oklahoma.

Karush, W. (1939). *Minima of functions of several variables with inequalities as side constraints.* M.Sc. Thesis. Department of Mathematics, Universtiy of Chicago.

Kuhn, H. W., & Tucker, A. (1951). Nonlinear programming. In *Proceedings of the Second Berkeley Symposium on Mathematical Statistics and Probability.* University of California Press.

Mistree, F., Hughes, O., & Phuoc, H. (1981). An optimization method for the design of large, highly constrained complex systems. *Engineering Optimization., 5*(3), 179–197.

Mistree, F., Hughes, O. F., & Bras, B. (1993). The compromise decision support problem and the adaptive linear programming algorithm. In M. P. Kamat (Ed.), *Structural optimization: Status and promise* (pp. 247–286). AIAA.

Mistree, F., Patel, B., & Vadde, S. (1994). On modeling multiple objectives and multi-level decisions in concurrent design. In *Advances in design automation* (DE-Vol. 69-2, pp. 151–161). ASME.

Nellippallil, A. B., Allen, J. K., Gautham, B., Singh, A. K., & Mistree, F. (2020). *Architecting robust co-design of materials, products, and manufacturing processes.* Springer.

Nellippallil, A. B., Mohan, P., Allen, J. K., & Mistree, F. (2018). Robust concept exploration of materials, products and associated manufacturing processes. In *ASME Design Automation Conference.* Quebec City, Quebec, Canada. Paper Number: DETC2018-85913.

Xu, Y., Zhang, H., Zeng, X., & Nojima, Y. (2022). An adaptive convergence enhanced evolutionary algorithm for many-objective optimization problems. *Swarm and Evolutionary Computation., 75*, Article 101180.

Chapter 3
Proposed Methods

The Design Evolution Loop

As discussed in Sect. 1.2.1, the rapidly changing nature of complex systems requires designers to evolve their designs continuously. Optimization methods often fall short in such dynamic environments. Therefore, we adopt a satisficing strategy that aims to achieve solutions that are good enough rather than optimal because we find it more practical.

In this chapter, we delve into the core of the design evolution cycle, with four stages in a loop—formulation, approximation, exploration, and evaluation; see Fig. 3.1 (also as Fig. 1.4 in Sect. 1.1). Using these methods, we connect multiple stages so that designers can iterate the cycle. We build upon our foundation for implementing a satisficing strategy using cDSP, ALP, and DSIDES. These methods collectively form a robust computational framework for addressing nonlinear, multi-goal problems where optimization techniques might fail.

We also outline our validation plan to ensure the proposed methods are effective and applicable across various domains. The validation of engineering research has historically been rooted in the principles of scientific inquiry, emphasizing formal, rigorous, and quantitative methods (Barlas et al. 1990). This approach, relying on logical induction and deduction, has proven highly effective due to the mathematical modeling that underpins much engineering research. By employing a mix of theoretical analysis and practical application, we aim to demonstrate the efficacy of the satisficing strategy in managing complex systems.

3.1 Organization of This Chapter

The logic flow between this chapter and previous chapters is illustrated in Fig. 3.2.

In this chapter is organized as follows:

- Sect. 3.1: We introduce the elements of model evolution, extending the tasks in the design evolution cycle described in Sect. 2.4.2.

L. Guo et al., *Architecting a Satisficing Strategy for Designing Complex Systems*,
https://doi.org/10.1007/978-3-032-04577-5_3

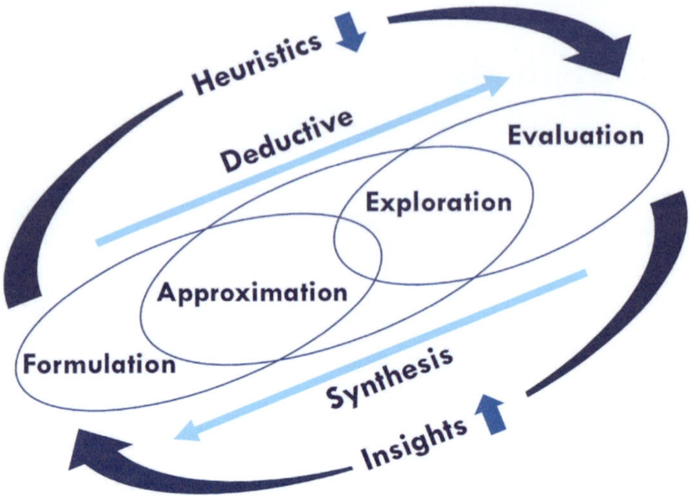

Fig. 3.1 The evolution cycle of complex systems realization

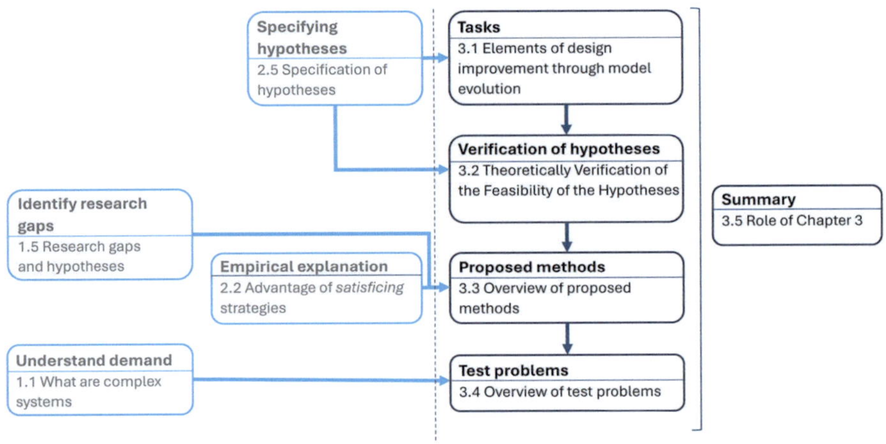

Fig. 3.2 Organization of this chapter

- Sect. 3.2: We state the hypotheses, which are tentative proposals for answering the research questions in Sect. 2.4.
- Sect. 3.3: We provide an overview of the proposed methods leveraging a satisficing strategy to fill research gaps identified in Sect. 1.4.
- Sect. 3.4: We introduce example problems for testing the proposed methods, representing typical characteristics of complex systems.
- Sect. 3.5: We summarize the role of Chapter 3 in this monograph.

Table 3.1 Summary interpretation of the problems investigated in this chapter

Elements	What?
Components	Stages of the design evolution cycle—formulation, approximation, exploration, and evaluation, and their connections
Connectors	Algorithms to pass information through the stages, learn knowledge to replace heuristics, and make cross-stage modification
Form	*How?*
Component roles	Provide the constructs for connecting stages and enabling loops using rule-based learning
Properties	Make improvements by design evolution—by replacing heuristics with learned knowledge
Relationship	Supporting selection, compromise, and coupled decisions while managing four types of uncertainties
Rationale	*Why?*
Motivation	Provide support to assist designers in design evolution when designing complex systems
Assumptions	Heuristics are used to model complex systems in the early stages of design. The mathematical relationship between model settings and system performances can be learned and implemented to replace the heuristics, thereby accomplishing design evolution and system improvement
Interpretation	Understand how rule-based learning algorithms help obtain insight into model behavior and how replacing heuristics with insight helps in design evolution

We summarize the interpretation of the problems investigated in this chapter in Table 3.1. The plan for addressing the research questions in this chapter by theoretically verifying the feasibility of the hypotheses and proposing methods is shown in Table 3.2 and Fig. 3.3.

3.2 Elements of Design Improvement Through Model Evolution, (Task 1–4)

The interactions among multiple stages in the evolution cycle of the realization of model-based complex systems are critical to exploring the solution space (ESS). In this monograph, the elements being explored are as follows.

3.2.1 Task 1: Formulation-Exploration

As the boundary formed by the constraints and bounds may vary due to (1) parameter variations, (2) variation in decision variables, and (3) variation in model structure,

Table 3.2 Plan for theoretically verifying the hypotheses and demonstrating the proposed methods in each chapter

Chapter	Chap. 1	Chap. 2		Chap. 3		Chap. 4	Chap. 5	Chap. 6	Chap. 7	Chap. 8	Chap. 9
Actions	RG H	RDI-II	RQ1 SH1	TVe1: Explore the boundary of the system by connecting formulation with exploration	M1: Formulation-Exploration framework	EVe SQT AQ				CQ EVa	TE
		RDIII	RQ2 SH2	TVe2: Improve the approximation by connecting approximation, exploration, and evaluation	M2: Adaptive Linear Programming Algorithm with Parameter Learning (ALPPL)		EVe SQT AQ				
		RDIV	RQ3 SH3	TVe3: Speed up learning the system nature by connecting formulation, exploration, and evaluation	M3: Adaptive Leveling-Weighting-Clustering algorithm (ALWC)			EVe SQT AQ			

(continued)

Table 3.2 (continued)

Chapter	Chap. 1	Chap. 2		Chap. 3		Chap. 4	Chap. 5	Chap. 6	Chap. 7	Chap. 8	Chap. 9
			RQ4 SH4	TVe4: Learn emergent property by connecting formulation, approximation, exploration, and evaluation	M4: Scenario planning in agent-based modeling				EVe SQT AQ		
Nomenclature	AQ—answer research questions CQ—closure of the answers to research questions EVa—empirically validate hypotheses EVe—empirically verify hypotheses H—give hypotheses RQ—pose a research question M—introduce methods RD—tie to robust design RG—give research gaps SH—specify hypotheses SQT—specify research questions in the context of test problems TE—theoretically extend the research TVe—theoretically verify hypotheses										

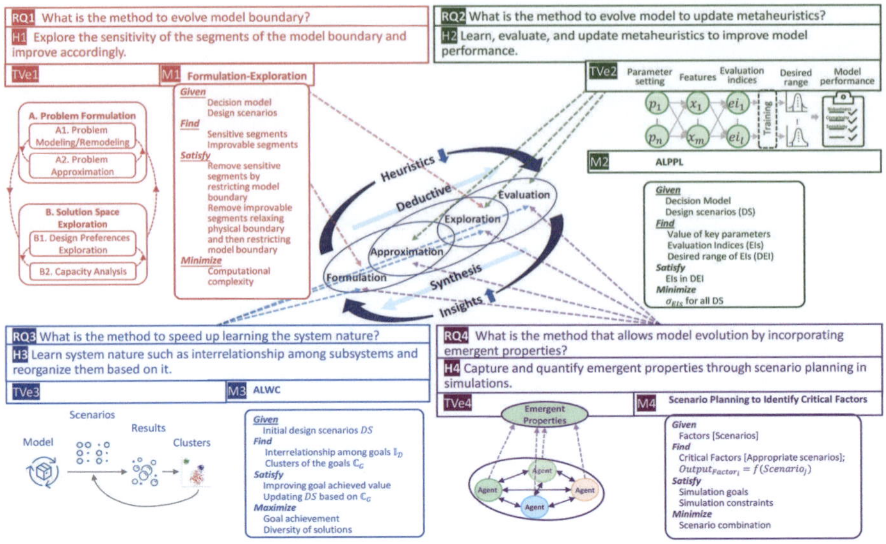

Fig. 3.3 Illustration of Research Questions (RQ1-RQ4), Hypotheses (H1-H4), Theoretical Verification of Hypotheses (TVe1-TVe4), and Methods (M1-M4) in the context of design evolution cycle

the exploration of the boundary and their reformulation allows a decision maker to improve model accuracy and robustness.

By establishing connections between model formulation and exploration, we aim to explore model performance and pass the information to reformulate the model to improve design accuracy and robustness. In Fig. 3.4, highlight the procedures and methods that are involved in doing this.

For example, in a dam network, the reservoir behind each dam has a lower bound—the water level of the inactive pool and an upper bound—the water level of the flood pool. However, some dams have more stringent bounds, which become bottlenecks for the entire dam network. To make the dam-network system more robust, the boundaries should be explored to pinpoint bottlenecks and add buffers to them and the appropriate mathematical model. The solution of the mathematical model can be away from the boundary of the physical system to ensure the system's robustness. A hypothesis for connecting formulation with exploration is shown in Sect. 3.2.1. The detailed description of the test problems and the method are in Sects. 4.2 and 4.3.

3.2.2 Task 2: Approximation-Exploration-Evaluation

As discussed in Sect. 1.2, design methods and solution algorithms for dealing with complex problems fall into two categories, namely, formulating a complex problem exactly and solving it approximately, or approximating a complex problem

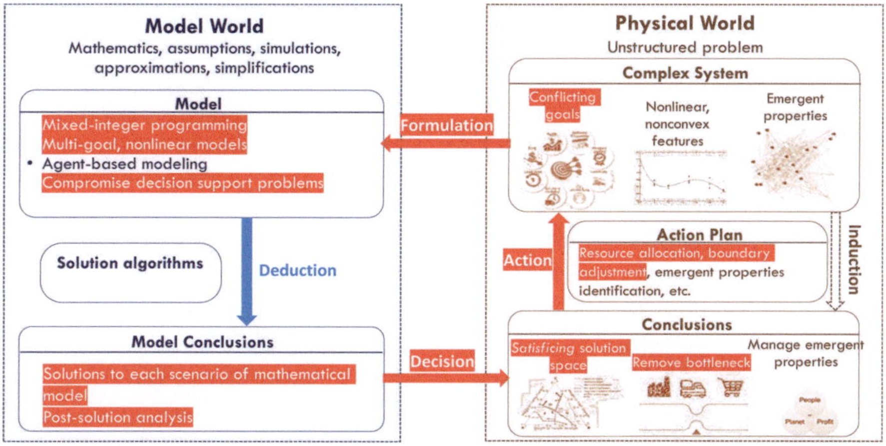

Fig. 3.4 The methods and procedures involved in formulation-exploration—realizing the model evolution using the underlined items

and solving it exactly. For both methods, designers must approximate the physical problem; what matters to the robustness of the result is when and how the approximation takes place. In this monograph, we assume that any approximation relies on heuristics or metaheuristics to make rules. A heuristic is a mental shortcut that allows a designer to solve problems quickly and efficiently. A metaheuristic is a higher-level procedure to find a heuristic that may provide a sufficiently good solution to a problem. There are no perfect heuristics or metaheuristics. However, given the evolving knowledge about the problem and the demands, designers can use this knowledge to update the metaheuristics or heuristics to improve the robustness of the approximation and hence improve the design.

The utility of ***evaluation*** lies in processing data analytics to learn patterns in the data generated by adopting various scenarios. Based on the evaluation, predictions on model performance regarding more scenarios can be made, allowing designers to evolve the model by improving the approximation.

In Fig. 3.5, we highlight the procedures and methods that are involved in understanding the connections. By establishing connections in model approximation, exploration, and evaluation, we aim to explore the model performance associated with the metaheuristics used in model approximation and use this information to update the metaheuristics to make the design satisficing and relatively insensitive to metaheuristics.

As the approximation of the model, by linearization for example, may impact model accuracy and solution quality, capturing the connections among the critical parameters in the approximation algorithm and the solution quality is critical to model evolution. Heuristics are used to determine the critical parameters of the approximation algorithm, such as the reduced move coefficient (RMC) in Adaptive Linear Programming (ALP). If the quality of the solution can be clearly defined, and the association between the quality of the solution and the RMC value can be

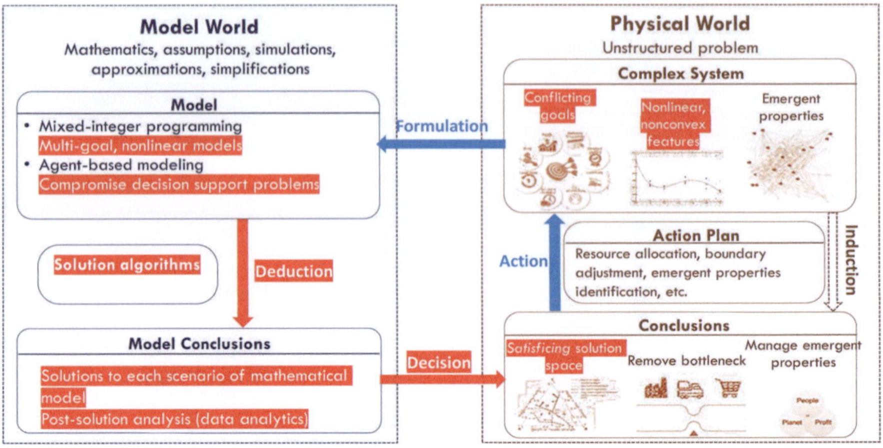

Fig. 3.5 The methods and procedures involved in approximation-exploration-evaluation—realizing the model evolution through the underlined items

established, then the solution quality as well as the robustness of the model can be improved by adjusting the RMC. The hypothesis on how to connect approximation with exploration and evaluation is in Sect. 3.2.2. The detailed description of the test problem and the method are presented in Chap. 5.

3.2.3 Task 3: Formulation-Exploration-Evaluation

The structure of the model, or any heuristics, assumptions, or simplifications applied consciously or subconsciously by the designers in model formulation can be evaluated and updated to improve design robustness. If a connection among formulation, exploration, and evaluation is established, this information can be interpreted as knowledge. Designers need to identify what knowledge obtained from the exploration and evaluation phase is useful. This includes knowledge about the improvement of model formulation that can benefit model accuracy and robustness. The improvement of model formulation is a broad topic, including the setting of parameters, variables, equations linking parameters and variables, the formulation of the goals, and the priority and combination of the multiple goals. A method of returning information from the solution space back to the design space regarding model reformulation is required.

In Fig. 3.6, we highlight the procedures and methods that are involved in the connections work. By establishing connections among model formulation, exploration, and evaluation, we aim to explore better model formulations based on information learned from post-solution analysis, such as, the interrelationship among subsystems.

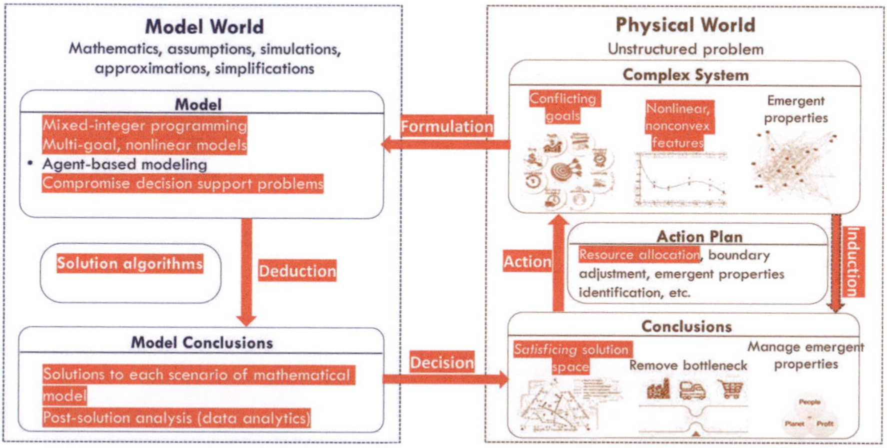

Fig. 3.6 The methods and procedures involved in formulation-exploration-evaluation—realizing model evolution through the items which are underlined

In a concurrent design problem with multiple subsystems, especially when the correlations and interactions among the subsystems are initially unknown to the designers, analyzing the solutions generated by varying scenarios can reveal interrelationships among the subsystems. This knowledge allows the model formulation to be adjusted accordingly. The method for correcting the formulation with exploration and evaluation is detailed in Sect. 3.2.3. The detailed description of the test problem and the method are presented in Chap. 6.

3.2.4 Task 4: Formulation-Approximation-Exploration-Evaluation

To manage the emergent properties of complex systems, it is important to consider the gap between the behaviors of individual components and the system behavior as a whole. Setting behavior rules for each individual or component of a complex system may not result in expected system behavior. Therefore, establishing mechanisms for capturing emergent properties and incorporating them into the procedures of the evolution cycle is a necessary element of model evolution.

In Fig. 3.7, we highlight the procedures and methods involved in establishing the connections among model formulation, approximation, exploration, and evaluation. We aim to capture emergent properties and incorporate them into the model. For example, in a social network, how can designers capture emergent properties in human behavior and promote a new technology by leveraging critical and sensitive factors? The hypothesis on how to connect all four stages in the model evolution

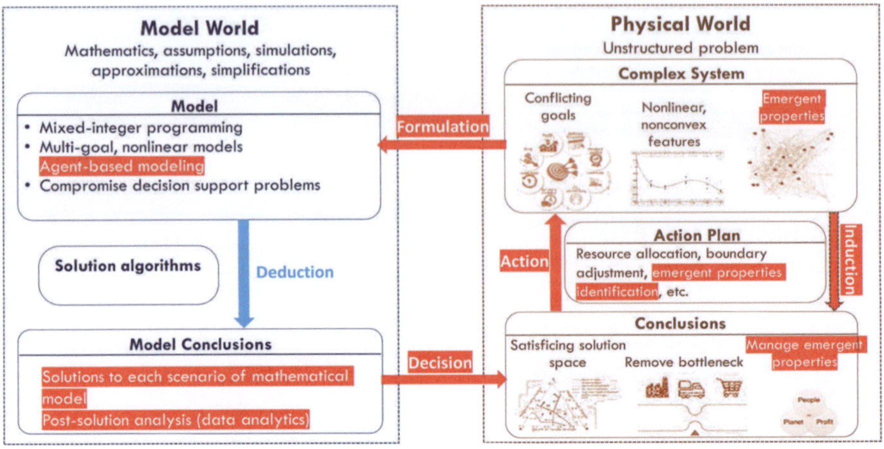

Fig. 3.7 The methods and procedures involved in approximation-formulation-exploration-evaluation—realizing the model evolution through the underlined items

is given in detail in Sect. 3.2.4. A detailed description of the test problem and the method are in Chap. 7.

3.2.5 The Design Evolution Cycle is Open and Extendable

There can be other elements in model evolution. The conception and procedure of model evolution are open and can be expanded to other relevant activities. In this monograph, the focus is on the elements mentioned above. Based on the conception and elements of the evolution cycle, hypotheses are proposed in Sect. 3.2.

3.3 Theoretical Verification of the Feasibility of the Hypotheses (TVe1-TVe4)

To realize the four elements of model evolution and answer the four research questions (RQ1-RQ4), there are four hypotheses (H1-H4[1]) (Table 2.17). The theoretical verification of the feasibility of these four hypotheses follows.

[1] *H1: Explore the sensitivity of the segments of the model boundary and improve accordingly. H2: Learn, evaluate, and update metaheuristics to improve model performance. H3: Learn system nature such as interrelationship among subsystems and reorganize them based on it. H4: Capture and quantify emergent properties through scenario planning in simulation.*

3.3.1 Theoretical Verification of Hypothesis 1 (TVe1)

Research Question 1 (RQ1)—What is the method to evolve the model boundary?

Hypothesis 1 (H1)—Explore the sensitivity of the segments of the model boundary and improve appropriately

Model formulation, especially for engineering-design problems, is usually based on designers' domain expertise, experience, some established conventions, assumptions, simplifications, or initial settings, which can be wrong or evolve. To refine the boundary and make the solution space relatively insensitive to the errors or uncertainties embodied in, or affected by, the model boundary, we hypothesize that exploring the sensitivity of the segments of the model boundary to multiple versions of extreme uncertainty is essential; see Fig. 3.8.

We ***hypothesize*** that through the outer cycle (Fig. 3.1)—the two large outer curved arrows marked with "Heuristics ↓" and "Insights ↑" which is to establish connections between problem formulation and solution space exploration, the forward information (on the next scenarios for solution space exploration) and backward information (on the next plan for model improvement) can help evolve the model boundary and improve design performance regarding the robustness, accuracy, and computational

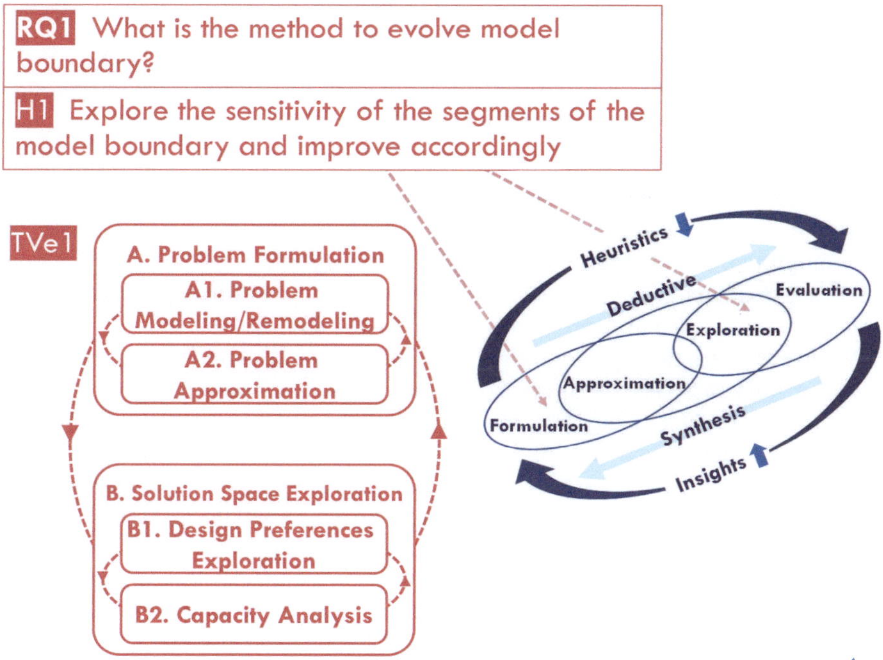

Fig. 3.8 Theoretical Verification of Hypothesis I (TVe1)—exploring the sensitivity of the segments of the model boundary and improve accordingly

complexity. In this monograph, design robustness means the capacity of the solution to be insensitive to model errors and variations, design accuracy means the degree of representativeness of the scenarios of the design, and computational complexity means the complexity of the calculations required to obtain *satisficing* results.

We ***hypothesize*** that through the inner cycles, the ones that circle two neighboring stages (formulation and approximation, approximation and exploration, and exploration and evaluation), to establish connections between different procedures in problem formulation or solution space exploration, that the two-way information transmission mechanisms can also gradually improve each procedure.

The rule-based connection mechanism is based on heuristics, but in Hypothesis 1 (H1), we do not evaluate mechanisms and update heuristics. In Hypotheses 2 and 3 (H2 and H3), we incorporate evaluation in the multi-stage connections and manage updating the heuristics.

Given the hypothesis, we propose the method, the Formulation-Exploration framework, which is briefly introduced in Sect. 3.3.1 and further introduced and tested using two test problems in Chap. 4.

3.3.2 Theoretical Verification of Hypothesis 2 (TVe2)

Research Question 2 (RQ2)—What is the method to evolve model to update metaheuristics?

Hypothesis 2 (H2)—Metaheuristics can be updated by learning, evaluating, and updating them to improve model performance

Model approximation especially for nonlinear, nonconvex, and multi-goal, engineering-design problems are usually based on rules which are made using designers' metaheuristics and experience, which can be wrong, inaccurate, or over-simplified. To improve the accuracy, efficiency and robustness of the rule-based approximation, we hypothesize that by learning, evaluating, and updating metaheuristics we can improve model performance; see Fig. 3.9.

We ***hypothesize*** that by identifying the metaheuristics that have a critical impact on the results in terms of solution robustness, such as, but are not limited to, settings of parameters, designers can learn the features that represent the performance of the approximation. From the features, evaluation indices can be created. Designers should train the metaheuristics to bring the evaluation indices into a desired range, which ensures that the approximation performance and the insensitivity of the solution space to model errors, uncertainties, and metaheuristics are improved.

Given the hypothesis, we propose a method, using parameter learning to improve approximation and solution algorithms, which is briefly introduced in Sect. 3.3.2 and further discussed and demonstrated using a test problem in Chap. 5.

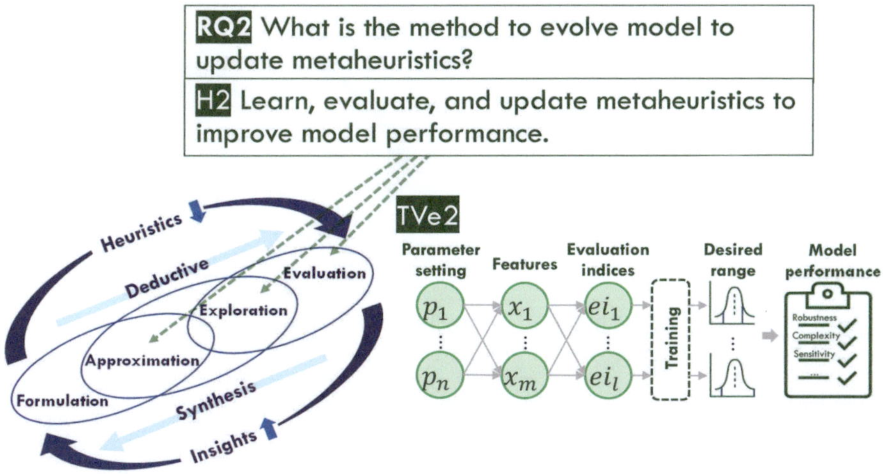

Fig. 3.9 Theoretical Verification of Hypothesis 2 (TVe2)—learn, evaluate, and update metaheuristics to improve model approximation

3.3.3 Theoretical Verification of Hypothesis 3 (TVe3)

<u>Research Question 3 (RQ3)</u>—What is a method to speed up learning the nature of the system?

<u>Hypothesis 3 (H3)</u>—Learn the nature of the system, the interrelationships among subsystems and reorganize them based on these interrelationships

For multi-stage, concurrent design problems, designers may initially lack profound knowledge about the nature of the system, such as the interrelationships among subsystems. For some problems, designers may not even be aware that subsystems exist. If there is insufficient domain expertise, or if designers mistakenly assume they possess domain knowledge or experience managing the subsystems, but their knowledge or experience is wrong or impractical, the design evolution loop can be used to verify or falsify their knowledge or experience.

We ***hypothesize*** that through post-solution analysis, designers may attain insight on how to design experiments and select the scenarios to better identify and organize the subsystems, so they can verify or falsify their assumptions. We ***hypothesize*** that such an analyzing-interpreting-selecting process can help speed up designers' learning about the system's nature (see Fig. 3.10).

The utility of adding an "evaluation" step to the formulation-exploration (Hypothesis 1) is extracting hidden information by learning from time-series data generated while iterating the design evolution cycle. We propose a method, using unsupervised learning to cluster the system into subsystems, learn their interrelationships, and update the design scenarios based on this. This method is briefly introduced in Sect. 3.3.3 and further discussed and demonstrated using a test problem in Chap. 6.

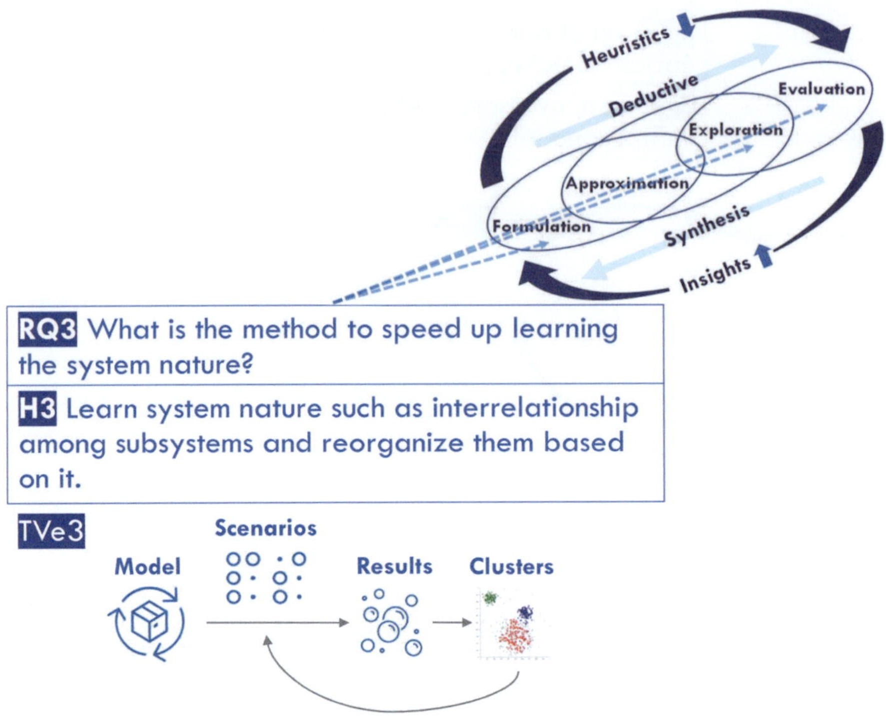

Fig. 3.10 Theoretical Verification of Hypothesis 3 (TVe3)—learn the system nature, such as, interrelationships among subsystems and reorganize the systems based on this knowledge

3.3.4 Theoretical Verification of Hypothesis 4 (TVe4)

Research Question 4 (RQ4)—What is the method that allows model evolution by incorporating emergent properties?

Hypothesis 4 (H4)—Emergent properties can be captured through scenario planning in simulation

"In planning and policy, a wicked problem is a problem that is difficult or impossible to solve because of incomplete, contradictory, and changing requirements that are often difficult to recognize."[2] We ***hypothesize*** that by capturing emergent properties and incorporating them into decision models as an element of model evolution, designers can convert wicked problems into regular complex-system design problems. We ***hypothesize*** that by using simulation tools such as agent-based modeling, emergent properties such as collective behavior, which is nonlinear with individual behaviors, can be observed, quantified, and predicted; see Fig. 3.11.

[2] This definition of "wicked problem" is from Wikipedia: https://en.wikipedia.org/wiki/Wicked_problem.

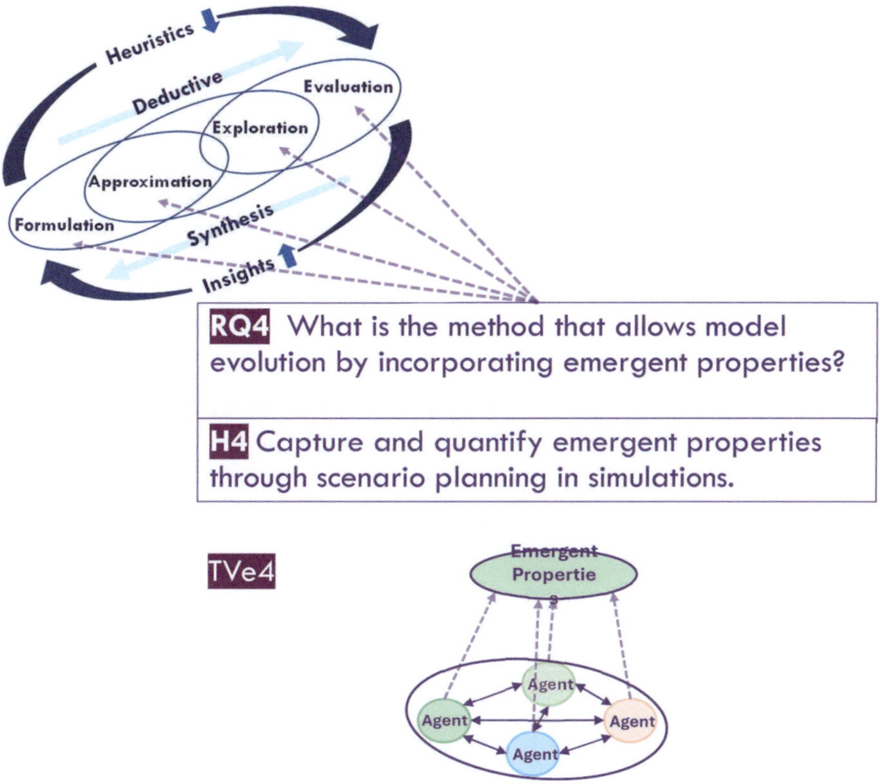

Fig. 3.11 Theoretical Verification of Hypothesis 4 (TVe4)—capture and quantify emergent properties through scenario planning in simulations

We ***hypothesize*** that learning and adding emergent properties to decision models can connect and affect all four stages of model evolution and result in design improvement. We propose a method using simulation tools to generate synthetic data under a variety of design scenarios for a wicked problem, clustering the system into subsystems, learning their interrelationships, and updating the design scenario based on this method. This method is briefly introduced in Sect. 3.3.4 and further discussed and demonstrated using a test problem in Chap. 7.

3.4 Overview of Proposed Methods (M1-M4)

To test the four hypotheses (H1-H4), four methods (M1-M4) are proposed. They are briefly introduced in this section one-by-one. Empirical verification using test problems is covered in Chaps. 4, 5, 6, and 7.

3.4.1 M1: Exploration of the Boundary Using the Formulation-Exploration Framework

To test Hypothesis 1, we explore the sensitivity of the segments of the model boundary and improve them accordingly. We propose the Formulation-Exploration framework so that we can establish connections between the two stages: formulation and exploration. The method in a word-form of the compromise Decision Support Problem (cDSP) is written as follows. Given the decision model and design scenarios, designers need to find the sensitive segments and improvable segments of the decision model, which are to be removed, while minimizing the computational complexity; see (Hajihashemi, 2023; Mistree et al., 1981, 1993).

Establishing connections between Formulation and Exploration

Given

Decision model

Design scenarios

Find

Sensitive segments

Improvable segments

Satisfy

Remove sensitive segments by restricting the model boundary

Remove improvable segments by relaxing the physical boundary and then restricting the model boundary

Minimize

Computational complexity

Sensitive segments refer to the parameters or equations, perhaps response surfaces, that embody Type I and Type II uncertainty, which are variation in parameters or unparameterizable factors (noise factors) or variables (control factors) (Nellippallil et al., 2020). When either Type I or Type II uncertainty occur, these segments change, and a feasible solution may be lost; see Fig. 3.12. For example, if an inequality constraint $g_i(x) \geq 0$ becomes active at a solution point $\boldsymbol{x^o}$, $g_i(\boldsymbol{x^o}) = 0$, i.e., when $\boldsymbol{x^o}$ is entered into an inequality constraint, the slack or surplus of the constraint is zero. Such a constraint is a sensitive segment of the decision model in the design scenario when $\boldsymbol{x^o}$ is a solution point. By changing design scenarios, designers obtain a set of solution points, X, and if $g_i(\boldsymbol{x^o}) = 0, \forall \boldsymbol{x^o} \in \boldsymbol{X}$, then constraint $g_i(x) \geq 0$ is a sensitive segment.

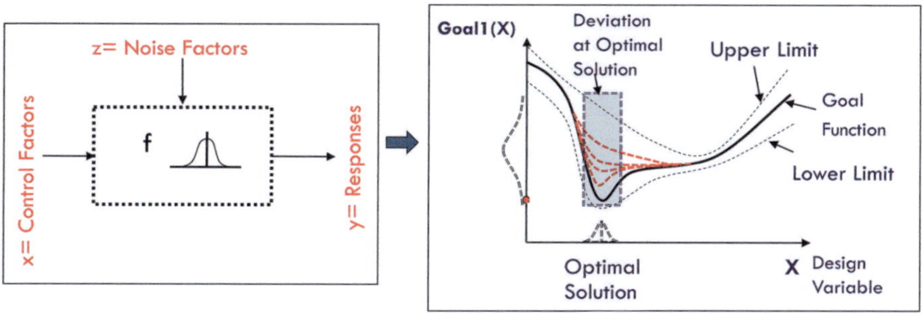

Fig. 3.12 The control factors and noise factors bring variation in goal function

Control factors are factors that are under designers' direct control. It is assumed that a control factor is accurately quantifiable and controllable, so these control factors can only be decision variables of the decision model. ***Noise factors***, on the other hand, cannot be controlled by designers directly. Sometimes noise factors are not even quantifiable or parameterizable, and they may not be incorporated into the decision model. However, initially, designers' understanding of the control factors and noise factors is insufficient to determine whether Type I and Type II uncertainty is present.. If this is the case, then actions to explore the solution space, such as design preferences exploration and capacity analysis, can help identify the sensitive segments.

Improvable segments are the parameters or equations that can be modified, and when modified in certain ways, the goal(s) can be better achieved. For example, a constraint with a large dual price is the amount that the goal (objective) could be improved (achieved more completely) as the right-hand side of the constraint is relaxed by a single unit (Diewert, 1974).

The method of each step in the framework and the rationale for each method is illustrated in Fig. 3.13. To realize a *satisficing* strategy in engineering design, we choose the compromise Decision Support Problem (cDSP) construct to formulate design problems and the ALP algorithm to approximate nonlinear, non-convex problems. The reason why using the cDSP and the ALP can identify *satisficing* solution space is in discussed Sect. 2.1. In solution space exploration, there can be additional activities. In this monograph, we discuss weight sensitivity analysis, which involves exploring weight vectors when combining multiple goals, and capacity analysis, to identify constraints with limited capacity.

In Chap. 4, we use two test problems—a continuous problem and a discrete problem—to demonstrate how to use the formulation-exploration framework to explore the boundary of a problem and make the design relatively insensitive to boundary variations.

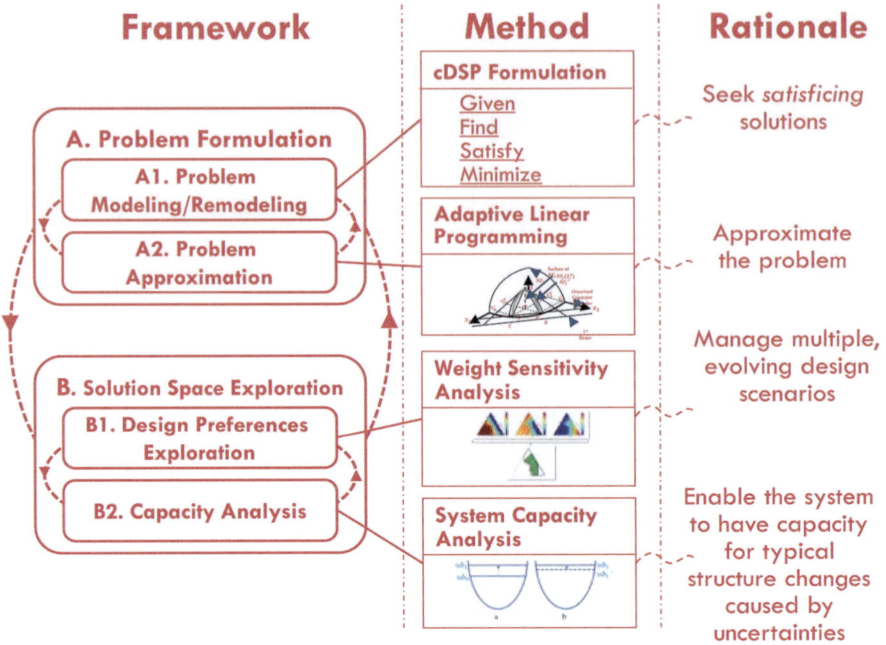

Fig. 3.13 Formulation-exploration framework

3.4.2 M2: Improving Algorithm Robustness Using Parameter Learning

To test Hypothesis 2, *learn, evaluate, and update metaheuristics to improve model performance*, we propose enabling information passing through the approximation, exploration, and evaluation by adopting the Adaptive Linear Programming algorithm with Parameter Learning (ALPPL) (Guo et al., 2024). The conception of the method in a word-form of the cDSP is as follows. Given the decision model and design scenarios (DS), designers need to find the appropriate values of key parameters by making the evaluation indices (EIs) fall into a desired range (DEI), while minimizing the standard deviation of the EIs under all DSs. We use the standard deviation as one of the statistics to represent solution robustness with respect to changes in the design scenario (DS) for the test problem, to improve ALP algorithm by updating the heuristics in the parameter settings.

Given

Decision Model

Design scenarios (DS)

Find

Value of key parameters

Evaluation Indices (EIs)

Desired range of EIs (DEI)

Satisfy

EIs in DEI

Minimize

σ_{EIs} For all DS

Evaluation indices (EIs) are the indices that can effectively reflect the approximation performance of a design regarding the criteria that designers desire, such as the approximation accuracy, the robustness of a solution with respect to approximation accuracy, and the computational complexity of the approximation. In this monograph, it is emphasized that evaluation indices can be customized for each problem. Evaluation indices are developed by designers based on (1) their domain knowledge or experience and (2) data generated when exploring the solution space. If designers adopt both methods, it means that they incorporate the evaluation procedure into the design cycle, which may make the evaluation indices less sensitive to human error. In this monograph, "evaluation" is defined as a procedure that requires using data analytics to analyze the data and gain knowledge about different factors from multiple stages in the design cycle.

The desired range of EIs (DEI) is a range in which, if the value of the evaluation index falls in the range, the approximation performance can be guaranteed to be at an acceptable level. By identifying the desired range of each evaluation index, designers can determine the association between the evaluation indices and the approximation performance. By bringing the evaluation indices into their desired ranges, the designers ensure that the approximation of the design is acceptable.

While changing design scenarios and parameter values, designers obtain results whose evaluation indices follow a distribution. Statistics are used to control the robustness and computational complexity of the solutions while choosing appropriate design scenarios.

The steps added to the ALP are illustrated in Fig. 3.14 (Guo et al., 2020). After initializing the decision model and the algorithm with specific parameter settings, designers can run the algorithm and obtain results. By interpreting the results to evaluate the approximate performance, insight on how to further update the parameters is obtained and used in the next iteration.

In Chap. 5, we present a test problem for designing the cooling stage of the hot rod rolling chain to demonstrate how parameter learning can improve an approximation and solution algorithm.

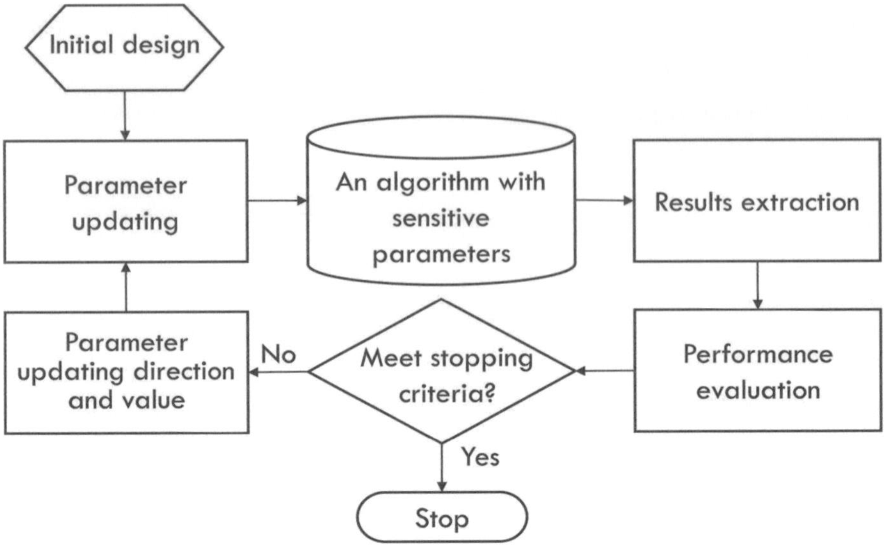

Fig. 3.14 Learn and update metaheuristics in an algorithm using parameter learning

3.4.3 M3: Exploring Interrelationships Among Subsystems Using Unsupervised Learning

To realize and test Hypothesis 3, *Learn the nature of the system such as the interrelationships among subsystems and adjust the model accordingly,* we propose the Adaptive-Leveling-Weighting-Clustering (ALWC) algorithm (Guo et al., 2023) to establish connections between formulation, exploration, and evaluation. The conception of the method in a word-form cDSP is as follows. Given the decision model and initial design scenarios, designers need to find interrelationship among goals, which they can use to update the format of the combination of the goals. In this way, it is expected that the achieved value of the goals will be improved and the most representative design scenarios can be identified and explored rather than enumerating all scenarios. Meanwhile, designers can maximize goal achievement and solution diversity.

Given

Decision model

Initial design scenarios (DS)

Find

Interrelationship among goals ($\mathbb{I}_{\mathcal{D}}$)

Clusters of the goals $\mathbb{C}_G$

Satisfy

Improving goal achieved value

Updating DS based on $\mathbb{C}_G$

Maximize

Goal achievement

Diversity of solutions

For this method, we assume that each goal represents a design requirement, which is the requirement of the performance of a subsystem. Exploring the interrelationships among the goals is a way of learning the interaction of subsystems.

The ***interrelationship*** of the goals can be one or more measurements, such as the correlation among the achieved values of the goals under multiple design scenarios, or the orthogonality among the deviations of the goals under multiple design scenarios. The choice of measurement depends on the nature of the problem or the designers' preferences.

The ***diversity of solutions*** is defined as how different the solutions are. The diversity of the solutions can reflect whether the solution space is explored sufficiently and whether there are adequate alternative solutions for designers to select in various situations.

The procedures of the ALWC algorithm are shown in Fig. 3.15. After initializing the design scenarios on subsystem organization, designers adopt the design scenarios and obtain a deviation matrix. The deviation is the distance between the achieved value of a goal and the target of the goal. A column of the deviation matrix is the deviation of all goals under a specific design scenario. By learning the interrelationship among the rows of the deviation matrix, designers obtain the correlation, orthogonality, or other types of interrelationship among the goals. Based on such interrelationships, the goals can be grouped into clusters, and then the goals can be reorganized for the next iteration by using the most representative design scenarios.

In Chap. 6, we use a concurrent engineering design test problem to demonstrate how to use the proposed method to learn and speed up learning the structure of the system.

3.4.4 M4: Exploring Critical Factors by Scenario Planning in Agent-Based Modeling

To realize and test Hypothesis 4, which involves *capturing and quantifying emergent properties by scenario planning in simulation,* we propose a framework to learn critical factors using simulation. In a word-form cDSP the concept of this method is as follows. Given all "m" factors that may or may not affect the simulation results and n scenarios of each factor, find the critical factors that severely impact the simulation

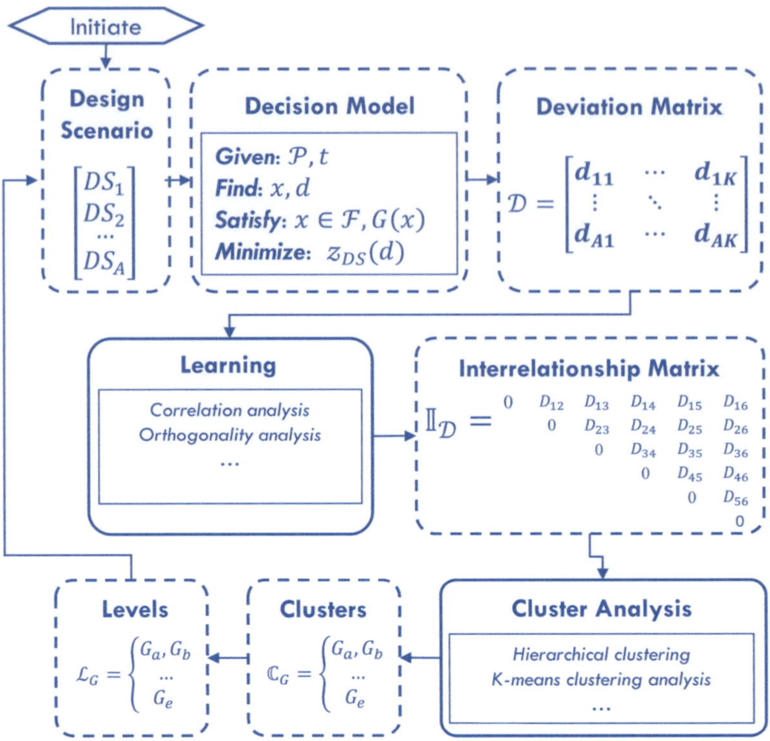

Fig. 3.15 Learn and speed up the learning of systems using machine learning

results and the appropriate scenarios for those factor that result in the desired output, while satisfying the simulation goals and constraints and minimizing number of the scenario combinations.

Given

Factors [Scenarios]

Find

Critical Factors [Appropriate scenarios]

$\text{Output}_{\text{Factor}_i} = \text{f}(\text{Scenario}_j)$, where $i = 1, \ldots m, j = 1, \ldots, n$

Satisfy

Simulation goals

Simulation constraints

Minimize

Scenario combinations

Factors are variables or parameters with a range, multiple scenarios, or uncertainties. Designers lack knowledge of the critical factor settings and the corresponding simulation results, especially the output of simulations under the influence of a combination of multiple scenarios of different factors. Among all the factors, some significantly impact the simulation results when changing scenarios, while others do not.

Critical factors have a relatively significant impact on the simulation results in different scenarios. The rationale for learning the emergent properties is to identify critical factors, quantify their impacts on the simulation results, and find specific scenarios that lead to the desired results.

The procedure of learning about critical factors in a simulation, such as agent-based modeling is shown in Fig. 3.16. After determining which factors have uncertainties, designers need to decide whether these critical factors are controllable or not. For the controllable factors, they must become decision variables. Then possible scenarios and the physical meanings for these scenarios and the ways of setting each scenario must be determined. Designers need to set the uncontrollable factors as parameters with uncertainties or unparameterizable noise factors and identify their possible scenarios as well. Then, by exploring the scenarios and the combination of scenarios, they obtain the simulation output of each scenario or combination of scenarios. For the factor(s) for which the simulation results do not vary significantly with changing scenarios, these are noncritical factors. Otherwise, a factor is a critical factor, and designers can learn the mathematical relation between the scenarios and the simulation outputs, for which decision support on scenario selection, setting, or timing is acquired.

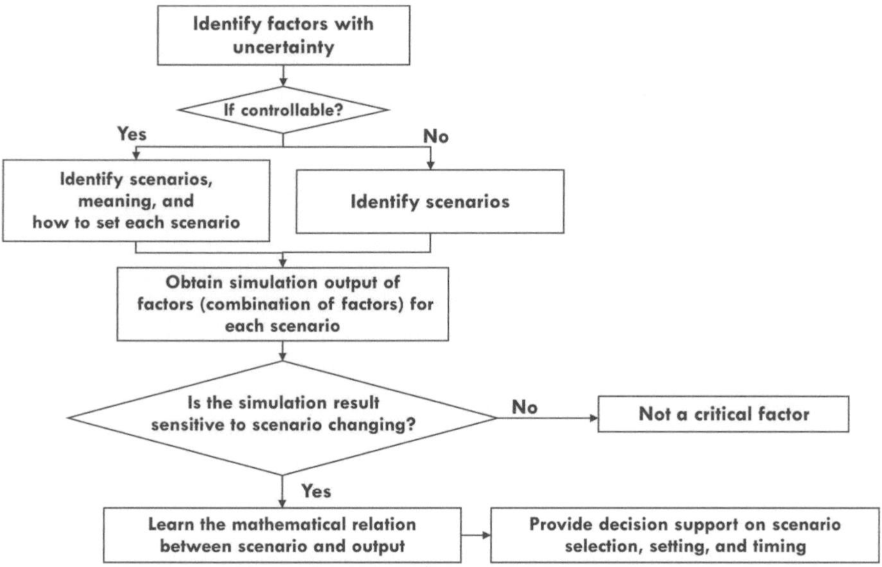

Fig. 3.16 The process of learning critical factors in a simulation

In Chap. 7, we use a test problem to demonstrate the process of identifying the critical factors to demonstrate the proposed method to learn and leverage critical factors.

3.5 Overview of Test Problems

In this section, the connections and transitions between the method's theoretical verification and the empirical verification are given, that is, the logic flow between the validation square's Quadrant 1 (Q1) and Quadrant 2 (Q2) (See Fig. 1.11 in Sect. 1.6).

3.5.1 Required Characteristics of the Test Problems

To realize robust design Type I and Type II (RDI-II) to identify a *satisficing* solution space that is relatively insensitive to variations in parameters and variables and causes segments of the boundary to be sensitive, we need to use test problems that have parameter and variable uncertainty. As shown in Fig. 3.17, Type I and Type II uncertainty may impact the boundary of the physical system, so designers should refine the mathematical model(s) accordingly, as shown by the bold dotted red lines in Fig. 3.17. By refining the model formulation, designers expect to identify the deviation at the optimal solution and *identify satisficing* solutions that are relatively insensitive to Type I and Type II uncertainty. This is Type I and Type II Robust Design.

Therefore, the features of the test problem(s) to realize RDI-II are:

- There are fewer factors to control but many requirements, or, in other words, the number of decision variables is fewer than the number of constraints. The model's coefficient matrix has more rows than columns. In the physical world, this means that the variables that the designers can control are limited, whereas there are many requirements that must be meet by controlling the limited variables.
- Multiple goals with no knowledge about ways of combining the goals associated with design preferences. When changing the way of combining the multiple goals to represent different design preferences, the solution may change as well. So, the solutions to the problem are usually sensitive to design preferences. In some cases, designers may not even have the knowledge of ways of combining the goals which are associated with the design preferences.
- Variation in parameters and decision variables can make the optimal solution infeasible.
- Parameter settings significantly impact the design results, however designers rely on metaheuristics to set parameters and there may be no mechanism to evaluate and update the metaheuristics.

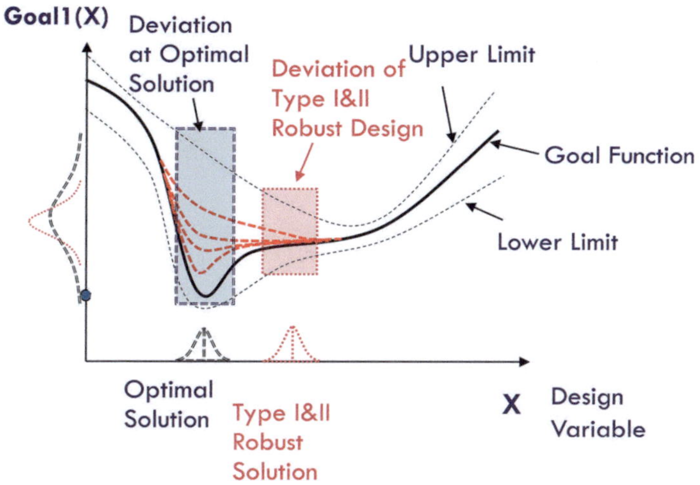

Fig. 3.17 Test Problems for RDI-II—refining the model formulation and identifying *satisficing* solutions relatively insensitive to the variation in parameters and decision variables (Choi, 2005; Nellippallil et al., 2020)

To realize the robust design Type III (RDIII) to identify a *satisficing* solution space that is relatively insensitive to variations in the mathematical model, we need to demonstrate the method with test problems with model structure uncertainty. As shown in Fig. 3.18, as Type III uncertainty impacts the structure of the model, designers must learn how the source of the uncertainty disturbs the model and then make decisions based on the learned information. These decisions lead to a range in the solution space with relatively narrow variation in model structure.

Therefore, the features required of test problem(s) for realizing RDIII are.

- Designers rely on domain expertise to make decisions on the approximation or simplification of the model formulation and solution. Domain-independent knowledge can be acquired through post-solution analyses, which can be used to improve the design..
- Designers rely on metaheuristics to make rules in one or more stages of the design—formulation, approximation, exploration, and evaluation, but there is no effective mechanism to update metaheuristics. However, through post-solution analyses, evaluation of metaheuristic performance can be carried out and the results used to improve metaheuristic updating.

To realize robust design Type IV (RDIV) to capture the system's emergent properties and incorporate them into the decision model, we need to use test problems with emergent properties, and learn and leverage critical factors to identify a *satisficing* solution space that is relatively insensitive to the uncertainties due to managing the first three types of uncertainty.

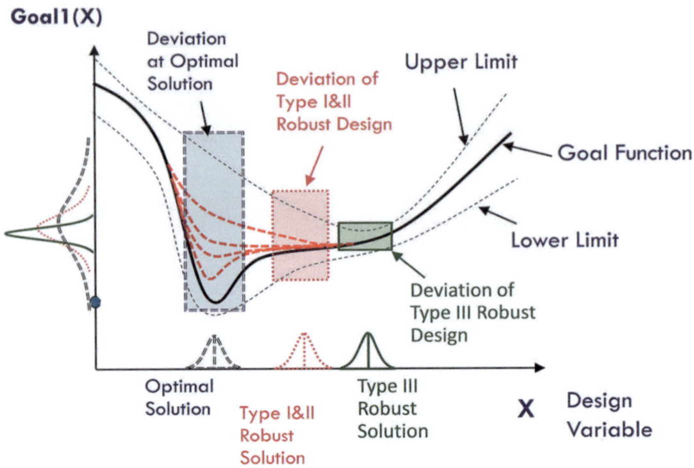

Fig. 3.18 Test Problems for RDI-II—refining the model formulation and identifying *satisficing* solutions relatively insensitive to the s and decision variables

There are different sources of emergent properties. First, some functional relationships, especially cause-and-effect relations, are hidden and designers may ignore these in the initial formulation. When gradually recognizing these functional relationships in the solution and post-solution analyses, designers may take them as emergent properties and incorporate them into the decision model during the model improvement process. By doing this, the decision model can become more accurate and complete as it evolves.

Second, as designers attempt to manage Type I, II, and III uncertainties, their actions induce emergent properties in the system. If they are unaware of the emergent properties and attempt to cope with them, the solutions may no longer be *satisficing*. So Type IV Robust Design allows designers to capture, quantify, and model emergent properties. In this way, the decision model can become more robust to multiple types of uncertainty as it evolves.

By managing emergent properties, designers can identify a solution space that is relatively insensitive to all four types of uncertainty. This is a dynamic process. How can designers manage emergent properties? By identifying the critical factors that can be controlled as inputs that can significantly affect the system's output, identifying the mathematical relations between the critical factors and the output, and leveraging the input to obtain the desired output.

Therefore, the features of the test problem(s) for realizing RDIV are.

- Designers' domain knowledge and experience may be insufficient to create an acceptably complete, accurate and high fidelity model. In this situation, decision support is useful. For example, designers' domain knowledge may not support the identification and definition of all necessary design variables and recognize all system requirements.
- There are controllable or uncontrollable factors, that may or may not impact the output of the system. Designers have no idea of what those factors are, how to control them, and how they affect the output. However, they can learn them using empirical methods such as simulation and scenario planning.
- When designers attempt to manage Types I, II, or III uncertainty, their actions may create new uncertainties.

Given the required features of the test problems, in Sect. 3.4.2, we briefly introduce the test problems—why they are appropriate for this monograph, how they can be used to demonstrate and verify the proposed methods, and what output is expected from the test problems.

3.5.2 Brief Introduction of Each Test Problem

In Table 3.3, the five test problems are summarized, including the robust design types they illustrate, what methods they are used to test, what uncertainties are contained in each problem, and what types of uncertainty they are.

Test Problem 1.1 (T1.1)—designing a dam network by controlling the water outflow. The only control factor of this problem is how much water each reservoir should release downstream each month, but multiple goals must be satisfied, and uncertainty in precipitation and the priority of various user groups needs to be managed. In this monograph, the uncertainty in water inflow is managed by exploring and refining the system's boundary using the formulation-exploration framework. The detailed introduction of T1.1 and the empirical verification of M1 is in Sect. 4.2.

Test Problems 1.2 (T1.2)—designing a supply chain by positioning the customer order decoupling point (CODP). Unlike the dam-network design problem, this problem has discrete variables, so it is a coupled decision problem, which includes two types of decisions, namely, selecting one scenario from several alternatives and compromising among multiple goals by determining the value of continuous decision variables. In this problem, uncertainties on the supply side and demand side are managed and multiple goals in different phases of the product life cycle are managed based on evolving preferences. The detailed introduction of T1.2 and the empirical verification of M1 is in Sect. 4.3.

Test Problem 1.2 (T1.2)—improving the parameter settings (setting the reduced move coefficient (RMC)) of the approximation algorithm (the Adaptive Linear

Table 3.3 Summary of test problems—the robust design type, testified methods, and uncertainties of each test problem. the uncertainties underlined in italic are managed in this monograph

RD Type	**RDI-II**				
			RDIII		
				RDIV	
Method	M1: Formulation-Exploration Framework		M2: Adaptive Linear Programming with Parameter Learning (ALPPL)	M3: Adaptive Leveling-Weighting-Clustering Algorithm (ALWC)	M4: Scenario Planning in Agent-Based Modeling
Chapter	Ch 4		Ch 5	Ch 6	Ch 7
Test Problem / Uncertainty	T1.1: Dam network	T1.2: Supply chain	T2: Hot rolling process chain	T3: Thermal system	T4: Promoting second season farming
Type I	*Uncertainty in timing and amount of inflow (precipitation) and outflow demand (user demand)*	*Uncertainty in demand (order fluctuation)*	*Uncertainty in hyper parameter setting (Parameters in approximation algorithm)*	*Uncertainty in parameter setting in solution algorithm (Starting point of searching)*	*Uncertainty in price (Price of agriculture products)*
Type II	*Uncertainty in outflow (water release)*	*Uncertainty in supply side (productivity fluctuation)*	*Uncertainty in user preferences*		*Promotion efforts and timing*
Type III			*Uncertainty in model approximation due to heuristics in approximation*	*Uncertainty in model approximation (ways of combining multiple goals)*	
Type IV				*Uncertainty in using domain knowledge to simplify the model (fixing decision variables and selecting design scenarios)*	*Interventions that change the mathematical relationships among promotion and results (developing local market)*

RD – robust design
M – method
EVe – empirical verification of the method
T – test problem

Programming algorithm (ALP)) to make the design (the cooling stage of hot-rod-rolling process chain) relatively insensitive to approximations. In the cooling stage of the hot-rod-rolling process chain problem, the value of the RMC can seriously influence the size and robustness of the solution space, especially the range of the weights that return solutions satisficing different design preferences. Uncertainties caused by using heuristics in parameter settings are managed by using parameter learning. A detailed introduction of T2 and the empirical verification of M2 is in Chap. 5.

Test Problem 3 (T3)—improving the understanding of the interrelationship among subsystems of a concurrent engineering design problem, the thermal system, by using unsupervised learning to analyze the correlation or orthogonality among the deviation matrix of the goals in multiple design scenarios. In this example, we demonstrate how subsystem awareness and reformulating the model by re-organizing the goals can help designers select the most representative design scenarios and boost system performance. The detailed introduction of T3 and the empirical verification of M3 is in Chap. 6.

Test Problem 4 (T4)—identifying controllable critical factors in a promotion by scenario planning in simulations and leveraging the critical factors to reach promotion goals. In a social system such as a rural community, in order to promote a new lifestyle based on the introduction of new technology, even if individuals' preferences and behaviors, are known, it is difficult to predict the collective behavior. Simulations using agent-based modeling can help designers understand collective behavior and the interventions that can be done to drive the system behavior in the desired direction with the right timing. The detailed introduction of T4 and the empirical verification of M4 is in Chap. 7.

3.6 Role of This Chapter in This Monograph

In this chapter, the elements of design improvement through model evolution, including the possible ways and tasks needed to answer the research questions, are discussed in Sect. 3.1. Theoretical verification of the feasibility of demonstrating the hypotheses by carrying out the tasks is presented in Sect. 3.2. Methods for completing the tasks are proposed in Sect. 3.3. The test problems used to test the hypotheses and methods are briefly introduced in Sect. 3.4. In summary, this chapter is foundational for Chaps. 4–7, where different test problems with desired characteristics are used to demonstrate the effectiveness of the proposed methods.

Chapters 1, 2 and 3 shown in Quadrant 1 of Fig. 3.19 constitute the Theoretical Structural Validity of what is embodied in this monograph; see Chap. 1, Fig. 1.2. In this chapter, we introduce the design evolution loop and the methods that allow designers to realize the loop; see Fig. 3.20.

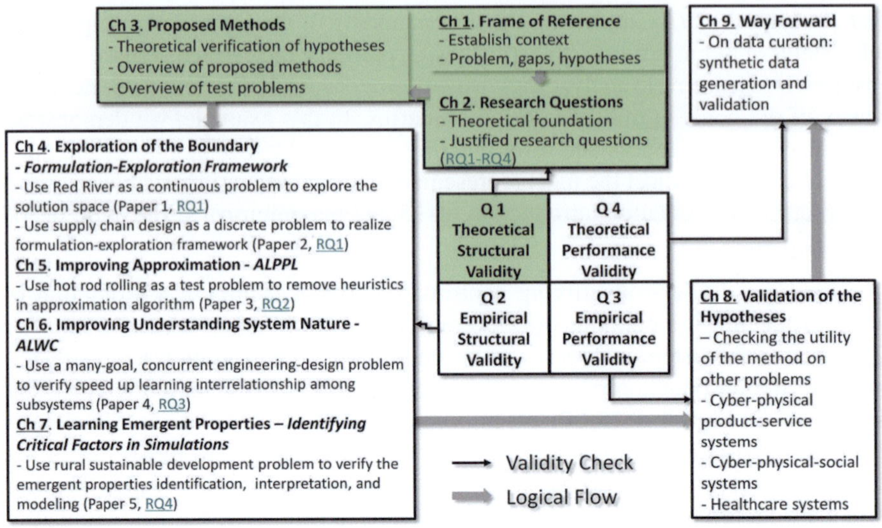

Fig. 3.19 Theoretical structural validity

Fig. 3.20 The role of this chapter in the knowledge foundation

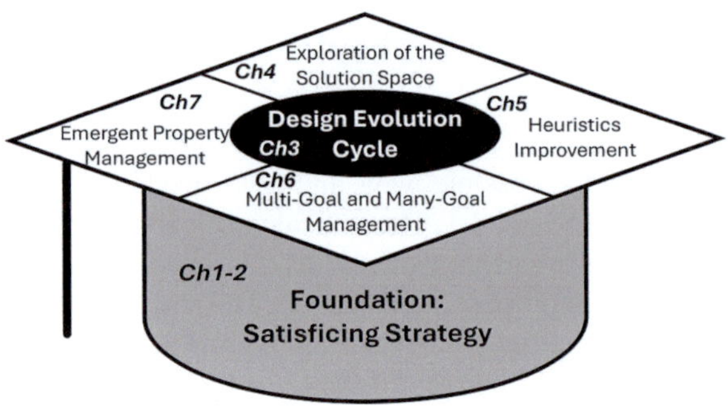

References

Barlas, Y., & Carpenter, S. (1990). Philosophical roots of model validation: Two paradigms. *System Dynamics Review, 6*(2), 148–166.

Choi, H.-J. (2005). A Robust Design Method for Model and Propagated Uncertainty, The George W. Woodruff School of Mechanical Eningeeering, Georgia Institute of Technology, Atlanta, GA, PhD Dissertation.

Diewert, E. (1974). Applications of duality theory. In M. D. Intriligator & D. A. Kendrick (Eds.), *Frontiers of quantitative economics* (Vol. 11, pp. 106–199).

Guo, L., Milisavljevic-Syed, J., Wang, R., Huang, Y., Allen, J. K., & Mistree, F. (2023). Managing multi-goal design problems using adaptive leveling-weighting-clustering algorithm. *Research in Engineering Design, 34*(1), 39–60.

Guo, L., Nellippallil, A. B., Smith, W. F., Allen, J. K., & Mistree, F. (2024). An adaptive linear programming algorithm with parameter learning. *Algorithms, 17*(2), 88.

Guo, L., Nellippallil, A.B., Smith, W. F., Allen, J. K., & Mistree, F. (2020). Adaptive linear programming algorithm with parameter learning for managing engineering-design problems. In *ASME design automation conference*. Virtual. Paper Number DETC2020–22602.

Hajihashemi, S. (2023). A User-friendly wrapper for DSIDES (Decision Support in the Design of Engineering Systems). MS Thesis. Industrial and Systems Engineering, University of Oklahoma.

Mistree, F., Hughes, O., & Phuoc, H. (1981). An optimization method for the design of large, highly constrained complex systems. *Engineering Optimization, 5*(3), 179–197.

Mistree, F., Hughes, O. F., & Bras, B. (1993). Compromise decision support problem and the adaptive linear programming algorithm. In: M. P. Kamat (Ed.), *Progress in astronautics and aeronautics: Structural optimization: Status and promise* (pp. 769–806). AIAA.

Nellippallil, A. B., Allen, J. K., Gautham, B., Singh, A. K., & Mistree, F. (2020). *Architecting robust co-design of materials, products, and manufacturing processes*. Springer.

Chapter 4
Types I and II Robust Design Through Formulation-Exploration Framework

Exploration of the Boundary of the System

Imagine you are the water resource manager of a dam network, tasked with managing water levels amidst uncertain precipitation and fluctuating water demand. How do you decide the amount of water each reservoir should store or release monthly? How do you assess whether the capacity of a reservoir is appropriate or whether the pool allocation within a reservoir is reasonable?

Now, consider managing a three-echelon, push–pull supply chain with significant fluctuations in supply and demand. How do you determine the best location for the customer-order decoupling point (CODP), where the push strategy transitions to a pull strategy? How do you choose the best trade-off scenarios among profit, service level, and reliability variance at each echelon? How do you adjust your decisions as a product progresses through different stages of its life cycle?

In both examples, decision models must account for uncertainties in parameters and variables, as well as the possibility of incomplete or inaccurate information. Some model elements may be more sensitive than others, potentially affecting the overall outcomes.

In this chapter, we introduce the Three-Step Exploration Method (Fig. 4.9*), which helps designers identify and improve the model's sensitive elements*

We then propose a framework incorporating the Three-Step Exploration Method-Method for managing mixed-variable, coupled decision models: the Formulation-Exploration framework (Fig. 4.22).

4.1 Organization of This Chapter

The logical flow between this chapter and the previous chapters is illustrated in Fig. 4.1. In Sect. 4.2, we frame the reference in the context of satisficing design strategies regardless of the specific field of applications or test problems. In Sect. 4.3, we use the dam-network problem to empirically verify the Three-Step Exploration

L. Guo et al., *Architecting a Satisficing Strategy for Designing Complex Systems*,
https://doi.org/10.1007/978-3-032-04577-5_4

Method. In Sect. 4.4, we use the supply chain problem to empirically verify the Formulation-Exploration framework. In Sect. 4.5, we describe the role of thus chapter in this monograph. In Table 4.1 we summarize the interpretation of the problems investigated in this chapter.

Let us recall Research Question 1 (RQ1): ***What is the method to evolve the model boundary?***

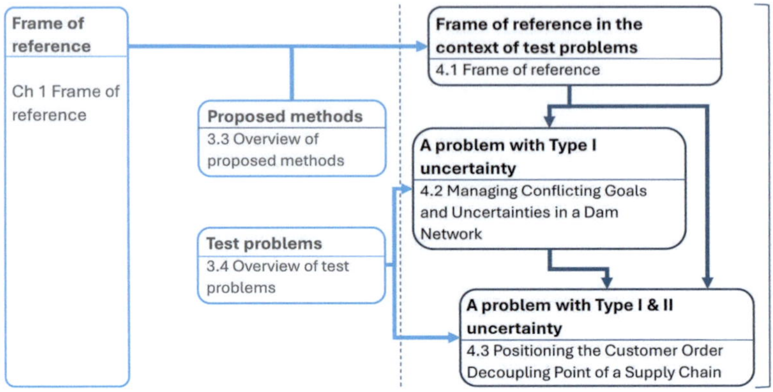

Fig. 4.1 Organization of this chapter

Table 4.1 Summary interpretation of the problems investigated in this chapter

Elements	What?
Components	The formulation and exploration stages of the design evolution cycle
Connectors	The Three-Step Exploration Method and the Formulation-Exploration framework pass information between the formulation and exploration stages, learn knowledge to replace heuristics on the boundary of a decision model and the corresponding physical systems, and make cross-stage modifications
Form	*How?*
Component roles	Provides the constructs for connecting the formulation and exploration stages and enabling model improvement
Properties	Attributes or features of the formulation and exploration stages—In the formulation stage, designers apply heuristics (e.g., the relation between model boundary and system performance) to proceed with the design in the early stage. In contrast, acquiring insight to improve the formulation is feasible in the exploration stage
Relationship	The relations among the four activities—heuristics application, exploration of the solution space, insight awareness, and heuristics replacement
Rationale	*Why?*
Motivation	Provide support to assist designers in improving model formulation to realize Type I and II Robust design through exploring the solution space
Assumptions	Vertex solutions on the boundary of the mathematical model but away from the boundary of the physical system are relatively insensitive to uncertainties that lead to boundary variations
Interpretation	Understanding how sensitivity analysis, weight scenario exploration, and physical boundary exploration help in solution space exploration

The plan of specifying RQ1 in the context of the test problems is shown in Table 4.2. In this chapter, the Proposed Method 1 (M1), Formulation-Exploration framework, is empirically verified (EVe1) using two test problems, designing a dam network by controlling the water outflow (T1.1) and designing a supply chain by positioning the customer order decoupling point (T1.2). Question 1 (RQ1) is specified in the context of the test problems (SQT1) and answered (AQ1) by testifying M1. The empirical validation and theoretical validation are in Chaps. 8 and 9.

In this monograph, we define the model boundary as the equations that bound the feasible solution space of a computational model. Therefore, we must explore the solution space to evolve the model boundary. To answer RQ1, we need to know the formulation and procedures that allow the exploration of the solution space boundary. When it is necessary to change the model boundary, we need to send that information to the formulation stage to update the model, establishing a connection between the formulation and exploration stages in the design evolution cycle (see Fig. 4.2).

4.2 Frame of References on Satisficing Strategy

George Box, a British mathematician and professor of statistics, wrote "essentially, all models are wrong, but some are useful" (Box and Draper, 1987). In keeping with George Box's observation, the decision maker must be able to work constructively with decision models that are typically incomplete and inaccurate (Simon,

Table 4.2 Plan of Specifying Question 1 (RQ1) and Empirically Verifying the Formulation-Exploration Framework (M1)

Chapters	Ch1	Ch2	Ch3	Ch4–7		Ch8	Ch9
				Ch4	Ch5–7		
Actions	RG H	RD RQ SH	TVe M	EVe1: use two test problems with uncertainties in parameters and variables to verify SH1 and demonstrate M1	EVe SQT AQ	CQ EVa	TE
				SQT1: Specify RQ1 in the context of the test problems—designing a dam network and designing a supply chain: Q1.1—What is needed for a designer to make a satisfactory decision concerning different design preferences and requirements in a changing design environment? Q1.2—What is needed for a designer to check the flexibility of the design in the face of errors in the model and variations in the environment? Q1.3—What modification can be made to the cDSP to improve feasibility robustness? Q1.4—How can designers improve the robustness (insensitivity) of the design of the customer order decoupling point without sacrificing the performance of the supply chain to facilitate mass customization?			

(continued)

Table 4.2 (continued)

Chapters	Ch1	Ch2	Ch3	Ch4–7			Ch8	Ch9
				Ch4		Ch5–7		
				AQ in T1.1 dam-network design: AQ1.1—With design scenarios representing different design preferences, designers identify the satisficing area of the weights of the multiple goals and provide their physical meanings AQ1.2—Using inflow scenarios considering different weather and climate conditions to identify the sensitive segments and exploring the practicality of removing the sensitive segments by modifying the mathematical model and adjusting the physical system AQ1.3—Using the information on the practicality of modifying the model and changing the system, improvements can be made, including changing the parameters' value to add buffers and reallocating water to different pools to change the physical boundary. Then, go through the three steps again. Running such a loop improves the model and provides satisficing solutions relatively insensitive to uncertainties	AQ in T1.2 Supply Chain design: AQ1.4—Using the Formulation-Exploration framework, designers can explore the solution space using representative design scenarios and the potential of boosting the system's performance by adopting physical means. By doing this in iterations, designers can exploit all means that lead to reaching the most desirable satisficing solution space given the resources			
Nomenclature	AQ—answer questions CQ—closure the answers to questions EVa—empirically validate hypotheses EVe—empirically verify hypotheses H—give hypotheses M—introduce methods RD—tie to roust design RG—give research gaps RQ—pose questions SH—specify hypotheses SQT—specify questions in the context of test problems TE—theoretically extend the research TVe—theoretically verify hypotheses							

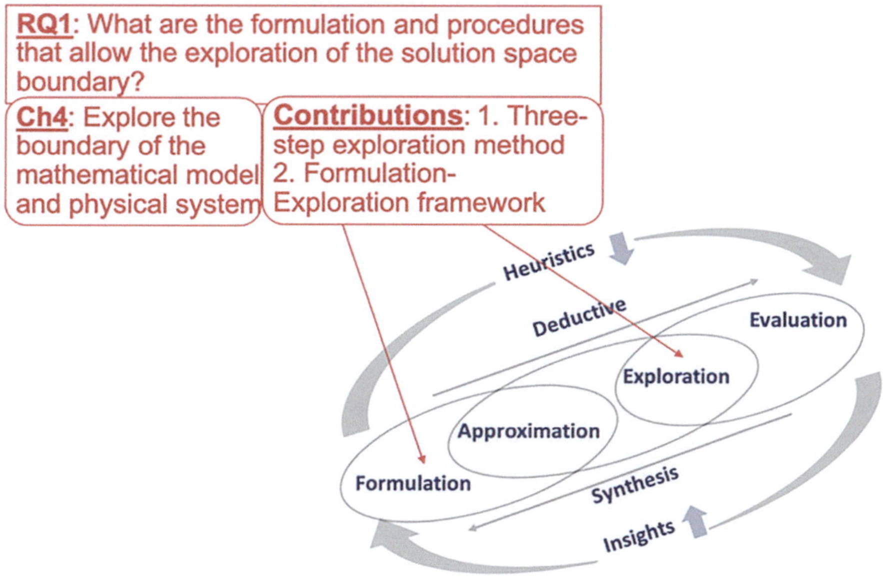

Fig. 4.2 Specified research question 1 and the relevant stages to be connected in the design evolution cycle

1996). The analysis embodied in a decision model does not represent the physical world completely and accurately, making it impossible to predict the future state exactly. A designer can work around this limitation by identifying relatively insensitive solutions to inaccuracies embodied in the analysis models; see (Triantaphyllou and Sánchez, 1997).

In this chapter, a construct to exercise a decision model, the compromise Decision Support Problem (cDSP), is introduced. Based on the cDSP construct, the exploration of the solution space is presented. The Three-Step Exploration Method and the Formulation-Exploration framework are presented to facilitate the exploration. The method and the framework allow a designer to ascertain to what extent the solution is insensitive to errors inherent in the modeling of the decision problem (Sabeghi et al., 2015), and answers to key questions (extended questions based on RQ1) such as:

Q1.1—*What is needed for a designer to make a satisficing decision concerning different design preferences related to the different requirements in a changing design environment?*

Q1.2—*What is needed for a designer to check the flexibility of the design in the face of errors in the model and variations in the environment?*

Q1.3—*What modification can be made to the cDSP to improve feasibility robustness?*

Moreover, suppose it is for a particular system, such as a push—pull supply chain with a customer order decoupling point (CODP). In that case, the research question can be further specified in the context of the problem:

Q1.4*—How can designers improve the robustness (insensitivity) of the CODP design without sacrificing the supply chain's performance to facilitate mass customization?*

The compromise Decision Support Problem (cDSP) is a multi-objective decision model (Mistree et al., 1993) which enables a designer to determine values of design variables that satisfy a set of constraints to achieve a set of goals (Chen et al., 1995). The objective is to minimize the deviations of different goals from target values using lexicographic minimization (Sabeghi et al., 2016).

In a decision problem, different stakeholders have dissimilar perspectives and must be accommodated (Aitken et al., 2003). The major challenge is determining the weights assigned to linearly combine the deviation variables (Neely et al., 1980). Understanding the inherent choices and risks within a design context leads to justifiable decisions (Smith et al., 2015).

In a multidisciplinary system, managing conflicting goals is critical to making the system sustainable despite considerable uncertainty. The priorities of multiple goals and resource availability vary depending on the external environment. Uncertainties such as parameter variations and consequent variations in constraint boundary can intensify the discrepancies between the designers' desire (the target of the goals) and what designers achieve (the completeness of the goals). To reduce such discrepancies, we seek to *satisfice* the goals, considering typical uncertainties.

4.3 Managing Conflicting Goals and Uncertainties in a Dam Network

–Test Problem 1.1 (T1.1): Apply ESS to a Continuous Problem

The new knowledge from managing Test Problem 1.1 is:

A method that allows us to identify the sensitive elements of the model and improve the model accordingly—the Three-Step Exploration Method.

In a multi-reservoir system, ensuring adequate water availability while managing conflicting goals is critical to making the social-ecological system sustainable despite considerable uncertainty. The priorities of multiple user groups and water resource availability vary with time, weather, and other factors. Uncertainties such as variations in precipitation can intensify the discrepancies between water supply and demand. To reduce such discrepancies, *satisficing* conflicting goals is desired.

It is observed that models are incomplete and inaccurate, which calls into question using a single-point solution and suggests the need for robust solutions to uncertainties. So, *satisficing* solutions are explored so that the solutions that are relatively insensitive to uncertainties are identified and associated with different design preferences. Sensitive elements of the model are identified, and the model is improved accordingly. In this section we present an example of the exploration of the solution space to enhance sustainability in multi-disciplinary systems, when goals conflict, preferences are evolving, and uncertainties add complexity. The proposed three-step method of solution space exploration can be applied in mechanical design.

Nomenclature for Sect. 4.3.

CCLP	Chance Constraint Linear Programming
cDSP	compromise Decision Support Problem
DP	Dynamic Programming
DSP	Decision Support Problem
GP	Goal Programming
IP	Interior Point
IS	Inflow Scenario
LHS	Left-hand-side
LP	Linear Programming
MILP	Mixed Integer Linear Programming
NFP	Network Flow Programming
RHS	Right-hand-side
SLP	Stochastic Linear Programming
WS	Weight Scenario

4.3.1 Problem Statement—Test Problem 1.1: Dam-Network Planning

The Red River is a tributary of the Atchafalaya River, a Mississippi River distributary, and flows separately into the Gulf of Mexico. We use a part of its dam-network system on the border of Oklahoma and Texas (Fig. 4.3). In 2015, the Red River basin experienced a severe drought followed by flooding; both impacted people, the planet, and profit. Therefore, managing fresh water supply and sensible distribution to support human activity while sustaining a vigorous, effective ecosystem is a major ecological challenge (Poff et al., 2016). We use data from a recent large-scale, comprehensive analysis of the hydrology, societal water usage, and water availability for the Red River by Xue and coauthors (2015) and McPherson and coauthors (McPherson et al., 2016).

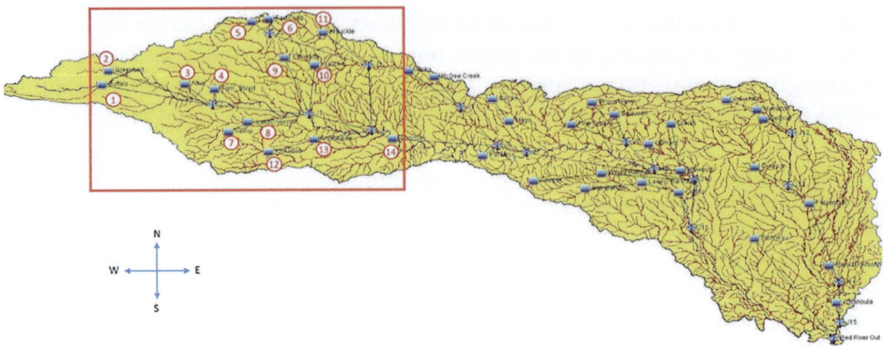

Fig. 4.3 Dams along the Red River basin

Goals: We have three user groups in the basin—people, fish in the reservoirs, and fish in the streams between reservoirs. To meet water demands, there are three goals:

- To reach the target for water storage in reservoirs.
- To meet people's demand for water—including agricultural and municipal demand.
- To meet the water requirements for the fish in streams.

We wish to minimize the difference between water supply and demand for each goal. Water volume is measured in cubic feet.

Problem Size: We consider a 14-dam network, shown in the rectangle in Figs. 4.3 and 4.4. We select these fourteen dams for three reasons:

1. Independence. This 14-dam network is a relatively independent sub-network in the 38 dams in the Red River basin, and its interaction with other dams is by managing Dam 14, Texoma.

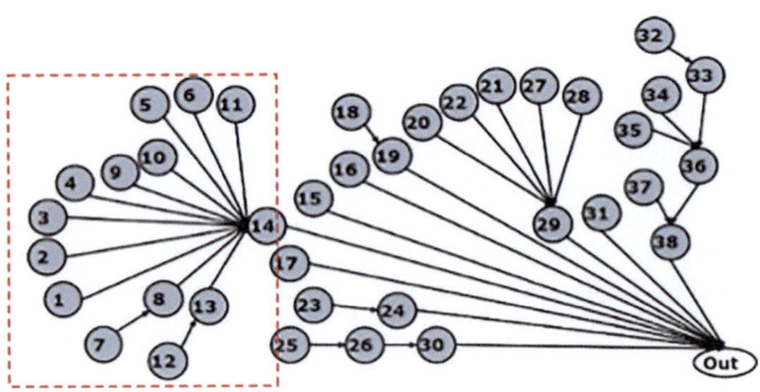

Fig. 4.4 The 14-Dam network

2. Representativeness. This 14-dam network is representative of a sub-network of the Red River basin because of its features, such as a hierarchical upstream–downstream structure and key nodes that connect with ordinary nodes.
3. Direct interactions with other dams. This 14-dam network is used to release water directly to other dams (not in the rectangle in Fig. 4.3) in the Red River basin. This release goes through a single dam—Texoma. The structure of the 14-dam network is shown in Fig. 4.3.[1] The physical features of the network are given in detail in Sect. 4.3.4.
4. Avoid repeated calculation and analysis. In this section, we focus on the method rather than the results. We use the 14-dam network to prove the utility of the proposed method. The size can be enlarged, and the time scale can be changed—we can apply our method in the 38-dam network, we can manage 12 months rather than 3 months, we can use a smaller time period such as one week or one day to further reduce the risk, and we can incorporate uncertainties in outflows.

Here, we consider three-month planning horizon and a one-month planning time unit. Each dam corresponds to a reservoir, and each node represents one dam and the reservoir behind it.

Variables and Parameters: The total water inflows of each reservoir have two parts:

- Water released from the direct upstream dams of Dam d ($d' \in D^{(d)}$) in the previous month (t'), $\sum_{d' \in D^{(d)}} F_d^{t'}$. This water conserves fish in the streams and becomes the inflow of Reservoir d. These inflows are crucial for planning and are controlled by designers, so they are decision variables.
- Natural water from the outside of Reservoir (d) in month t includes tributary inflow (I_d^t) and precipitation (Pr_d^t), $I_d^t + Pr_d^t$. These are crucial to the results but cannot be controlled, so they are uncertain parameters.
- The total water outflows of each reservoir have three parts:
- Water released for people by Dam (d) in Month t, A_d^t, includes agricultural and municipal use. They are decision variables.
- Water released by Dam (d) in Month t, F_d^t F_d^t, for the fish living between Dam d and its downstream dams. They are critical and can be controlled, so they are decision variables.
- Natural water loss in Reservoir (d) in Month t includes evaporation (E_d^t) and seepage loss (P_d^t), $E_d^t + P_d^t$. From the historical data, the natural loss of the 14 dams is only 2% the total water storage volume, and the natural loss in 12 months is only 11% of the variance of theinflows. So, the natural loss and uncertainty are ignorable compared to water storage. Therefore, they are defined as constant parameters. The water stored in Reservoir d in Month t, S_d^t, is used to conserve the fish in the reservoir. They are important to the planning and can be controlled, so they are decision variables.

[1] The nodes' position and the arches' length in Fig. 4.4 are different from those in Fig. 4.3. They are changed for the convenience of visualization. The network structure in the two figures is the same.

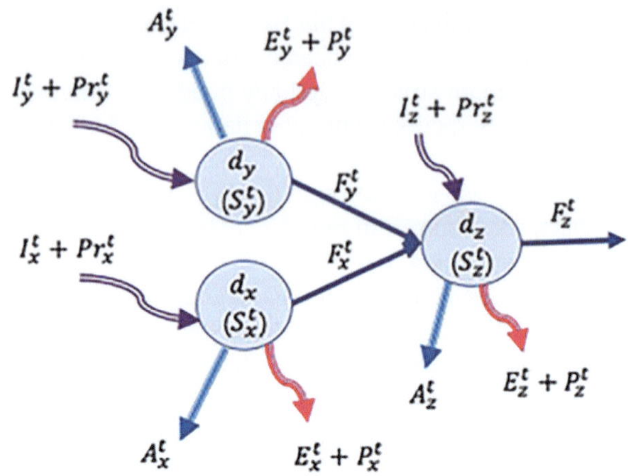

Fig. 4.5 A small part of the dam-network in the Red River Basin

In Fig. 4.5, we illustrate a part of the dam network with three dams—Dam x, y, and z—and show how these decision variables and parameters affect the system.

In addition to the variables and parameters of inflows and outflows, each goal has a target value.

- The target value of water stored (S) in Reservoir (d)in Month t, ST_d^t
- The target value of water released for people (A) by Dam (d) in Month t, AT_d^t
- The target value of water released for fish (F) by Dam (d) in Month t, FT_d^t

With the parameters and variables, here is the mathematical form of the goals—Eqs. 4.1, 4.2 and 4.3. To avoid over-supply or under-supply, we use the sum of the squares of the difference between water supply and water demand in each month so that each goal indicates the achievement of water demand for one user group.

$$\sum_{d\epsilon D}\sum_{t\in T}\left(1-\frac{S_d^t}{ST_d^t}\right)^2+v_1^- - v_1^+ = 0 \tag{4.1}$$

The goal for water stored in Reservoir (d) in Month t

$$\sum_{d\epsilon D}\sum_{t\in T}\left(1-\frac{A_d^t}{AT_d^t}\right)^2+v_2^- - v_2^+ = 0 \tag{4.2}$$

The goal is for water to be released to people by Dam (d) in Month t

$$\sum_{d\epsilon D}\sum_{t\in T}\left(1-\frac{F_d^t}{FT_d^t}\right)^2+v_3^- - v_3^+ = 0 \tag{4.3}$$

The goal for water released to fish by Dam (d) in Month t.

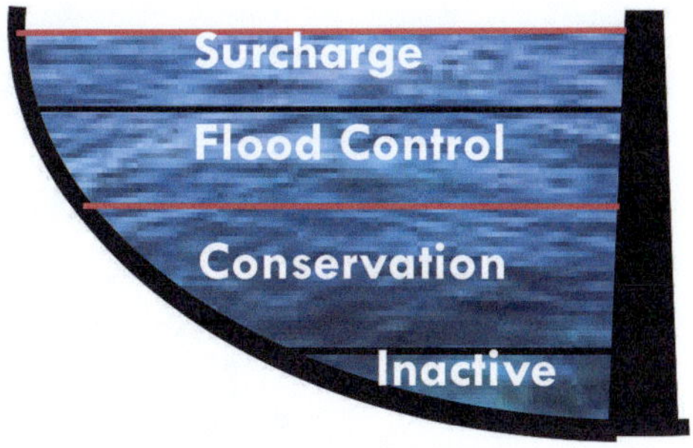

Fig. 4.6 The pools of a reservoir

Bounds: As shown in Fig. 4.6, each reservoir has several pools for different functionalities. The water level in the inactive pool never changes; the conservation pool is to conserve fish. Within the conservation pool, a certain volume is necessary to contain enough microorganisms and nutrients for the fish to live and reproduce healthily. The flood control pool is reserved for flood runoff and must be evacuated immediately to prepare the space for the next flood. On the top of the reservoir, there is a surcharge pool. We do not plan to use the surcharge pool because it is the last backup space.

For Reservoir d, there is a lower bound of water storage volume, CM_d, to guarantee that the amount of water in the conservation pool is acceptable and does not cause system failures. Its upper bound, CF_d, is the capacity of the inactive, conservation, and flood control pools.

The lower bound of water release volume, both for fish (F_d^t) and for people (A_d^t), is zero since water outflows subtract backflows are always positive. We do not have an upper bound of F_d^t and A_d^t. Goal 2 and Goal 3 are set to restrain the overachievement of water released for people and fish.

In summary, the bounds are the capacity of reservoirs and the upper/lower limit of water release volume and water storage volume—Eqs. 4.4–4.7.

$$\mathrm{S_d^t} \leq CF_d \tag{4.4}$$

The upper bound of the water released by Dam (d) in Month t

$$\mathrm{S_d^t} \geq CM_d \tag{4.5}$$

The lower bound of the water released by Dam (d) in Month t

$$\mathrm{F_d^t} \geq 0 \tag{4.6}$$

The lower bound of the water released to fish by Dam (d) in Month t

$$\mathbf{A_d^t} \geq 0 \tag{4.7}$$

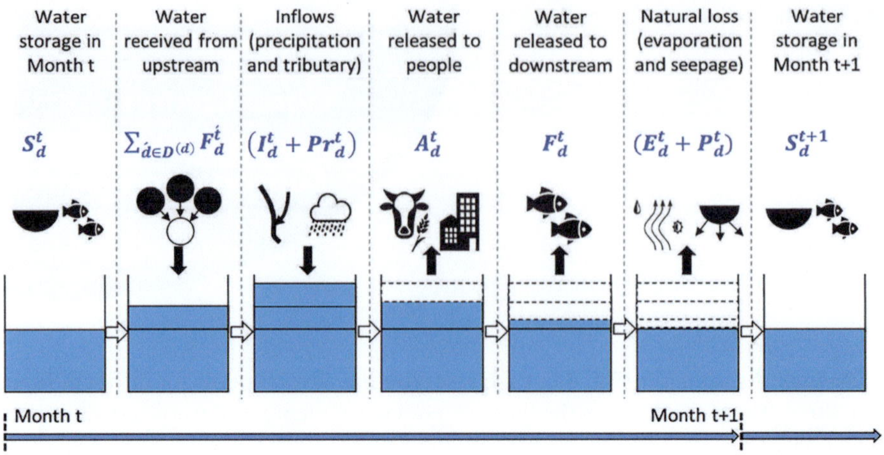

Fig. 4.7 Illustration of the equality constraints for dam (Reservoir) d

The lower bound of the water released to people by Dam (d) in Month t.

Constraints: to the water storage of Reservoir d at the beginning of Month t, S_d^t, the total water inflow is added, $\sum_{d'\in D(d)} F_d^{t'} + I_d^t + \mathrm{Pr}_d^t$ and the total water outflow is subtracted, $-F_d^t - A_d^t - E_d^t - P_d^t$, so the reservoir's water storage at the beginning of the next month, S_d^{t+1}, is obtained. (Eq. 4.8 and Fig. 4.7)

$$S_d^t + \sum_{\forall d' \in UD_d} F_d^t + I_d^t + \mathop{\mathrm{Pr}}_d^t - F_d^t - A_d^t - E_d^t - P_d^t = S_d^{t+1}, \text{ where } t = 1, 2, 3 \quad (4.8)$$

In this model, we have water storage volume as a goal. Meanwhile, we have upper bounds and lower bounds of water storage volume as bounds. The meaning of the goal and bounds are different. The water stored in each reservoir ensures the healthiness of the conservation pool to conserve the fish. The upper and lower bounds of water storage guarantee operable conditions, ensuring the reservoirs can run smoothly with a low risk of system breakdown. However, the water demand of the fish in reservoirs is more specific. In other words, the boundaries of the water storage in a reservoir allow operators to have a larger range of operations. In contrast, the target for water volume in the reservoir is an ideal value that boosts the environmental condition.

In this section, only the quantity of the water is considered without considering the water quality. The flows for each month are planned, and the daily differences are ignored.

Uncertainty: Based on our data, we simultaneously manage two uncertainties in water inflows—the variations in precipitation and tributary inflow under present-day conditions. The uncertainties are modeled by considering different inflow scenarios (ISs) of precipitation and tributary inflow. Each IS represents a typical weather and climate condition, from extreme drought to flood and uneven cases such as drought followed by flood. We use ISs instead of stochastic variables to incorporate the uncertainty to avoid making assumptions about the uncertainty distribution. This is explained further in Sect. 4.3.3.

4.3.2 Critical Review of the Literature on Dam-Network Water Resource Management

Challenges such as high dimensionality and computational complexity are encountered by dam-network designers (Reddy and Kumar, 2007)—multiple user groups compete for a limited resource, and designers must deal with complicated variables. These variables include precipitation, tributary inflow, dam storage, irrigation, and municipal water demands (Rani and Moreira, 2010). They are complicated because they often bring uncertainties into the system. Once a dam network is planned, operational plans and evaluations should make the system's performance meet the water demand of all user groups as closely as possible.

Scholars have tried many methods to plan dams to make the reservoirs serve people's goals. Popular methods of dam water planning include Linear Programming (LP), Mixed Integer Linear Programming (MILP), Stochastic Linear Programming (SLP), Chance Constrained Linear Programming (CCLP), Network Flow Programming (NFP), Interior Point (IP), Nonlinear Programming (NLP), Dynamic Programming (DP), and Goal Programming (GP) (Wurbs). However, there are limitations to each of these methods. See Table 4.3.

Using deterministic LP, Crawley and Dandy, Crawley and Dandy (1993) determine optimal operating policies using an LP model, Loucks (2000) sizes reservoir capacities, Dahe and Srivastava (2002) apply yield models to design an eight-reservoir system, Needham and coauthors (2000) offer methods for flood control, and Vedula and coauthors (2005) make an optimal conjunctive use policy for irrigation in a reservoir-canal–aquifer system. Nonlinear features or uncertainties in the system cannot be captured and managed in LP, such as uncertainties in water inflow and outflow.

Stochastic Linear Programming (SLP) and Chance Constrained Linear Programming (CCLP) are widely used to manage uncertainties, such as variations in parameters. Loucks and coauthors (1981) use an SLP to get optimal steady-state probabilities for dam outflows and storage, assuming the inflows follow a single Markov chain. Sreenivasan and Vedula (1996) apply a CCLP to determine the optimal hydropower production while satisfying irrigation demands at a specified level of reliability. Van Ackooij and coauthors (2014) present a cascaded reservoir optimization problem with the uncertainty of inflows in a joint CCLP setting and present an iterative algorithm for solving similarly structured problems that require a Slater point and the computation of gradients. SLP has been used with assumptions about the distribution of stochastic variables, which may cause errors. Another problem with SLP and CCLP is that they only guarantee a small probability of system failure but cannot ensure that for each failure, the severity is within a manageable level.

Table 4.3 Gaps and limitations of the methods in the literature

Methods in literature	Manage nonlinear/nonconvex features	Manage uncertainties in inflows/outflows	Manageable computational complexity	Relatively stable performances as size increases	Relatively stable performances as dimensionality changes	Gaps and limitations
Deterministic Linear Programming (LP)			*	*	*	Cannot manage nonlinearity, nonconvexity, discontinuity, or variations in parameters
Mixed Integer Linear Programming (MILP)	*					Computational complexity increases exponentially as the number of integer variables increases
Stochastic Linear Programming (SLP)	*	*	*	*	*	The assumptions of the distribution of stochastic variables may be incorrect
Chance Constrained Linear Programming (CCLP)	*	*	*	*	*	It can be used to decrease the frequency of system failures but cannot manage the severity of each system failure
Network Flow Programming (NFP)			*	*		It cannot be used to evaluate the structure of the network and output improvement suggestions
Interior Point Methods (IP)	*			*	*	IP methods are efficient only when the problem is large-scale and relatively hard to implement

(continued)

Table 4.3 (continued)

Methods in literature	Manage nonlinear/ nonconvex features	Manage uncertainties in inflows/outflows	Manageable computational complexity	Relatively stable performances as size increases	Relatively stable performances as dimensionality changes	Gaps and limitations
Nonlinear Programming (NLP)	*	*				It could be more computationally efficient for engineering design purposes and may trap the designer in a local optimum
Dynamic Programming (DP)			*			Curse of dimensionality—adding extra dimensions in Euclidean space causes exponential increases in calculation volume
Goal Programming (GP)			*	*	*	*Preemptive GP:* An error in the choice of a primary goal can prevent a possible large improvement in a secondary goal, yet setting a hierarchy of the goals is based on the designers' domain knowledge and intuition rather than general quantified methods; *Weighted GP*: It is difficult to find the appropriate values of the goal weights and evaluate their rationality

Since the reservoirs work as a network, Network Flow Programming (NFP) deals with multi-reservoir system planning. It is considered to be a computationally efficient form of LP. Ford and Fulkerson (1962) treat reservoir systems as general configurations of capacitated networks, and they maximize the flow while minimizing the cost. Kuczera and Diment (1988) discuss the principles of formulating network LPs and apply these principles to develop a simulation model, WASP (Water Assignment Simulation Package). Hsu and Cheng (2002) apply a generalized NFP model for long-term supply–demand analysis for basin-wide water resources planning. NFP allows us to manage the amount and timing of the flow in the network as a whole, but one of its limitations is that we cannot determine the potential for improvement of the network structure. Interior Point (IP) Methods are efficient in solving large problems. Ponnambalam and coauthors (1989) introduce Karmarkar's interior point LP (IPLP) approach to reservoir operation, showing that the IPLP algorithm is capable of solving large multi-reservoir operation problems faster than the simplex method. Seifi and Hipel (2001) apply an improved IP method to multi-reservoir operation planning and show how exploiting the problem structure enhances algorithm performance. Mousavi and coauthors (2004) use a primal–dual IP algorithm to resolve the dimensionality of a multi-reservoir, multi-functional system and demonstrate the computational efficiency of the proposed method using historical data. However, IP only works better than the Simplex algorithm for large-scale problems, and many IP algorithms are relatively hard to implement.

Managing the multiple functions of reservoir systems is computationally difficult due to the nonlinearity in the complex relationships among physical and hydrological variables. This problem is often solved by approximation (linearization) or successive applications of LP, dynamic programming, or algorithms such as sequential quadratic programming (SQP) and generalized reduced gradients (GRG). Zhou and coauthors (2007) use the combination of entropy and fuzzy optimization to improve reservoir operation. Teegavarapu and Simonovic (2000) use a mixed integer NLP formulation with binary variables to study the daily hydropower operation of four cascading reservoirs. Barros and coauthors (2003) demonstrate that NLP is particularly fit for setting up guidelines for real-time operations using inflow prediction with frequent updating. A well-known drawback of NLP is its computational complexity. Although the NLPs have good fidelity and are suited for real-time operations, they could be more computationally efficient for engineering design purposes. In addition, using NLP may trap a designer in a local optimum without sufficiently exploring the design space. Interior Point (IP) Methods are efficient in solving large problems. Ponnambalam and coauthors (1989) introduce Karmarkar's interior point LP (IPLP) approach to reservoir operation, showing that the IPLP algorithm is capable of solving large multi-reservoir operation problems faster than the simplex method. Seifi and Hipel (2001) apply an improved IP method to multi-reservoir operation planning and show how exploiting the problem structure enhances algorithm performance. Mousavi and coauthors (2004) use a primal–dual IP algorithm to resolve the dimensionality of a multi-reservoir, multi-functional system and demonstrate the computational efficiency of the proposed method using historical data. However, IP only works better than the Simplex algorithm for large-scale problems, and many IP algorithms are relatively hard to implement.

Managing the multiple functions of reservoir systems is computationally difficult due to the nonlinearity in the complex relationships among physical and hydrological variables. This problem is often solved by approximation (linearization) or successive applications of LP, dynamic programming, or algorithms such as sequential quadratic programming (SQP) and generalized reduced gradients (GRG). Zhou and coauthors (2007) use the combination of entropy and fuzzy optimization to improve reservoir operation. Teegavarapu and Simonovic (2000) use a mixed integer NLP formulation with binary variables to study the daily hydropower operation of four cascading reservoirs. Barros and coauthors (2003) demonstrate that NLP is particularly fit for setting up guidelines for real-time operations using inflow prediction with frequent updating. A well-known drawback of NLP is its computational complexity. Although the NLPs have good fidelity and are suited for real-time operations, they could be more computationally efficient for engineering design purposes. In addition, using NLP may trap a designer in a local optimum without sufficiently exploring the design space.

To obtain acceptable computational complexity while managing multi-period plans, Dynamic Programming (DP) is introduced (Rani and Moreira, 2010). Yakowitz (1982) reviews DP models and concludes that computational considerations severely limit the scale of DP problems. Nandalal and Bogardi (2007) discuss the applicability and limits of DP methods, specifically for reservoir operation problems. The well-known "curse of dimensionality," however, is the main drawback of DP (Dreyfus, 1957) because adding extra dimensions in Euclidean space may cause exponential increases in volume (Bellman and Kalaba, 1965).

Goal Programming (GP) is an efficient method for managing multiple conflicting goals (Ignizio, 1982). GP is a multi-objective optimization method that allows the flexible expression of policy constraints as objectives. Clayton and coauthors (1982) developed an approach to goal programming using a modified pattern search routine. Loganathan and Bhattacharya (1990) apply five GP schemes (preemptive, weighted, min-max, fuzzy, and interval), minimizing deviations from a set of preferred target storage and flow values and finding efficient alternative optima. Eschenbach and coauthors (2001) use preemptive GP and combined detailed system representations, policy expression flexibility, and computational speed for routine daily scheduling of large complex multi-objective reservoir systems. Changchit and Terrell (1993) present an application of chance-constrained GP (CCGP) to a system of multipurpose reservoirs for planning rather than real-time operation. The problem with GP is that designers cannot separate the goals into "rigid" and "soft" goals to manage them differently. Moreover, the problem with preemptive GP is that even a tiny difficulty of a primary goal may block a large improvement in a secondary goal. The challenge with weighted GP is determining the appropriate weights to represent the goal's priority.

The limitations of these methods are summarized in Table 4.1. Besides, there is another difficulty. All of these methods are used to find optimal solutions, which are boundary solutions. As models are incomplete and inaccurate, boundary solutions are sensitive to uncertainties in the constraints and bounds. According to George Box (Box and Draper, 1987), “All models are wrong, but some are useful.” In other words, models are typically incomplete, inaccurate, and of different fidelities. Hence, using a multi-objective optimization model to obtain a single optimal solution is questionable, especially in a dam network with various uncertainties, such as variations in precipitation, uncertain tributary inflow, and climate change.

Therefore, instead of searching for optimal solutions, we search for *satisficing* solutions, as described by Simon and Kadane (1975). *Satisficing* solutions are “good enough” solutions. A designer can work around this research gap by providing insensitive solutions that are relatively insensitive to inaccuracies embodied in the models (Triantaphyllou and Sánchez, 1997) and uncertainties that cannot be well captured in the models. The method descriptions are given in Sect. 4.3.3.

4.3.3 Proposed Methods—The Three-Step Exploration Method

As the total water resource is limited, the designer may only meet some user groups' water demands at all times. Hence, we work to minimize the discrepancies between water supply and demand for each user group. We use a method that is a hybrid of mathematical programming and GP—the compromise Decision Support Problem (cDSP) because we can manage “rigid” goals (such as the physical capacity of the reservoirs) as constraints and manage “soft” goals (such as the water demand of different user groups) as goals. Mistree and coauthors (1993) developed the cDSP. It has the following advantages:

1. Generic. Any optimization model has an equivalent representation in the cDSP.
2. Appropriate accuracy and manageable computational complexity. Not only does using the cDSP offer a method for an alternative representation, but it also provides a representation for effectively capturing the nature of real problems by incorporating nonlinear constraints, goals, and mixed variables. After approximating the problem as a linear one, the accuracy of the approximated problem is maintained at a “good enough” level, and its computational complexity is reduced to a manageable level (Mistree et al., 1981).
3. Providing *satisficing* solutions. Using the cDSP construct, designers can explore the solution space to *satisfice* different design preferences, manage multiple types of uncertainties in the system, and improve the model and the physical system.

By exploring the solution space, a designer can identify a space containing *satisficing* solutions that are relatively insensitive to (1) errors anchored in incomplete and possibly inaccurate information embodied in the analysis models used to define the behavior of the dam network and (2) unpredictable uncertainties in parameters. Applying the method may allow a designer to answer three key questions in the context of network planning:

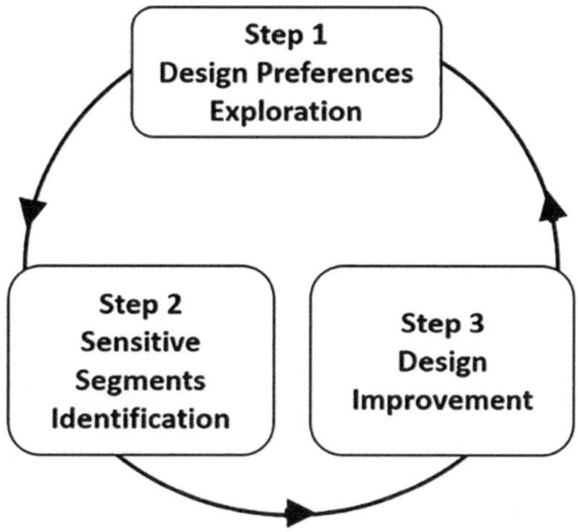

Fig. 4.8 Three steps for the exploration of the solution space

Q1.1-1. How can a designer manage conflicting water resource goals with reasonable priorities under different precipitation scenarios?

Q1.2-1. How can a designer manage uncertainties in the inflows (precipitation and inflows from upstream) by making water flow plans that are relatively insensitive to these uncertainties?

Q1.3-1. How can a designer improve the mathematical model and give recommendations on the physical system improvement so that the system can be relatively insensitive to uncertainties?

To answer the three key questions, we hypothesize three steps to be accomplished—design preference exploration, sensitive segment identification, and design improvement, which are formed in a loop (Fig. 4.8).

The detailed procedural steps for exploring the solution space are given in Fok and Chopra (1986), with extensions and applications in Sabeghi et al., (2016). In this section, the method is extended; see Fig. 4.9.

Step 1 (Fig. 4.9**): Design preference exploration**. In this step, the precipitation and tributary forecasts are treated as deterministic values to obtain an overview of the feasible space for the goal weights, ignoring any uncertainty.

- **Step 1.1:** A group of weight scenarios (WSs) representing different goal priorities is used to obtain solutions.
- **Step 1.2:** Using ternary plots to identify the range of weights that satisfice different design preferences.

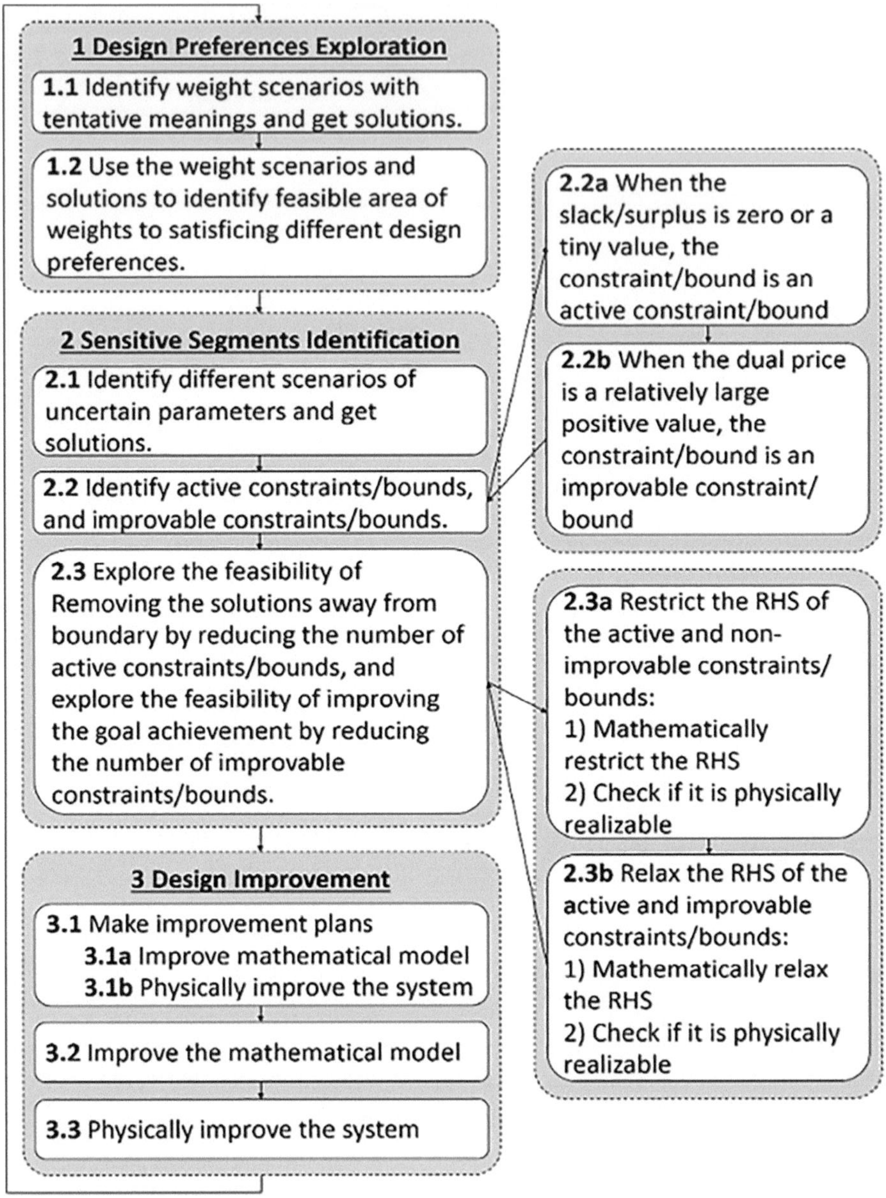

Fig. 4.9 Method for exploration of the solution space

Step 2 (Fig. 4.9**): Sensitive segment identification**. There are two types of sensitive segments:

- **Active constraints/bounds:** Constraints or bounds with zero or very small slack/surplus. We define "very small slack/surplus" as "less than 1% of the larger RHS and LHS of a constraint/bound."
- **Improvable constraints/bounds:** Constraints and bounds with relatively large positive dual prices. We define a dual price greater than "0.1% of the achieved value of the goals" as a relatively large positive dual price.

Active constraints and improvable constraints are sensitive segments because:

- The solutions are always on the boundary formed by active constraints/bounds, so they are relatively sensitive to changes to the active constraints/bounds. When their RHS values change, we may lose the solutions. Therefore, we explore the practicality of moving the solutions away from the boundary to make them relatively insensitive to uncertainties.
- Relaxing the improvable constraints/bounds can help achieve the goals more completely. Hence, we explore the practicality of relaxing them.

In this step, we consider the uncertainties using different parameter scenarios (inflow scenarios—ISs) representing different weather and climate conditions. By analyzing the practicality of moving the boundary solutions and improving goal achievement, we suggest model modifications for system improvement.

- **Step 2.1:** Consider possible weather and climate conditions using representative inflow scenarios (ISs) and obtain solutions for different inflow scenarios.
- **Step 2.2:** With the solutions, identify the sensitive segments.
- **Step 2.3:** Explore the practicality of bringing the solutions away from the boundary by reducing the number of active constraints/bounds and the practicality of improving the goal achievement by reducing the number of improvable constraints/bounds.

Step 3 (Fig. 4.9**): Design improvement.** Using the outcomes from Steps 2.2 and 2.3, we improve the model mathematically or the system physically. We improve the design by modifying the mathematical model or adjusting the physical system.

- **Step 3.1:** Improvements are planned based on Steps 2.2 and 2.3. Determine which improvement actions we shall apply.
  - The sensitive segments can be divided into active and improvable constraints/bounds or active and non-improvable constraints. In other words, all improvable constraints/bounds are active constraints/bounds, but not all active constraints/bounds are improvable constraints/bounds.
  - There are two active and improvable constraints/bounds characteristics: (1) they are too restrictive. Thus, if they are relaxed, the goals will be achieved more completely, and (2) the solutions are on the boundary, so they are sensitive to uncertainties. We need to deal with each type separately. The system is

adjusted physically (Step 3.1b), and for the second type, the mathematical model is refined (Step 3.1a).
  - The active and non-improvable constraints/bounds are of the second type, so we modify the mathematical model (Step 3.1a).

- **Step 3.2:** The mathematical model is improved by adjusting the relevant parameters.
  - Options for modifying the mathematical model:
  - Create a more conservative mathematical boundary to move the solution from the physical boundary. Since the solution is always on the boundary, we "add a buffer" by creating a more conservative mathematical boundary. The solution will remain on the mathematical boundary while avoiding the physical boundary.
  - Manage the uncertainty by obtaining a more accurate assessment of the parameters. The parameters that cause uncertainty make the solution infeasible or close to the boundary, so we can avoid some of such uncertainties through parameter assessment.

- **Step 3.3:** The physical system is improved, and the mathematical model is improved correspondingly.
  - We use the results of Step 2.3 here—if the improvable constraints/bounds can be relaxed by making a physical change to the system, then we make the physical change and then the corresponding mathematical change so that the goals can be achieved more completely. If we cannot make any physical change, we accept the current physical boundary.

With the proposed method, we formulate the model in Sect. 4.3.4.

4.3.4 Formulation of Compromise DSP (cDSP)

In this section, we demonstrate the use of Step 1.1 (Fig. 4.9). Based on the network structure in Fig. 4.4, we summarize the information on each dam in Table 4.4. For example, for Dam 14, Texoma, 11 upstream dams (1, 2, 3, 4, 5, 6, 8, 9, 10, 11, 13) release water to it.

In this step, we use a deterministic precipitation value based on historical data. The cDSP for the 14-dam network is as follows.[2]

[2] In this problem, to separate the Dam "d" from the deviation variable "d^-" and "d^+", the deviation variables are represented by "v^-" and "v^+". For the other problems in this monograph, the deviation variables are all represented using "d^-" and "d^+".

Table 4.4 Features of 14 Dams in Red River Basin

Number	Dams	Set of upstream dams (UD_d)
1	Buffalo	
2	Greenbelt	
3	Altus	
4	Tom Steed	
5	Foss	
6	Fort Cobb	
7	Kemp	
8	Lake Diversion	7
9	Ellsworth	
10	Waurika	
11	Arbuckle	
12	Kickapoo	
13	Arrowhead	12
14	Texoma	1, 2, 3, 4, 5, 6, 8, 9, 10, 11, 13
Out	To the other part of the network	14

Given

System parameters

$D = \{d\} = \{1,2, \ldots, 14\}$//The set of 14 dams (reservoirs)

$T = \{t\} = \{1,2, 3\}$//Planning period—three months

$UD_d = \{d\prime\}$//The set of upstream dams (Table 4.2)

ST_d^t//Target of water storage volume for Reservoir d at the beginning of Month t

FT_d^t//Target of water release volume to downstream fish from Dam d in month t

AT_d^t//Target of water release volume to people from Dam d in month t

$(E_d^t + P_d^t)$//Natural loss (evaporation and seepage) of Reservoir d in Month t

CF_d^t//Flood capacity of Reservoir d

CM_d^t//Minimum storage volume of Reservoir d

$(I_d^t + Pr_d^t)$//Anticipated tributary inflow and precipitation for Reservoir d

w_i, where $i = 1,2, 3$//Weight of Goal i

Find

System variables

S_d^t//The volume of water stored in Reservoir d at the beginning of Month t

A_dd//The volume of water released from Dam d to people in Month t

F_d^t //The volume of water released from Dam d to downstream fish in Month t

Deviation variables

v_i^+, v_i^-, where $i = 1,2, 3$ //Over-achievement and underachievement of Goal i

Satisfy

System constraint

$S_d^t + \sum_{\forall d' \in UD_d} F_{d'}^t + I_d^t + Pr_d^t - F_d^t - A_d^t - E_d^t - P_d^t = S_d^{t+1}$, where $t = 1,2,3$

System goals

$\sum_{d\epsilon D}\sum_{t\in T}(1-\frac{S_d^t}{ST_d^t})^2 + \sum_{d\in D}(1-\frac{S_d^4}{ST_d^4})^2 + v_1^- - v_1^+ = 0$

//Goal 1—Reservoir: Reach the target of water storage in each reservoir at the beginning of each month and the end of the last month

$\sum_{d\epsilon D}\sum_{t\in T}(1-\frac{A_d^t}{AT_d^t})^2 + v_2^- - v_2^+ = 0$

//Goal 2—People: reach the target of water released to people by each dam in each month

$\sum_{d\epsilon D}\sum_{t\in T}(1-\frac{F_d^t}{FT_d^t})^2 + v_3^- - v_3^+ = 0$

//Goal 3—Fish: Reach the target of water released to fish in streams by each dam each month

Bounds

$\mathrm{S_d^t} \le CF_d$

$\mathrm{S_d^t} \ge CM_d$

$\mathrm{F_d^t} \ge 0$

$\mathrm{A_d^t} \ge 0$//Bounds of system variables

$v_i^- \ge 0,$

$v_i^+ \ge 0,$

$v_i^- \cdot v_i^+ = 0$

$where i = 1, 2, 3$ //Bounds of deviation variables

Minimize

The deviation function

$z = \sum_{i=1}^{3} w_i \cdot v_i^-$, where $0 \le w_i \le 1, and \sum_{i=1}^{3} w_i = 1$

//The weighted sum of deviation variables

Explanation of the Goal Formulation: To scale the three goals, we use the variable divided by its target value as the achievement rate of each variable ($\frac{S_d^t}{ST_d^t}$, $\frac{A_d^t}{AT_d^t}$, and $\frac{F_d^t}{FT_d^t}$), and then use "One subtracts the achievement rate" as the difference between supply and demand. In this way, we scale the water demand of different user groups. We calculate the percentage by which we do not meet a user group's demand, instead of the quantity of water (cubic feet). According to our data, the average target for water storage (ST_d^t) is around 390,000 cubic feet. In contrast, the average target for water release for people (AT_d^t) and water release for fish (FT_d^t) are around 30,000 and 51,000 cubic feet, respectively. If we do not scale them, the water storage goal will always dominate the other two goals.

The Utility and Meaning of the Weights: Sub-networks along the Red River basin serve different conditions and circumstances—different populations, crops, wild fish species, and water storage flexibilities; thus, the priority of the goals may vary from dam to dam. Even for a single area, the priority for meeting different users' demands may vary with seasons. To empower decision-makers to have the flexibility of managing goals associated with different design preferences, we assign different weight scenarios (WSs) to the goals and make rules for them to follow. This process is defined as Design Preference Exploration (Fig. 4.9, Step 1).

4.3.5 Water Resource Planning Results and Discussion

Step 1.1 Identify Weight Scenarios and Obtain Solutions: We obtain the results (Table 4.5) by assigning equal weights to the three goals and solving the problem using Lingo 17.0. In this scenario, the three weights are 0.333, 0.333, and 0.333, respectively. The value under each goal is the value of the deviation variable, the sum of the squares of the discrepancies between supply and demand; hence, we prefer small values.

We use more weight scenarios (WSs) to generate and identify *satisfying* solutions. These WSs represent different considerations of the goals' priorities.

Block 1 (Fig. 4.9) **Design Preference Exploration**

Step 1.2 Use Weight Scenarios and Solutions to Identify Feasible Areas of Weights that Satisfice Different Design Preferences. As we have three goals, we use ternary plots to visualize the achievement of each goal corresponding to different weight scenarios (WSs) and identify the meaning of each WS. A ternary plot can expand discrete results to the whole design space (weight space) and visualize the *satisficing* space.

We visualize the location of 8 WSs in a ternary (Fig. 4.10) and give their tentative physical meanings (Table 4.6). The three axes (sides of the triangle) represent the weights of the three goals, so each point in the ternary is a unique WS. The sum of the coordinates along the three axes of any point in the ternary plot is always one. To ensure WSs are representative and cover a variety of design preferences, we solved the problem with 34 WSs and we present the results in Appendix 4.1.

Data Normalization: We normalize the deviations of each goal within the range of [0, 1] for visualization. For each goal, we convert the minimum deviation value to 0, the maximum deviation to 1, and other values between 0 and 1.

Table 4.5 Results for equal weights on preferences

W1	W2	W3	Goal 1 (v_1^+)	Goal 2 (v_2^+)	Goal 3 (v_3^+)
0.333	0.333	0.333	0.49	0.53	0.31

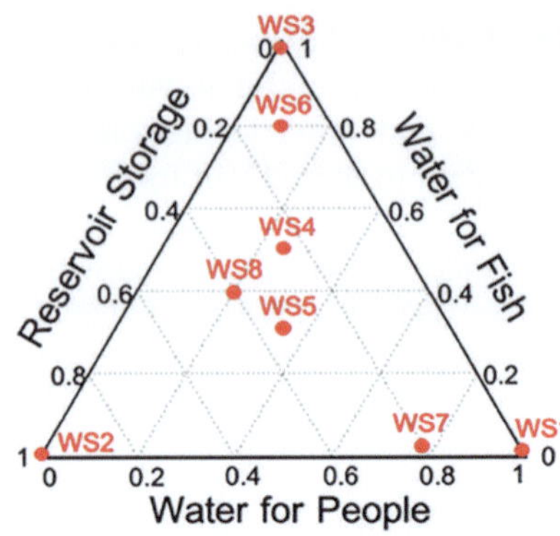

Fig. 4.10 Visualization of the eight WSs in the ternary plot

Table 4.6 Physical Meaning of the eight WSs—Type II uncertainty

WS	W1	W2	W3	Tentative physical meaning
1	0	1	0	People are the only important goal
2	1	0	0	Reservoir is the only important goal
3	0	0	1	Stream fish is the only important goal
4	0.25	0.25	0.5	Reservoirs and people are of equal importance, whereas stream fish are more important than the former two
5	0.33	0.33	0.33	All three goals are equally important
6	0.1	0.1	0.8	Reservoirs and people are not the priority, whereas stream fish are much more important than the former two
7	0.2	0.79	0.01	People are the most important, followed by the reservoir, whereas stream fish have a much lower priority[a]
8	0.4	0.2	0.4	Reservoir and stream fish have equal importance, whereas people are less important than the former two

[a]According to experts in water resources, WS1 and WS7 are often used in reality

Results Visualization: We plotted the results of each goal using color contours in ternary plots (Figs. 4.11, 4.12, and 4.13 for Goals 1, 2, and 3, respectively) The WS with the smallest deviation value we seek is shown in dark blue, with intermediate values represented by colors between dark blue and dark red. We can identify the value of each point by looking up the color bar on the right.

Identifying the Feasible Area of the Weight for Each Goal: The aspiration may not be met for each goal, but with expertise in water resource management,[3] an acceptable value of achieving the goal is identified as a threshold, converted to normalized deviation values of 0.29, 0.32, and 0.3 for reservoir, people, and fish, respectively.

[3] The data and relevant knowledge are from Xue, X., K. Zhang, Y. Hong, J. J. Gourley, W. Kellogg, R. A. McPherson, Z. Wan and B. N. Austin (2015). "New multisite cascading calibration approach for hydrological models: Case study in the Red River basin using the VIC model." Journal of Hydrologic Engineering **21**(2): 05,015,019. and McPherson, R., et al. (2016). "Impacts of climate change on flows in the Red River Basin." Final report to the South Central Climate Science Center.

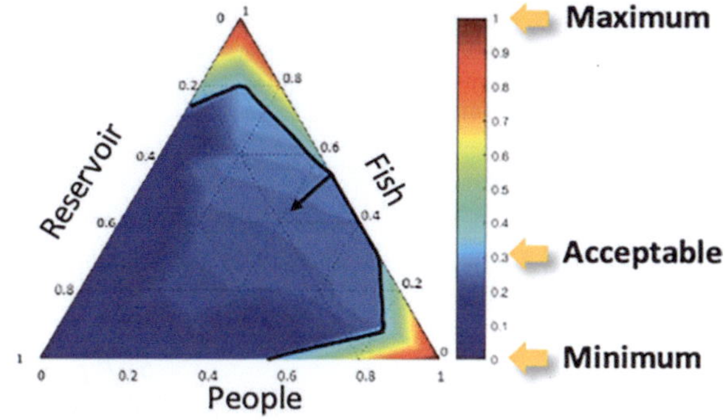

Fig. 4.11 Feasible weight area of goal 1—Reservoir

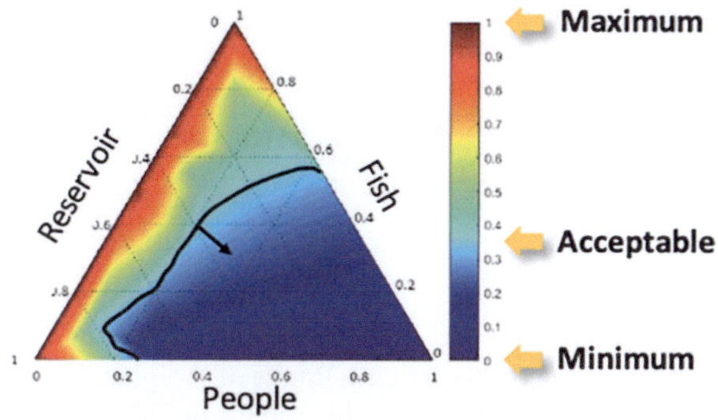

Fig. 4.12 Feasible weight area of goal 2—People

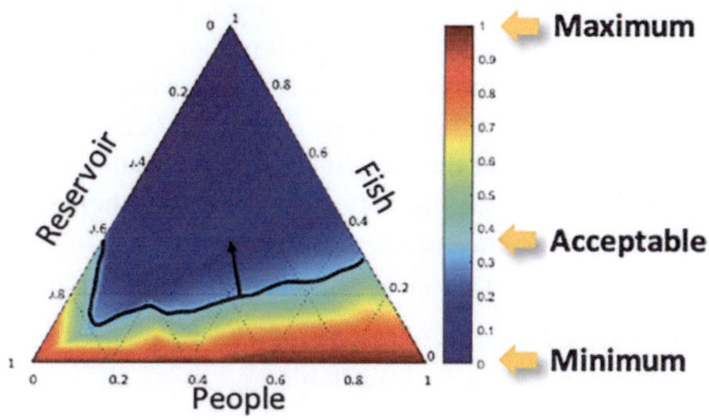

Fig. 4.13 Feasible weight area of goal 3—Fish

For each goal, the weights yielding results that are less than or equal to the threshold are feasible, and the area containing all feasible weights is the feasible area of the weights. In each ternary plot, we illustrate the threshold with a line (the more WSs we use, the smoother the line), and arrows show the region of feasibility for the weights.

Identifying the Satisficing Area of Weights: When we overlap the feasible area for the three weights, we obtain the superimposed area as the *satisficing* area of weights (shaded area in Fig. 4.14).

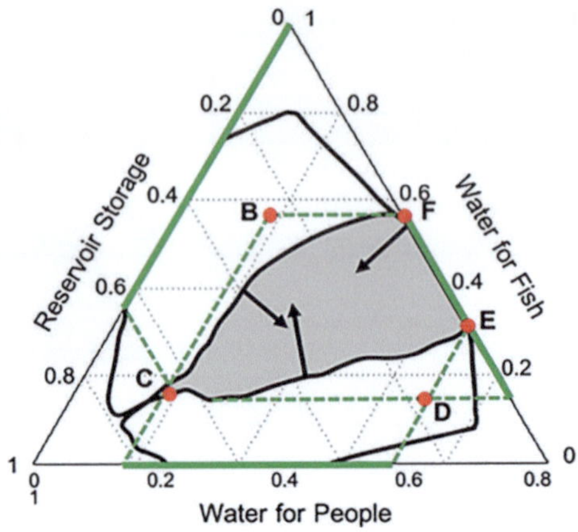

Fig. 4.14 Satisficing weight area for three goals[4]

Table 4.7 Range of weights of the satisficing space

Weight	Range
W1	0–0.65
W2	0.18–0.7
W3	0.15–0.56

Identifying the Satisficing Range of Weights: We identify the range of the weights in the satisficing area and show them in Table 4.7, using the ternary plot. We first identify the vertices of the area (C, F, and E). Then, we make lines parallel to the axes crossing the vertices. The range covered by the intersection of these parallel lines and the axes is the range of the satisficing weights. However, such ranges do not guarantee satisficing solutions, as we have enlarged the irregular satisficing area (the shaded area in Fig. 4.16) to the Pentagon BCDEF. For example, Spot A is within the identified range of the weights (Table 4.5) but not in the satisficing area. Therefore, when selecting weights, we need to visualize them in a ternary plot to avoid choosing a point outside the *satisficing* area as a *satisficing* weight. The ranges in Table 4.7 provide only us a rough idea, but they do not ensure satisficing solutions.

WS-Design Preference Look-Up Table: Designers can use the WSs and results in Appendix 4.1 as a look-up table and meet different design preferences by setting appropriate weights.

Block 2 (Fig. 4.9) **Sensitive Segment Identification**

Step 2.1 Identify Different Scenarios of Uncertain Parameters and Obtain Solutions: We use ten inflow scenarios (ISs) to obtain solutions and identify the sensitive segments. The ISs are generated as follows:.

[4] In Fig. 4.14, the solid lines show the *satisficing* weight range of each goal, and the dashed lines are the auxiliary lines that facilitate identifying the *satisficing* weight range. Every auxiliary line is parallel with one of the axes.

Using Inflow Scenarios: To ensure the frequency and severity of system failures are manageable, we test multiple parameters with uncertainties (inflow scenarios—ISs).

In our ISs, we consider extremely dry, flooding, and uneven precipitation scenarios to study the model's sensitivity. The ten ISs are listed in Table 4.8. The "No Rain" scenario means no inflow in three months. The "Flood" scenario indicates that the inflow is four times the forecasted value in three months.

We used the inflow scenarios (ISs) in Table 4.8 and weight scenarios (WSs) in Appendix 4.1 to obtain solutions based on which we identified the sensitive segments.

Step 2.2 Identify Active Constraints/Bounds and Improvable Constraints/Bounds and Obtain Solutions: *Active Constraints/Bounds:* In Table 4.9, we list the active constraints/bounds and their active ISs and WSs. From this, we observe that in all ISs, the lower bound of storage in Reservoir 7 has limited capacity. In the flood scenario, the upper bound of storage in Reservoir 6 also has limited capacity.

Table 4.8 Inflow Scenarios (ISs)—Type I uncertainty

#	Inflow scenarios	The percentage of the inflow based on the forecast value in each month		
		M1 (%)	M2 (%)	M3 (%)
1	No Rain	0	0	0
2	Extremely Dry	20	20	20
3	Dry	60	60	60
4	Normal	100	100	100
5	Rainy	150	150	150
6	Extremely Rainy	200	200	200
7	Flood	400	400	400
8	Rain Unevenly I	0	200	100
9	Rain Unevenly II	400	0	0
10	Rain Unevenly III	400	0	400

Table 4.9 Active bounds in each IS and WS

Active bounds	Inflow scenarios	Weight scenarios	Physical meaning of the bounds
CM_7^2	1–10	All except 2, 4	The lower bound of storage volume in Reservoir 7 in Month 2 is relatively high
CM_7^4	1–10	All except 3	The lower bound of storage volume in Reservoir 7 in Month 4 is relatively high
CF_6^3	7	All except 2, 3, 4	The upper bound of storage volume in Reservoir 6 in Month 3 is relatively low
CF_6^4	7	All except 2, 3, 4	The upper bound of storage volume in Reservoir 6 in Month 4 is relatively low

Table 4.10 Improvable bounds in each IS and WS

Improvable bounds	Inflow scenarios	Weight scenarios	Physical meaning of the constraints of bounds
CF_6^3	1–10	1, 3, 26–28	The upper bound of storage volume in Reservoir 6 in Month 3 is relatively low
CM_6^4	1–10	1, 26, 28	The upper bound of storage volume in Reservoir 6 in Month 4 is relatively low

Improvable Constraints/Bounds: The constraints and bounds with relatively large positive dual prices are identified and shown in Table 4.10. From Table 4.10, we conclude that the upper bound of the storage volume in Reservoir 6 affects the achievement of the goals-if the upper bound is increased (relaxed), a better water flow plan can be obtained.

Improvement Indication: In conclusion, the lower bound of storage volume in Reservoir 7 is the active and non-improvable bound, and we can use the method in Step 2.3a to improve it; the upper bound of Reservoir 6 is the active and improvable bound, and we can use the method in Step 2.3b to improve it.

Model Improvement

In this section, we carry out Steps 2.3 and 3 using the improvement suggestions in Sect. 4.3.

First, we explore the practicality of adding a buffer to the active and non-improvable constraints/bounds.[5]

Step 2.3a, 3.1a, and 3.2, Active and Non-Improvable Constraints/Bounds Improvement—Restricting: First, we explore the practicality of restricting the RHS of CM_7^t, which is to increase the lower bound of Reservoir 7 in the mathematical model without changing the physical level of the conservation pool. This way, we add a "buffer" to the active bound by making its mathematical model more restrictive than its physical condition. We illustrate this concept in Fig. 4.17. Solution A is at the lower bound (lb) of Reservoir 7, which is sensitive to uncertainty in inflows. To bring it away from the boundary, we need to add a buffer to the RHS of the bound by changing the bound from lb to lb' to bring Solution A to Solution A'. Solution A' away from the physical bound lb.

Then, we explore the practicality of this change. By redesigning each pool's capacity, we ensure the reservoir meets the requirements of multiple functions. Increasing the lower bound of a reservoir means increasing the conservation pool's water level. We can realize this by reallocating the water volume of the pools–we

[5] One way of improving the mathematical model is to obtain a better assessment of the precipitation to reduce uncertainty. However, as experts in water resource management provide our precipitation data, we do not tackle precipitation improvement in this case. Therefore, we focus on the capacity of the reservoirs' pools.

increase the water level requirement of the conservation pool by 1% and reduce the water level requirement of the flood pool to offset the change because historical data and the physical condition of the reservoir inform us that such reduction is safe and attainable. As we make the physical change, we change the corresponding mathematical bounds (Fig. 4.15).

Original bound: $S_7^t \geq CM_7^t$

Improved bound: $S_7^t \geq CM_7^t * 1.01$, where t = 2, 4

Step 2.3b, Step 3.1b and Step 3.3, Active and Improvable Constraints/Bounds Improvement—Relaxing. First, we explore the practicality of relaxing the upper bound of Reservoir 6 by increasing the upper bound of Reservoir 6. This action will benefit the achievement of the goals. We run this for several iterations until we get a solution with a dual price that is not relatively large; then, we start treating it as an active and non-improvable constraint/bound and move the solution away from the boundary. However, if we continue iterating until we physically cannot relax the RHS further, the constraint/bound still has a positive dual price. We stop and deal with it as an active and non-improvable constraint/bound. This procedure explores the physical boundary of the system that best achieves the goals. Based on this physical boundary, we "add a buffer" to drive the solutions away from the physical boundary. In Fig. 4.16a, we illustrate this procedure. We explore the upper physical bound and bring the bound from ub_0 to ub_1 (Fig. 4.16a) to make the RHS of this bound reasonable and facilitate obtaining the best achievement of the goals. Then, we add a buffer to ub_1 by moving it to ub_1', as shown in Fig. 4.18b.

Then, we explore the practicality of physically increasing the upper bound of the water storage volume in Reservoir 6 by 1%. Our data indicate that the maximum storage of Reservoir 6 is 75,573 cubic feet, which is 1.03% larger than the upper bound of 74,799 cubic feet, so we can relax the RHS of this bound. Hence, we can increase the RHS of the upper bound of Reservoir 6 by 1%.

$$S_6^2 \leq CF_6^2$$
$$S_6^2 \leq CF_6^2 * 1.01$$

Table 4.11 displays the actions of model improvement. Thus, new solutions are obtained (Step 3) for the improved model (Appendix 4.2). After iterating, solutions

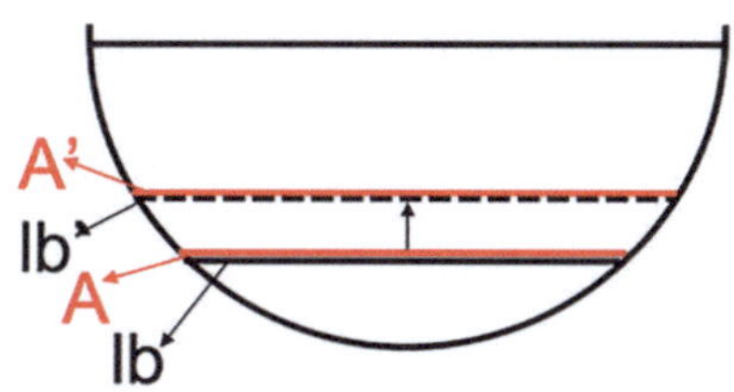

Fig. 4.15 Bring the Solution Away from the Boundary by Restricting the RHS

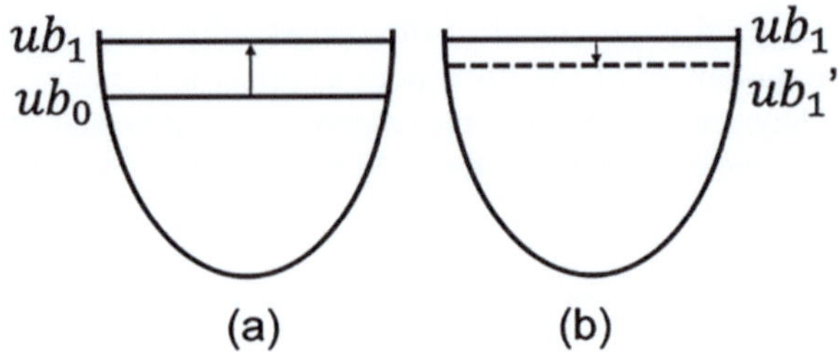

Fig. 4.16 Applying the physical boundary by relaxing RHS and then bringing the solution away from the physical boundary by restricting RHS

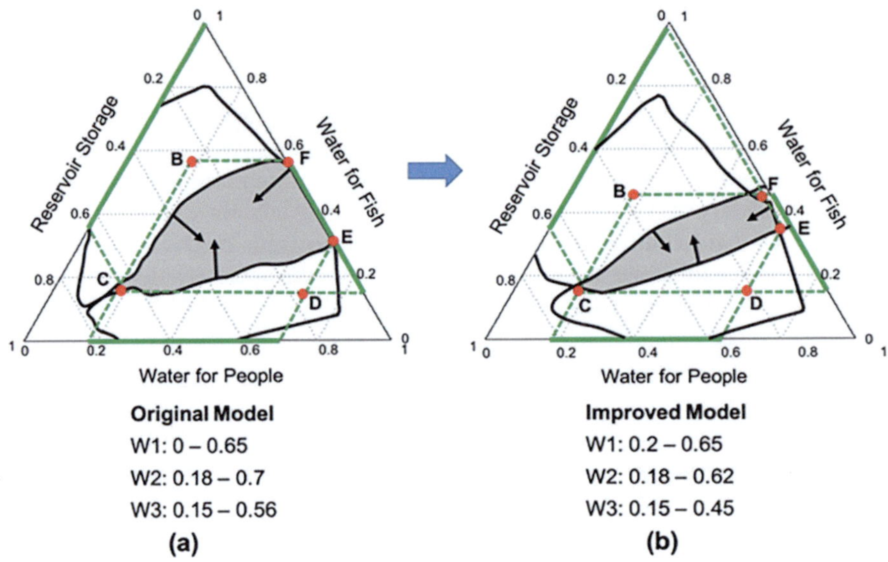

Fig. 4.17 The satisficing area of the weights of the original model (**a**) and improved model (**b**)

away from the boundary and yield a relatively better achievement of the goals are obtained.

Block 3 (Fig. 4.9) **Design Improvement Validation**

Design Preferences of Improved Model: As we go through Steps 1 to 3 for three iterations, we have no active constraints/bounds or improvable constraints/ bounds, so we stop the exploration of the solution space. We obtain a new satisficing area of the weights with the improved model in the final iteration. We compare the satisficing area of the original model and the improved model in Fig. 4.17. The satisficing area and the corresponding range of weights become smaller after model improvement because we eliminate the solutions sensitive to uncertainties. In the perspective of robust design, we seek an insensitive solution and corresponding weight range, so

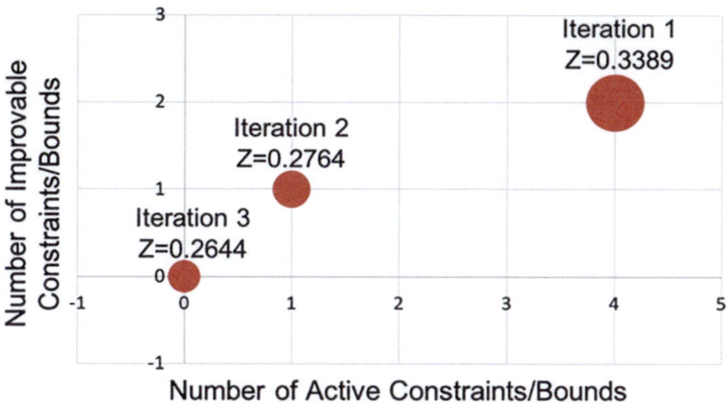

Fig. 4.18 Improvement through Iterating

Table 4.11 Suggestions for model improvement

#	Improvement	Constraints or bounds function changes, i = 1, 2, 3
1	Raise the lower bound of storage volume in Reservoir 7 in Months 2 and 4 by 1%	$S_7^t \geq CM_7^t$ $S_7^t \geq CM_7^t$*1.01, where t = 2, 4
2	Raise the upper bound of storage volume in Reservoir 6 by 1%	$S_6^2 \leq CF_6^2$ $S_6^2 \leq CF_6^2 * 1.01$

the improvement is not to enlarge the satisficing area, but to refine the satisficing area to obtain relatively insensitive solutions.

Sensitive Segments of Improved Model: As we go through Steps 2.1–2.3 with the improved model in each iteration, we identify new sensitive segments. The solutions are calculated using the improved mathematical model. However, the active constraints/bounds should be identified using the original model because we need to know whether the solution is close to the physical boundary. As to the improvable constraints/bounds, we need to use the improved model in each iteration because we update the physical boundary through iterating. In Table 4.12, we list the sensitive segments in the second iteration, and their physical meanings, with suggestions for improvement.

By carrying out the improvement suggestions in Table 4.12, in the third iteration, we have no sensitive segments, which means all the solutions are not at or highly close to the physical boundary, and there is no potential to further improve the achievement of the goals.

The evolution of the performance and the number of sensitive segments of the model, along with iterating, are illustrated in Fig. 4.18. The horizontal and vertical axes represent the number of active constraints/bounds and the number of improvable constraints/bounds. The size of the bubbles shows the average value of deviation z, of the 34 WSs and 10 ISs. In this case, improvement means driving the bubble from

Table 4.12 Sensitive segments of the model of the second iteration

Active bounds	Physical meaning of the bounds	Further improvement suggestions
S7M3L	The lower bound of storage volume in Reservoir 7 in Month 3 is relatively high	Raise the lower bound of storage volume in Reservoir 7 by 1%
Improvable Bounds	Physical Meaning of the Constraints or Bounds	Further Improvement Suggestions
S4M1	The storage volume in Reservoir 4 at the beginning of Month 1 is relatively low	Raise the lower bound of storage volume in Reservoir 4 by 1%

the up-right corner to the down-left corner, while making it smaller. As shown in Fig. 4.18, by iterating, we improve the insensitivity of the model from (4, 2) to (0, 0), and we improve its performance by reducing z from 0.3389 to 0.2644.

After three iterations, our model is neither sensitive to the uncertainties in inflows nor has the potential to improve the completeness of the goals. It is guaranteed that any extreme weather will not cause any system failure because now, all the solutions in all circumstances are away from the physical boundary. The Three-Step Method of exploration of the solution space is summarized as an algorithm in Table 4.13.

Table 4.13 The algorithm for model improvement

a. Identify n scenarios of parameters with uncertainties—ISs
Begin Iterations of Exploration of the Solution Space
b. Use the latest model to identify the feasible area of weights and identify m weight scenarios within the feasible area of weights representing different design preferences—WSs
c. Plug n ISs and m WSs into the latest model to get x solutions
d. Plug x solutions into the model in the first iteration and into the model in the current iteration
e. Identify active constraints/bounds using the model in the first iteration
f. Identify improvable constraints/bounds using the model in the current iteration
g. if no sensitive segment (active or improvable constraints/bounds)
Go to l
else
Continue with h
h. For each active and non-improvable constraint/bound
Explore the practicality of restricting their RHSs
i. For each active and improvable constraint/bound
Explore the practicality of relaxing their RHSs to the physical bound
j. Make model improvement plans based on the conclusion in h and i
k. Improve the model based on the improvement plans in g and go to b
l. The latest model is relatively insensitive to uncertainties and has no potential for improvement
End the iteration

4.3.6 Closing Remarks—Test Problem 1.1

In Sect. 4.3, we formulate a 14-dam-network problem using the compromise Decision Support Problem. We explore the solution space to improve the model and obtain *satisf*actory solutions relatively insensitive to the uncertainties in water inflows. We identify key questions and provide answers using a Three-Step Method to explore the solution.

Answers to the Research Questions

Q1.1 How can designers manage conflicting goals with reasonable priorities under different circumstances?

Step 1—Using design scenarios representing different design preferences, designers identify the satisficing area of the weights of the multiple goals and provide their physical meanings.

Q1.2 How can designers manage uncertainties in the inflows by making water flow plans that are relatively insensitive to these uncertainties?

Step 2: Use inflow scenarios considering different weather and climate conditions to identify the sensitive segments and explore the practicality of removing them by modifying the mathematical model and adjusting the physical system.

Q1.3 How can designers improve the mathematical model and the physical system to be relatively insensitive to uncertainties?

Step 3—Using the information about the practicality of modifying the model and changing the system, make improvements, including changing the parameters' values to add buffers and reallocating water to different pools to change the physical boundary. Then, go through the three steps again. Running such a loop improves the model and provides satisficing solutions that are relatively insensitive to uncertainties.

With satisficing solutions, we reduce the frequency and severity of discrepancies between water supply and water demand. The advantage of our method is that we boost the potential of the physical system while improving its robustness. Hence, we neither sacrifice system robustness for better performance nor do the opposite.

The dam-network planning problem only incorporates continuous variables and requires no coupling decisions. In Sect. 4.4, a supply chain design problem with mixed variables and coupling decisions is discussed, and the Three-Step Exploration Method is advanced to a decision framework—the Formulation-Exploration framework.

4.4 Positioning the Customer Order Decoupling Point of a Supply Chain

– Test Problem 1.2: Apply a Formulation-Exploration framework to a discrete problem

The new knowledge from managing Test Problem 1.2 is

A framework that incorporates the Three-Step Exploration Method allows the identification of the sensitive elements of a mixed-variable, coupled decision model and improves the model accordingly – the Formulation-Exploration framework.

As globalization continues in Industry 4.0, manufacturing enterprises must quickly perform mass customization (MC) to satisfy evolving market demands in different regions. One challenge of MC is to fulfill orders swiftly at an acceptable cost while maintaining service quality. To do this, the customer order decoupling point (CODP), where the value-adding activities occur, should be designed and adapted to the changing market demands.

In this section, a Formulation-Exploration framework is proposed to make decisions on CODP positioning and improve the SC to support MC. A test problem in auto parts manufacturing is used to establish the efficacy of our method. The Formulation-Exploration framework can be used to design SCs to facilitate product MC, especially when information is incomplete and inaccurate, goals conflict and multiple types of uncertainty add complexity.

cDSP	compromise Decision Support Problem
CODP	Customer order decoupling point
DCI	Design capability index
EMI	Error margin index
ESS	Exploration of the solution space
MTS	Make-to-stock production mode
MTO	Make-to-order production mode
PLC	Product life cycle

4.4.1 Problem Statement—Test Problem 1.2: CODP and the Challenges in Supply Chains

Mass customization (MC) dramatically enhances the emotional interaction between designers and the customers (Parker, 2016). MC is proposed by Naylor and coauthors (1999) as to combine agile and lean as a new strategy–*agile*. It is defined and broadly accepted as “postponing the task of differentiating a product for a specific customer until the latest possible point in the supply network" (Jacobs and Chase, 2004). Accordingly, determining the differentiation point of a product to adapt to the rapidly changing market is of great importance in supply chain (SC) design. This point is normally defined as the customer order decoupling point (CODP). The CODP is also known as the boundary between make-to-stock (MTS) production and make-to-order (MTO) production (Hajfathaliha et al., 2011), where the value-adding activities take place (Rudberg and Wikner, 2004).

In Fig. 4.19, we illustrate different production characteristics before (upstream from) and after (downstream from) the CODP of an SC. In MTS production (before CODP), manufacturers make production plans based on demand forecasts and pursue physical efficiency. MTS is often used in upstream SCs in labor-intensive industries, prioritizing low-cost and on-time delivery. On the contrary, in MTO production, manufacturers produce goods to meet customers' orders and emphasize market response and capacity flexibility. MTO is often used in the downstream SC of high-tech industries, or after-market items (Iravani et al., 2012), where differentiation and personalization are valued.

Theoretically, the CODP of a product can be anywhere in the supply value chain. Figure 4.20 shows some typical CODPs in different industries. In the coffee industry, the CODP is the retailer's raw material. When an end customer orders a cup of coffee, the retailer starts making it. In the electronic cabinet SC, MTS starts from the raw materials and moves to the regional distribution center (RDC), which stores the standard components. After receiving orders, the assembly plant determines the customers' needs regarding the exact models and the corresponding parts, then starts assembling and shipping. As for the shipbuilding industry, e.g., cargo ships, the whole SC is MTO due to its highly customized nature. There are common features of the CODP for different goods: convenient storage, commonality, and short lead-time requirement of the goods upstream from the CODP, whereas multiple consumption channels, affordable production shifting cost, and relatively long but acceptable lead-time of the goods downstream from CODP, etc.

MC aims to provide products that satisfy individual customers' requirements with near-mass-production efficiency (Salvendy, 2001). Usually, the efficiency of a SC varies significantly as the CODP changes. The CODP should vary in different phases of its life cycle for a single product to support MC. The rapid product upgrades and replacements require fast CODP switching. From the demand side, the users continuously put forward new needs, which force the CODP to go upwards along the SC. Maintaining a near-mass production efficiency in an SC with fast CODP switching is a challenge because the low-cost and product differentiation are contradictory. Therefore, a method that facilitates the exploration of the tradeoffs between cost and flexibility is needed.

Challenges in SCs

In the age of Industry 4.0, multiple types of workflows exist in large-scale supply networks that involve many entities of different types (Velasquez et al., 2019). Many goals that may conflict with each other must be managed (Fig. 4.21). Their priorities may evolve as circumstances change. Since mathematical models used to define SCs

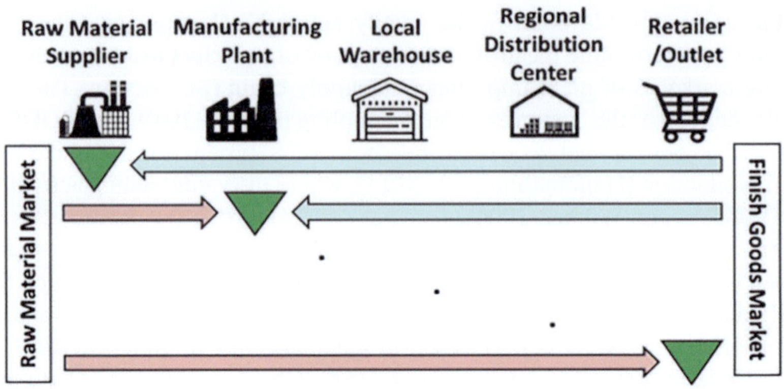

Legend

- CODP
- Possible locations for CODP omitted in the figure
- MTS production mode
- MTO production mode

Characteristics of the MTS and MTO production		
Production mode	Make-to-Stock (MTS)	Make-to-Order (MTO)
Strategy	Push	Pull
Driven type	Forecast-driven	Order-driven
Key strength	Physically efficient	Market responsive
Key performance	Cost-minimized	Capacity-flexible
Practice solution	Lean	Agile

Fig. 4.19 Possible location of CODP in a SC

are typically abstractions of reality, we are guided by Simon's maxim of looking for *satisficing* (good enough – not optimum) solutions (Simon, 1956, 1976).

According to George Box, "all models are wrong, but some are useful" (Box and Draper, 1987), the mathematical models that constitute a SC are incomplete and inaccurate (Talvitie, 1981) and of different fidelity. The decision model does not represent the physical world perfectly. Hence, designers should work with decision models that embody incompleteness and errors (Simon, 1996).

Multiple types of uncertainty in decision models affect the performance of the SC (Nellippallil et al., 2018). In Table 4.14, we illustrate the four types of robust design in association with the four types of uncertainty, which are defined, extended, and summarized by Taguchi (1993), Taguchi and Clausing (1990), Chen et al. (1996), Isukapalli et al. (1998), Choi et al. (2004, 2005, 2008a, 2008b) and well interpreted by Nellippallil et al. (2017). We tie this section to the SC. The four types of uncertainty in mechanical engineering have equivalents in the SC (Table 4.14).

Type I robust design is relatively insensitive to noise factors that are uncontrollable parameters such as uncertain demand. MTS production is planned based on the demand forecast, which may differ from the actual demand. MTO production is planned based on the orders received, which may be more urgent than usual or may

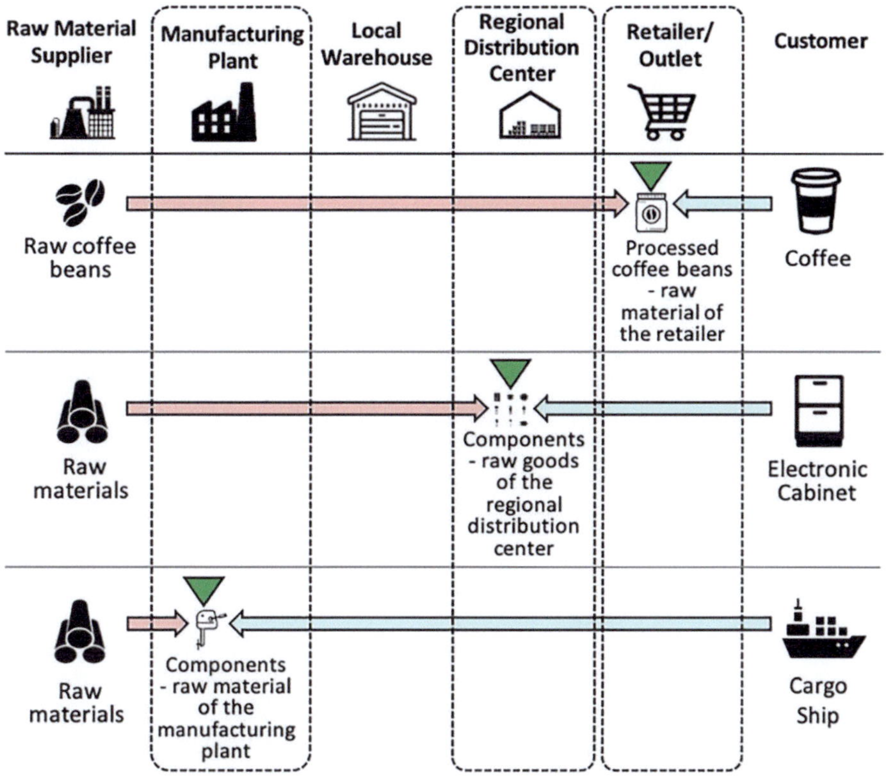

Fig. 4.20 CODP of different industries

Fig. 4.21 Multiple conflicting goals in SCs

Table 4.14 Four types of robust design and the interpretations in SC

Types of robust design	Types of uncertainty	Example in SC	Quantification
I	Uncertainty in parameter	Uncertainty on the demand side, such as unpredictable order	Type I, II: EMI, Monte Carlo simulation, Latin hypercube, First/Second moment method, etc.
II	Variable uncertainty	Uncertainty on the supply side, such as variation in productivity	
III	Uncertainty in model structure	The number of decision variables, constraints, or mathematical relations between variable changes due to machine failure, customer loss, impatient customer, or natural disaster	Type III: Design Capability Index (DCI), Variance Function Estimation, Prediction Interval Approach, etc.
IV	Uncertainty is created in the process chain	The bullwhip effect	Type IV: *Satisficing*, Exploring the Solution Space (ESS), etc.

be withdrawn later. Hence, demand is an uncertain parameter that the decision-maker cannot control.

Type II robust design is relatively insensitive to variations in design variables such as uncertain supply. The actual output of production can be different from the expected volume because of overuse of the machines, overtime of the labor, etc. The decision-maker may set the production volume to a certain level, but the output can differ. The error margin index (EMI) deals with Type I and Type II uncertainty by bringing the mean and minimizing the variance of the performances. EMI is a mathematical construct indicating the location of mean system performance and the spread of the performance considering variability in design variables (Choi et al., 2004). Type III robust design is relatively insensitive to uncertainties embedded within the model structure, such as dimensionality, constraints, or the relation between variables due to unexpected events, such as overcapacity, infrastructure damage, and political environment changes. The design capability index (DCI) handles Type III uncertainty. DCI is a mathematical construct for efficiently determining whether a ranged design specification can satisfy a range set of design requirements (Choi et al., 2005). Using a DCI, the designer identifies a range of design variables with minimized design specification variation and an acceptable performance variation. Within this range, the impact of the model's structural change on the model performance is minimized. Meanwhile, the performance is good enough and has a relatively small variance. Details of using EMI and DCI can be found in Chen et al., (1996), Choi et al., (2005, 2008a, 2008b), Nellippallil et al. (2018).

In Type IV robust design, the propagated uncertainty in process chains and the interaction among the three types of uncertainty are managed, for example, the bullwhip effect. We propose exploring the solution space (ESS) to improve the design's insensitivity to various situations. Using ESS, we explore design preferences, improve design capacity, and boost the system's potential. The method is introduced in detail in Sect. 4.4.3.

4.4.2 Literature Gap Analysis—In the Domain of Customer Order Decoupling Point Determination

There are two main approaches to locating the CODPs in SC. One is the qualitative approach, through which researchers make general rules on locations of CODP for different types of products. For example, CODP should be closer to the designer/manufacturer for personalized products, luxury goods, or services. The other approach involves quantitative methods–designers position the CODP by applying mathematical modeling. The latter has been mainstream in recent years. We focus on the quantitative approach in the literature review.

Queuing model. In a series of articles (Teimoury and Fathi, 2013; Teimoury et al., 2012), Teimoury et al. use queuing models to position CODP and make relevant decisions in SCs. Teimoury and coauthors propose a queuing and matrix-geometric method to determine CODP in a multi-product SC. Customer orders are assumed to arrive according to the Poisson process; see Teimoury et al., (2012). The objective is to minimize total cost with constraints of warehouse capacity and service level. The effect of impatient customer arrival on costs is considered. The assumption of the Poisson customer-order arrival may cause errors since the customer-order arrival varies a lot for different products or in different phases of the product life cycle. In Teimoury and Fathi (2013), Teimoury and Fathi model a two-echelon SC offering multiple products to customers with different preferences, considering both shared and unshared capacity models. They develop an integrated operations-marketing perspective while considering price-sensitive demand for CODP positioning based on a multi-class queueing model. Additionally, Zhou and coauthors (2014) apply a two-stage queuing model with customer orders modeled as Markov processes and find that the variance and correlation of the demand increase the total operation cost. The effect of the CODP's position on the unit inventory holding cost, the lead-time quotation policy, and the penalty cost for tardiness are considered. However, the single objective is simplified to minimize the total cost.

Time scheduling model. Liu et al. (2018) propose a time scheduling model to examine the dynamic positioning of CODP. Minimizing total cost and maximizing customer satisfaction are the two objectives, while the incremental cost of a new order, lead-time, and changing objectives values are constraints. However, they do not scale the two objectives with different units–scaling is necessary because the importance and utility of the two objectives may change with circumstances. Jeong (2011) proposes a dynamic model based on product life cycle theory to position the CODP and make a production-inventory plan. The author applies optimal control theory to a quadratic objective function. However, the author does not consider factors such as time-varying penalty costs and the delivery lead time.

Stochastic programming, multi-criteria optimization, dynamic programming, etc. Some authors apply programming methods with stochasticity or multiple criteria. Ghalehkhondabi and coauthors () leverage a stochastic programming model to identify CODP by minimizing total cost or maximizing total profit. Working time constraints, satisfied order percentage, and manufacturing capacity are considered. Customer orders are assumed to follow Poisson distribution. Shidpour et al. (2014) propose a multi-objective programming model based on the manufacturer's profit and customer perceived value to analyze the impacts of single-CODP and multi-CODP while considering service time constraints. They conclude that multi-CODP is preferred to single-CODP for a product portfolio, benefiting both the manufacturer and the customers. Ahmadi et al. (2008) developed a dynamic programming model to optimize CODP by minimizing holding costs, delivery delays, and the number of modules that are not ready to be assigned. They assume that customer demand is deterministic. Liu et al. (2015) construct a linear programming model with discrete variables in the logistics industry to position CODP by minimizing the total cost of a logistics service integrator. Procedure constraints and lead-time constraints are considered in the model. They conclude that the optimal CODP is not affected by decreasing order-transferring and waiting costs but is driven to the last procedure when order processing cost decreases. Other methods such as the analytic hierarchy process (AHP) model (Xu and Liang, 2011), value network simulation model (Daaboul et al., 2010, 2015), and tandem forecast-driven and order-driven simulation model (Wikner et al., 2017) are also explored to identify the optimal CODP.

In summary, recent researchers in CODP positioning have addressed the issue by constructing models with multiple objectives and constraints. There have employed traditional optimization methods such as stochastic programming, dynamic programming, and selection models such as AHP, and methods "borrowed" from other domains such as value networks, tandem circuit design, and polychromatic set theory. Researchers find various metaphors for CODP and SC design, to come up with interesting concepts and methods—such as viewing the MTS and MTO production mode as two circuits and looking for an appropriate way to connect them. This phenomenon implies that CODP design is a field combining science and art. With different metaphors, designers can have diverse perspectives and foci. Their observations are interesting and somehow useful to an extent in some situations. However, there are some limitations, and we summarize them as follows.

Limitation 1–There are assumptions in the distribution of non-deterministic methods (Ghalehkhondabi et al., 2016, 2017; Teimoury et al., 2012; Zhou et al., 2014). They assume that enough data supports making decisions on the distribution of customer-order arrivals, demand forecast accuracy, customer satisfaction quantification, etc. These decisions or assumptions can be wrong and irreversible once the design stage is over, and no adjustments can be made during operations.

Limitation 2–The optimization methods (Ahmadi and Teimouri, 2008; Ghalehkhondabi et al., 2016, 2017; Jeong, 2011; Liu et al., 2015; Shidpour et al., 2014; Teimoury et al., 2012; Zhou et al., 2014) are used to find optimal solutions, that are solutions on the boundary of the feasible solution space (for deterministic approach) or distributed close to the boundary of the fussy feasible solution space (for stochastic approach). As models are incomplete, inaccurate, and embody different levels of fidelity, the solution may be optimal for the model. However, it may not be optimal for the real problem, which is way more complicated than the model.

Facing the challenges in the SCs and MC, we need to overcome the limitations. The assumptions on data adequacy can be replaced with the information we obtain from post-solution analysis. Typical uncertainties need to be managed in the design and operation stage. Positioning strategies regarding the CODP should be tailored for different market environments or phases in the product life cycle.

Based on the limitations in the literature and challenges in CODP design and RQ1, we pose a question and hypothesis specified in the domain of SC. By answering the question and validating the hypothesis, we bridge the gap.

Question*—How can designers improve the robustness (insensitivity) of the CODP design without sacrificing the performance of the SC?*

Hypothesis*–We propose a Formulation-Exploration framework to explore the solution space concerning the design capacity under typical uncertainties. With insight from solution space exploration, we remove some errors and heuristics in the initial design and boost the system's potential.*

4.4.3 Proposed Methods—The Formulation-Exploration Framework

To obtain *satisficing* solutions insensitive to multiple types of uncertainty the enterprises need to apply some systematic methods to make timely decisions supporting MC. We propose the Formulation-Exploration framework (Fig. 4.22). The method consists of two stages: problem formulation and solution space exploration. We need to iterate between the two stages to converge on a satisfactory solution.

Compromise Decision Support Problem (cDSP). In the problem formulation stage, the problem is formulated as a cDSP and then solved using the Adaptive Linear Programming algorithm (Mistree et al., 1993). The mathematical form of a cDSP and why it delivers *satisficing* solutions are stated in Sect. 2.3. Due to the complexity of the SC with MC, the cDSP usually contains non-convexity and nonlinearity features, so one way of managing the complexity is to approximate the model to a classic linear model and solve the problem using a simplex algorithm. The model structural uncertainty (Type III) and process chain added uncertainty (Type IV) are tackled in the problem formulation stage.

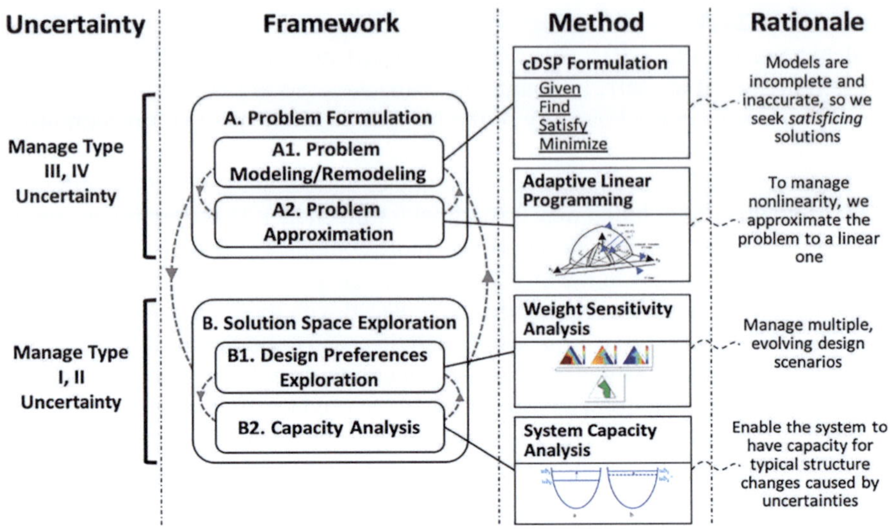

Fig. 4.22 Formulation-exploration framework

In the second stage, the solution space is explored to find *satisficing* solutions associated with each design preference (scenario) in different phases of the product life cycle. Type I and II uncertainty are managed by exploring the solution space (ESS). In this section, when referring to design preferences, we particularly focus on the importance of the different goals.

Weight sensitivity analysis–an exploration of the design preferences. We use weight sensitivity analysis to explore how the different weights affect the system's performance in achieving the goals. See Fig. 4.22, "weight sensitivity analysis" (Ahmed et al., 2014).

System capacity analysis–identification and management of the sensitive segment and bottleneck. To overcome the capacity limitation of constraints or bounds, we use the "system capacity analysis" method to identify the sensitive segment and bottleneck (Table 4.15). If an inequality constraint has zero or tiny surplus or slack compared with its right-hand-side value, we define it as an active constraint. The solution is on or close to the boundary of the active constraint, so the solution is sensitive to the uncertainty of the active constraint. By relaxing such an active constraint, suppose the shadow price of an active constraint is lower than other active constraints. In that case, we may not greatly improve the fulfillment of the goals, and we define such a constraint as a "sensitive segment." We then bring the solution away from the sensitive segment by restricting the active constraint, which means adding a buffer to the mathematical model to prevent the solution from reaching the physical boundary. If the shadow price of an active constraint is the largest value compared with that of other constraints, relaxing the constraint can result in the most improvement in fulfilling the goals, and we define such an active constraint as a "bottleneck." Also, we need to find ways of relaxing the bottleneck in

Table 4.15 Algorithm for system capacity analysis

Step 1. Identify n scenarios of parameters/bounds with uncertainties—ISs
Begin Iterations of Exploration of the Solution Space
Step 2. Use the latest model to identify the feasible area of weights and identify m weight scenarios within the feasible area of weights representing different design preferences—WSs
Step 3. Plug n ISs and m WSs into the latest model to get x solutions
Step 4. Plug x solutions into the model in the first iteration and into the model in the current iteration
Step 5. Identify active constraints/bounds and sensitive segments using the model in the first iteration
Step 6. Identify bottlenecks using the model in the current iteration
Step 7. **If** no sensitive segments or bottleneck
Go to 12
else
Continue with h
Step 8. For each sensitive segment
Explore the practicality of restricting their right-hand-side values (RHSs) in the model
Step 9. **For** each bottleneck
Explore the practicality of relaxing it in the physical system
Step 10. Make model improvement plans (restricting model or relaxing physical system) based on the conclusions in 8 and 9
Step 11. Improve the model based on the improvement plans in 7 and **go to** 2
Step 12. The latest model is relatively insensitive to uncertainties and has no potential to achieve a better solution
End the iteration

the physical system to boost the system's potential. Once there is no more potential for physical relaxation, we bring the solution away from the newly relaxed boundary by restricting the constraint in the mathematical model– adding a buffer to the physical boundary. In this way, we boost the potential of the physical system while improving its robustness; hence, we neither sacrifice system robustness for better fulfillment of the goals nor do the opposite.

We hypothesize that using the Formulation-Exploration framework, the CODP can be designed or redesigned to obtain satisficing solutions that support MC. A successful company should be able to perform repeated design based on the information attained from the previous cycle, so its SC can be flexible for MC. The Formulation-Exploration framework allows us to answer the research question in this section and validate the hypothesis.

In Sects. 4.4.4 and 4.4.5, we describe the application of the Formulation-Exploration framework to an industry inspired problem.

4.4.4 Model Formulation

The test problem is a three-echelon SC in the automobile industry with three players—a supplier, a manufacturer, and a retailer as shown in Fig. 4.23. We studied

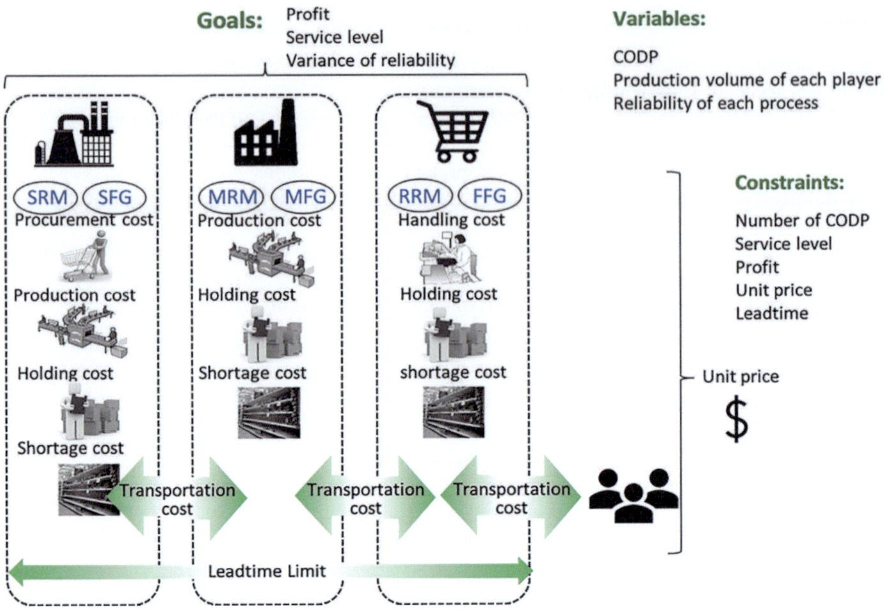

Fig. 4.23 A three-Echelon SC

the problem for one month. Our goals are to position the CODP and determine the appropriate reliability of each process, including the operation for each player and the transportation between two players. The concept of reliability comes from lean process management and the six-sigma quality control principle, which states that the processes in the automobile industry range from three-sigma (99.7%) to six-sigma (99.99997%). The constraints are cost and capacity in each process. We formulate the problem in a cDSP (See Appendix 4.3). The goals and constraints of the model are visualized in Fig. 4.23.

Goals. We have three goals: profit, service level, and variance in the reliability of processes. We have a target for each goal. The profit target is an ideal value that not only makes each player in this SC sustainable but also makes them extremely competitive by providing the stakeholders the greatest return—the biggest in the industry. We solve the problem by assuming the product's unit price is fixed within the modeling time, and the costs of all the processes affect the profit. In this way, we focus on improving the process regarding CODP positioning.

As the second goal, the service level is represented by multiplying the reliability of all processes in the SC, including the procurement, manufacturing, holding, handling (the display, movement, and maintenance in the retailer), and transportation. This service level definition is based on the reliability of a mechanical system with multiple components. In this sense, only the reliability of all processes maintains a high level and ensures an acceptable service level for the entire SC. We set the service level target as the six-sigma, 97.99997%.

The variance of reliability of processes is the third goal. The aim of setting such a goal is to enable all players of the SC to share the duty and pressure of reliability without over-relying on certain processes to maintain high reliability while ignoring the reliability management of some other processes. This prevents strong players from dominating the other players by squeezing their profit margins. For example, strong retailers have very strict requirements on the manufacturer's unit price, product quality, delivery time, etc., which can be reflected in the extremely high reliability of the manufacturer. If we minimize the differences between the reliability of different players, the responsibility shirking and domination between players can be avoided. The overall performance and profit of the entire SC can be boosted, and the sustainability of the SC, as well as each player, can be maximized.

System variables. We have two types of system variables, binary variables and continuous variables. The CODP of each candidate location is a binary variable. Each player has two candidate positions that may store the goods, raw materials, or finished goods, so there are six candidate locations for CODP in this SC, namely, supplier's raw material (SRM), supplier's finished goods (SFG), manufacturer's raw material (MRM), manufacturer's finished goods (MFG), retailer's raw material (RRM), and retailer's finished goods (RFG). In each design scenario representing a certain design preference, one of the six candidate locations should be the CODP. The production upstream to the CODP is make-to-stock, and the production downstream to the CODP is make-to-order.

The reliability of the processes is a continuous variable. We define reliability as a value that each process owner can determine—high reliability results in a high service level but low profit. By exploring the tradeoffs between profit and the other two goals, we can set appropriate values for the reliability of each process in different situations, such as different stages of the product life cycle.

The production volume of the supplier, the manufacturing volume, and the purchasing volume of the retailer are either equal to the forecast (if the production at the point is make-to-stock) or equal to the real demand (if the production at the point is make-to-order), so the production volumes are defined as dependent parameters rather than system variables.

Constraints. The cost of every process, the total number of CODPs, and the service level are the constraints. Holding costs: In the processes of make-to-stock production, when the real demand is less than the forecast, the extra part of the forecast becomes the remaining stock, which causes holding costs until the next time period; the remaining stock offsets the production or procurement volume of the next time period. Shortage cost: In the processes of make-to-stock production, when the real demand exceeds the forecast, we assume the shortage amount can be outsourced urgently, but urgent outsourcing causes extra cost, and we define such extra cost as the shortage cost.

In this problem, we define that there is always one CODP in this SC. The CODP can be changed as the product life cycle evolves, but at any specific time, it is required that all raw materials, semi-finished goods, and finished goods be stored at the single CODP.

The ideal value of the SC service level is defined as the target of the service level goal, but we also have a lower bound of the service level, which is the three Sigma, 99.7%. SC is only sustainable if the service level is at least 99.7%.

Different design scenarios result in different product life cycle stages (PLC). As the product enters different phases of its life cycle—introduction, growth, maturity, and decline—preferences on the three goals vary, and the CODP may move up and down accordingly.

4.4.5 CODP Results and Discussion

—Applying the Formulation-Exploration framework to improve the model and obtain satisficing solutions.

Weight sensitivity analysis. We use several weight scenarios (WSs) to represent typical design preferences in design preferences exploration. We apply the method (Seada and Deb, 2014) to determine the WSs. For an M-goal problem, if the designer wants to divide the weight range for each goal—we define the whole range as [0, 1]—into p pieces, we will have $H = \binom{M+p-1}{p}$ number of WSs. In this problem, since there are three goals, $M = 3$. We set $p = 2$, so $H = 6$. We want one more scenario assigning equal weights (1/3, 1/3, 1/3) to three goals; thereby, there are altogether seven WSs (Table 4.16). Plugging each of the seven WSs into our cDSP, we have seven sub-problems. Solving them individually, we obtain the results listed in Table 4.17.

We use Fig. 4.24 to show the achievement of the goals associated with the CODP results. From Fig. 4.24, we conclude that when profit is less important than service level and variance of variability, the CODP should be at the retailer's finished goods (RFG); if profit is important, the CODP should be at the manufacturer's raw materials (MRM).

System capacity analysis. We identify that the holding costs of the manufacturer and retailer have limited capacity but a relatively large shadow price, which means

Table 4.16 Seven scenarios—Type II uncertainty

WS	w1	w2	w3
1	1	0	0
2	0	1	0
3	0	0	1
4	0.5	0.5	0
5	0.5	0	0.5
6	0	0.5	0.5
7	0.33	0.33	0.33

Table 4.17 Results of the first iteration

WS	CODP	Profit	Service level (%)	Variance of reliability (%)
1	MRM	30,489	99.13	0.483
2	RFG	33,835	99.26	0.304
3	RFG	33,569	98.33	0.045
4	MRM	30,610	99.40	0.549
5	MRM	30,233	98.17	0.055
6	RFG	33,824	99.26	0.279
7	MRM	30,543	99.28	0.249

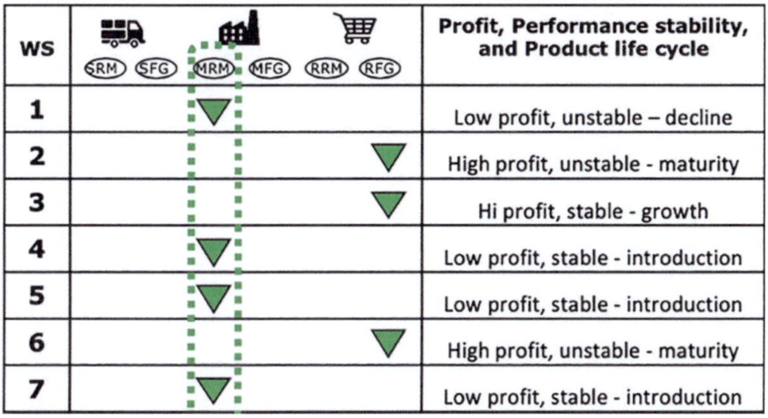

Fig. 4.24 CODP of different weight scenarios

this constraint is a bottleneck. If we can reduce the holding costs physically by some means (e.g., applying RFID technology or using a third-party vendor with lower unit cost), the achieved value of the goals can be improved. Assuming that the operational team takes action on the cost reduction, we reformulate the problem accordingly using the algorithm in Table 4.15. Then we run the second iteration of the Formulation-Exploration. After three iterations of formulation exploration, we get the results in Table 4.17. The final results are better than the first-iteration results (Table 4.18) regarding the achieved value for the goals.

In contrast, the service level and variance of reliability in some scenarios worsened, but only slightly. This means we saved the cost of achieving an unnecessarily high service level and equity between players to obtain higher profits. After three iterations, we boosted the system potential, so we stopped iterating.

In this case, we observe that in WS 1, 4, 5 and 7, when the profit goal (Goal 1) is achieved not so good,), the CODP is at MRM, the raw material of the manufacturer, which means if the decision maker allows the SC to give up some of its profit margins, (in the introduction phase, or decline phase of the product life cycle—PLC), In such

Table 4.18 Results of the seven design scenarios in the third iteration

WS	CODP	Profit	Service level (%)	Variance of reliability (%)
1	MRM	31,584	99.13	0.483
2	RFG	36,946	99.21	0.483
3	RFG	36,657	98.25	0.025
4	MRM	31,584	99.13	0.483
5	MRM	31,584	99.13	0.483
6	RFG	33,813	99.23	0.315
7	MRM	31,584	99.13	0.483

a situation, from SRM to MRM (from the raw material of the supplier to the raw material of the manufacturer), the productions are all based on the demand forecast; from MRM to RFG (from the raw material of the manufacturer to the finished goods at the retailer), the production or goods preparation are all based on customers' order. Standard components are stored at MRM and all the locations upstream from it, and customization takes place right after MRM. The profit is not at a high level, whereas the achievement of service level (the higher, the better) and the variation of the reliability (the lower, the better) are not always at a high level either. This indicates that in a product's introduction or decline phase, we need to sacrifice the performance of the SC for a better-customized product and a relatively more flexible SC.

On the contrary, if the profit is the priority for the SC, which usually happens in the growth or maturity phase of the PLC, the CODP should be at the last candidate node of the SC, that is, RFG, the finished goods of the retailer. In this way, the whole SC is an MTS one. The relatively stable and strong customer demand makes the demand forecast more reliable so that the MC can be done via the production through the manufacturers to the retailers based on relatively accurate forecasts due to market feedback. The whole SC is now for a product family. In such a way, the MC is done with a mass production cost. We tie the results in seven WSs with the different phases in PLC (Fig. 4.25). This indicates that even for a new, customized product, as it enters the growth phase when the demand is relatively stable and predictable, the customization can be done in advance to the customers' order intake until the product goes to the decline phase. Therefore, even for a product with customized solutions, we only sometimes have to adopt the MTO production mode because, in the growth and maturity phase, we can finish the production of the mass-customized product ahead of the orders with relatively low risk.

Verification and Discussion

To verify our conclusions and attain insight into the relation between CODP candidate variables and the achieved value of the goals, we perform a sensitivity analysis of the six CODP candidate locations. In each weight scenario, we fix the CODP in each of the six candidate locations and obtain the achieved value of the goals. We compare the results with CODP fixed at each point with the *satisficing* results (Table 4.18) in

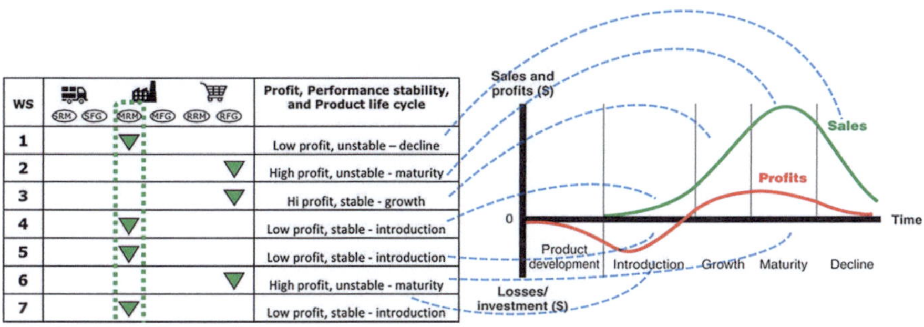

Fig. 4.25 CODP, achieved value of goals in different phases of a product life cycle—managing Type I uncertainty

Table 4.19. Under each WS, we show the *satisficing* result with an asterisk "*" in the column named "*satisficing.*" The percentage numbers in those *non-satisficing* lines are compared with those in *satisficing* ones. "–24%" means that it is 24% worse than the *satisficing* result under the same WS. In some scenarios, if we assign a candidate location as CODP, we can only get a feasible solution if the constraints are always enforced. For example, in WS 1, if we set SRM as the CODP, which means the whole SC is in MTO production mode, we cannot satisfy all constraints. So, given the resources that can be acquired in this SC, a whole MTO production mode is not recommended for this product if we set profit as the only goal.

From Table 4.19, we observe that we can only position CODP at one location, SRM, and SFG because we cannot obtain any feasible solutions. Therefore, the product is not fit for a pour pull strategy. In WS 1, 4, 5, and 7, RFG as CODP cannot give feasible solutions, whereas in all the other WSs—WS 2, 3, and 6—RFG is the *satisficing* solution. This means RFG is a sensitive node that gives the best or loses it all. This proves that weight sensitivity analysis is necessary so that we only pick appointments as CODP when it gives qualified results in certain design preferences.

Among the three goals, profit is relatively sensitive to CODP migration because when we move CODP to *non-satisficing* nodes, the achieved profit values decrease the most. Service level does not change much as CODP migrates, nor does it change much as WS changes. In this sense, we can remove the service level goal to reduce the computational complexity.

Among the six candidate locations, only two can be CODP in all seven WSs—MRM and RFG, so it simplifies the decision-makers' tasks. They need to focus on these two locations instead of all six. The responses to preference change are all about switching CODP between MRM and RFG. Decision makers do not need to make one-time investments in fixed assets to make other locations as CODP.

Using the Formulation-Exploration framework, when changes occur in the SC, such as design preferences unexpectedly evolving (Type III uncertainty), cost budget cutting due to crises (Type II uncertainty), production/transportation cost or capacity changes due to natural disasters (Type I or IV uncertainty), etc., we can reformulate the problem and output *satisfying* solutions in real-time. The switching of the

Table 4.19 Comparison of satisficing results with results from other CODP candidate locations

WS	Satisficing	CODP	Profit	Service level	Variance of reliability
1		SRM	Infeasible		
1		SFG	Infeasible		
1	*	MRM	31,584	99.13%	48.82%
1		MFG	–24%	0%	+0.25%
1		RRM	–13%	+0.1%	+85%
1		RFG	Infeasible		
2		SRM	Infeasible		
2		SFG	Infeasible		
2		MRM	Infeasible		
2		MFG	Infeasible		
2		RRM	Infeasible		
2	*	RFG	36,946	99.21%	0.483%
3		SRM	Infeasible		
3		SFG	Infeasible		
3		MRM	Infeasible		
3		MFG	Infeasible		
3		RRM	Infeasible		
3	*	RFG	36,657	98.25%	0.025%
4		SRM	Infeasible		
4		SFG	Infeasible		
4	*	MRM	31,584	31,584	31,584
4		MFG	–24%	–24%	–24%
4		RRM	–13%	–13%	–13%
4		RFG	Infeasible		
5		SRM	Infeasible		
5		SFG	Infeasible		
5	*	MRM	31,584	31,584	31,584
5		MFG	–24%	–24%	–24%
5		RRM	–13%	–13%	–13%
5		RFG	Infeasible		
6		SRM	Infeasible		
6		SFG	Infeasible		
6		MRM	Infeasible		
6		MFG	Infeasible		
6		RRM	Infeasible		
6	*	RFG	33,813	99.23%	0.315%
7		SRM	Infeasible		

(continued)

Table 4.19 (continued)

WS	Satisficing	CODP	Profit	Service level	Variance of reliability
7		SFG	Infeasible		
7	*	MRM	31,584	31,584	31,584
7		MFG	–24%	–24%	–24%
7		RRM	–13%	–13%	–13%
7		FRG	Infeasible		

PLC phases can be captured in time, and the production strategy can be updated accordingly.

4.4.6 Closing Remarks—Test Problem 1.2

This section proposes a practical framework, the Formulation-Exploration framework, for positioning CODP in SC networks to facilitate MC. We review the literature on CODP positioning, identify two major limitations—problematic assumptions on stochastic parameters and reliance on model accuracy—and analyze the challenges in SC regarding CODP positioning. We also define the concept of robust SC design and categorize the uncertainties in SC into four types, as shown in Table 4.20. We raise a question and hypothesis based on the limitations and research gaps in the literature and the challenges in the CODP design. To answer the question, we propose the Formulation-Exploration framework to explore the solution space to exploit the potential of the SC. Using a test problem in the automobile industry, we answer the question.

Question (Q1.4)*—How can designers improve the robustness (insensitivity) of the CODP design without sacrificing the performance of the SC to facilitate MC?*

Using the Formulation-Exploration framework, we explore the solution space using representative design scenarios and the potential of boosting the system's performance by adopting physical means, such as applying a new technology to cut down a certain cost. By doing this in iterations, we exploit all means that lead us to reach the most desirable *satisficing* given the resources. With the Formulation-Exploration framework, we can iteratively determine and update the CODP, which facilitates providing products to satisfy individual customers' requirements with near-mass production efficiency. We define the Formulation-Exploration framework as a dynamic, open meta-design scaffold on which new, customized modules can be added to manage domain-dependent and domain-independent design problems. The four current incorporated methods—cDSP, adaptive linear programming, weight sensitivity analysis, and system capacity analysis- are examples of how we launch our practice for *satisficing* design.

Table 4.20 Summary of Test Problems 1.1 and 1.2 regarding Type I and II uncertainty management

RD Type	**RDI-II**				
			RDIII		
				RDIV	
Method	M1: Formulation-Exploration framework		M2: Adaptive Linear Programming with Parameter Learning (ALPPL)	M3: Adaptive Leveling-Weighting-Clustering Algorithm (ALWC)	M4: Scenario Planning in Agent-Based Modeling
Chapter	Ch 4		Ch 5	Ch 6	Ch 7
Testt Problem / Uncertainty	T1: Dam network	T2: Supply chain	T3: Hot rolling process chain	T4: Thermal system	T5: Promoting second-season farming
Type I	***Uncertainty in timing and amount of inflow – Table 4.8***	***Uncertainty in demand side – Figure 4.25***	*Uncertainty in hyperparameter setting (Parameters in approximation algorithm)*	*Uncertainty in parameter setting in solution algorithm (Starting point of searching)*	*Uncertainty in price (Price of agriculture products)*
Type II	***Uncertainty in outflow (water release target) – Table 4.6***	***Uncertainty on the supply side - Table 4.18***	*Uncertainty in user preferences*		*Promotion effort and timing*
Type III			*Uncertainty in model approximation due to heuristics in approximation*	*Uncertainty in model approximation (ways of combining multiple goals)*	
Type IV				*Uncertainty in using domain knowledge to simplify the model (fixing decision variables and selecting design scenarios)*	*Interventions that change the mathematical relation between promotion and result (developing local market)*

RD – robust design
M – method
EVe – empirical verification of the method
T-test problem

There are two major contributions to this work. First, the proposed Formulation-Exploration framework allows designers to manage multiple conflicting goals by minimizing the distance between their targets and the achieved values. As different design scenarios that represent evolving preferences in various stages of a product life cycle (PLC) are explored, one can design the MC in different situations to have various flexibility. By exploring the tradeoffs between goals, an appropriate design scenario can be adopted for each stage of a PLC. As the product enters different stages, a real-time switching of CODP facilitates mass customization.

Second, we provide a new perspective on SC design—improving the sustainability of an SC by enabling players to share the duty and risk in reliability. Instead of modeling the SC conventionally by determining the production volume or inventory level, we determine the reliability of the processes in the SC and minimize their difference. We view the SC as a system and encourage the players to share responsibility and boost the system's sustainability by balancing the duty and the gains of each player. Multiple types of uncertainty are managed in this section by adding a buffer to the physical model and removing bottlenecks of the system.

There are two other contributions in this section. We apply the robust design concept in engineering to SC design and categorize the typical uncertainties in an SC into four types. Different modules in FME enable designers to manage different types of uncertainty in a SC. In addition, the concept of CODP can be expanded in other complex systems, such as decision workflow design. The essential idea of adopting the CODP in complex-system design and operation is positioning the key node or several key nodes as the system's CODP(s), where the system activity's characteristics change from active preparation to order-driven. By identifying the requirements at the CODP, a designer can proceed with the upstream and downstream processes. The benefit of having a CODP in a complex system is that by positioning and switching the CODP, one can ensure that the entire system maintains a sustainable and robust state while maximizing the reconfigurability and enabling mass customization.

The Formulation-Exploration framework can be applied to engineering design problems. Designers can improve the model and manage uncertainties by removing the sensitive and improvable segments of the engineering-design system. By viewing the different segments of a system as players collaborating with one another while competing for resources, we can improve system performance by letting them share responsibilities and risks. The Formulation-Exploration framework is especially useful for continuous improvement and sustainability improvement of a system with conflicting goals, evolving design preferences, and multiple players who seek equality and common interest.

4.5 Role of This Chapter in This Monograph

4.5.1 Summarizing How We Finish Task 1—Connecting Formulation and Exploration

When designing a complex system, especially in the early stages, designers may need more information and knowledge on the requirements or uncertainties in and around the system. Some mathematical relationships among parameters and variables are based on designers' experience, heuristics, or preferences.

The essence of Specific Hypothesis 3, "*explore the sensitivity of the segments of the model boundary and improve accordingly,*" is to refine the model formulation,

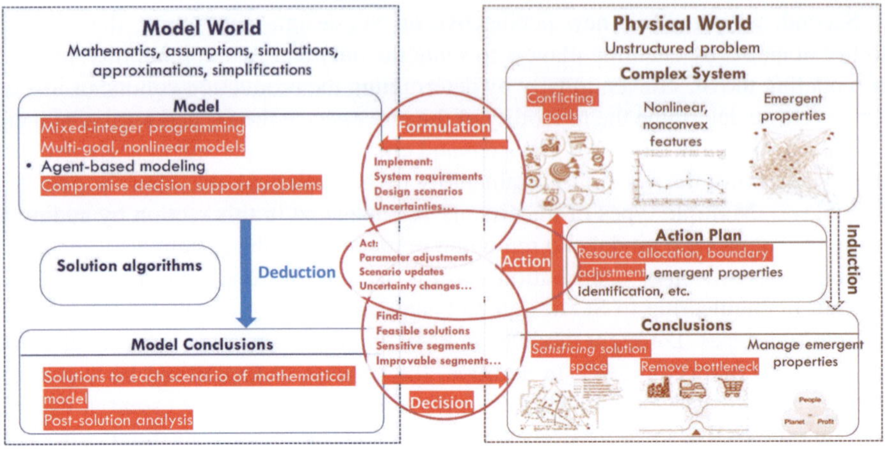

Fig. 4.26 The methods and procedures involved in formulation-exploration—establish the information exchange, knowledge awareness, and instructions sharing among the three highlighted processes

use appropriate design scenarios, and manage typical uncertainties. Designers often need sufficient exploration and experimentation to acquire data and interpret it into knowledge for model improvement actions.. In this chapter, we standardize such exploration-interpretation-improvement procedures by incorporating typical uncertainties, such as the variations in the water inflow of the dam network and the uncertainties in product demand and supply in a supply chain, then identifying the sensitive and improvable segments. By exploring how to remove the sensitive and improvable segments of the model, we ensure the model becomes less sensitive to scenario changes and uncertainties.

In other words, we strengthen the connections among the three processes shown in Fig. 4.26: formulation, decision, and action. Without using the Formulation-Exploration framework, the three processes are relatively isolated, without information exchange. Even with designers' efforts to improve design by passing information among them, the operations need to be standardized and generalized. In this monograph, we establish information exchange, knowledge awareness, and instructions sharing among the three processes and standardize them; see Tables 4.12 and 4.14.

4.5.2 *Summarizing How We Realize Type I and II Robust Design*

For the dam network problem, Type I uncertainty is identified as the variation in water inflow, which is managed by concretizing the typical cases as inflow scenarios (Table 4.8); Type II uncertainty is concretized as the design scenarios that represent the different priorities of different users' water demand (Table 4.6).

For the supply chain problem, Type I uncertainty is represented as the variation in product demand in different product life cycle phases (Fig. 4.25); Type II uncertainty is reflected as the evolving preference of different system goals (Table 4.18).

Implementing the scenarios, identifying the segments to be improved, and improving the model formulation manage Type I and Type II uncertainty. In this way, we realize Type I and II robust design; see the summary in Table 4.20 as the closing remarks of Table 3.3 regarding the robust design realization and uncertainty management for Test Problems 1.1 and 1.2.

4.5.3 Role of This Chapter

In Chapter 4, we are given the frame of reference for *satisficing* strategy, especially focusing on exploring the model's boundary, which is an extension of the frame of reference in Chap. 2. A method, the Three-Step Exploration Method, and a framework encompassing the method, the Formulation-Exploration framework, are proposed to explore the model's boundary. The "Exploration of the Solution Space" piece of the "Design Evolution Cycle" is completed (see Fig. 4.27). Two test problems, a continuous system–dam network and a coupled decision-making problem–designed customer-order-decoupling-point (CODP) of a supply chain, are used to verify the proposed methods. It is proven that using the Three-Step Exploration Method and its advanced framework, the Formulation-Exploration framework, the system can evolve to be relatively insensitive to the uncertainties that occur at the boundary of the system without sacrificing system performance (the achievement of the goals). Research Question 1 is addressed.

In the introductory story, we highlighted the challenges water resource and supply chain managers face in making critical decisions amidst uncertainty. These decisions involve determining optimal water storage and release levels for dams and selecting the best location for the customer-order decoupling point in a supply chain.

To address these challenges, we introduced the Three-Step Exploration Method, which enables designers to identify sensitive elements within decision models and

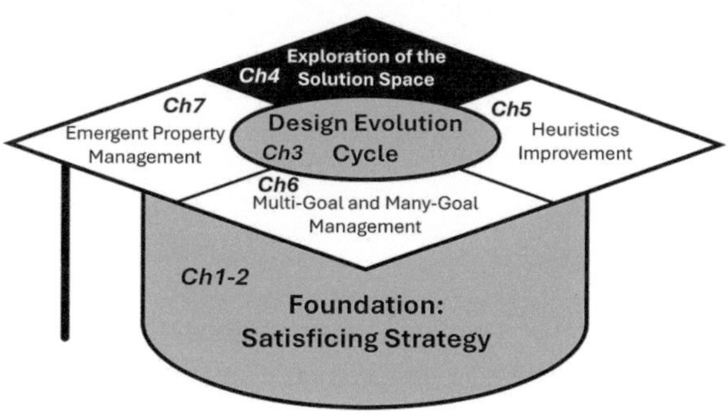

Fig. 4.27 The role of this chapter in the knowledge foundation

enhance them accordingly. Building upon this method, we proposed the Formulation-Exploration framework, specifically tailored for managing mixed-variable, coupled decision models.

Water resource and supply chain managers can use these methods to make informed decisions that account for uncertainties and optimize outcomes, ensuring efficient water management and supply chain operations despite fluctuating conditions.

Glossary

Acceptability threshold A value a designer identifies as the acceptable value of a goal that can be achieved.

Active constraints/bounds The constraints or bounds with zero or a very small slack or surplus are defined as active constraints/bounds. In this section, we define "very small" as less than 1% of the larger of the LHS and the RHS of the constraint/bound.

Boundary of the solution space The constraints and/or bounds of the model that bound the feasible solution space.

Dual price (shadow price) The dual price is the amount by which the goal (objective) would be improved (achieved more completely) as the RHS of the constraint/bound is relaxed by one unit (Lindo Online Help, 2024).

Improvable constraints/bounds Designers identify these constraints or bounds when they have relatively large positive dual prices. In this chapter, we treat any dual price above 0.1% of the achieved goal value as large enough to be improvable.

Improvement In Sect. 4.2, we define model and system improvement as the increasing of insensitivity to uncertainties.

Insensitive solutions Insensitive solutions are away from the physical boundary and relatively insensitive to uncertainty.

Key Nodes Nodes in a network that receive flows or information from multiple upstream nodes.

Ordinary Nodes Nodes in a network that receive flows or information from at most one upstream node. Designers ensure that every node meets one of these definitions.

Physical boundary Designers recognize a physical boundary as the limit that a system can reach physically. It may differ from the mathematical boundary.

	To make the mathematical model less sensitive to uncertainties, we add a buffer to that physical boundary. As a result, the mathematical boundary becomes more restrictive than the physical boundary.
Relaxing RHS	It is the opposite of restricting the RHS. On less-than-or-equal-to (≤) constraints/bounds, relaxing the RHS means increasing the RHS. On the greater-than-or-equal-to (≥) constraints/bounds, relaxing RHS means decreasing the RHS.
Restricting RHS	On less-than-or-equal-to (≤) constraints/bounds, restricting the RHS means decreasing the RHS. On the greater-than-or-equal-to (≥) constraints/bounds, restricting RHS means increasing the RHS.
Satisficing	*Satisficing* is a decision-making strategy or a cognitive heuristic that entails searching through available alternatives until they met an acceptability threshold rather than maximizing any specific objective (Byron, 1998).
Satisficing solutions	solutions for which the acceptability threshold for all goals is simultaneously met.
Sensitive segments	There are two types of sensitive segments: active constraints/bounds and improvable constraints or bounds.
Slack or surplus	Designers use the slack or surplus variable to gauge how close a solution comes to meeting a constraint or bound exactly. For a ' ≤ ' constraint, we call that variable slack. For a ' ≥ ' constraint, we call it surplus. If a solution meets a constraint exactly, its slack or surplus value becomes zero (Lindo Online Help, 2024).
Satisficing weight range	Designers identify an area in a ternary plot where goal weights yield satisficing solutions.

Appendix 4.1: The 34 Weight Scenarios (WSS) and the Corresponding Achievement of the Goals

We use W1, W2, and W3 as the three weights of the goals, and obtain the achievement values of the three goals: Goal 1, Goal 2 and Goal 3. The "Total" is the value of "z," the weighted sum of the three goals. These are the results of the original model. The results facilitate us in obtaining the ternary plots (Figs. 4.11, 4.12 and 4.13) and the *satisficing* area (Fig. 4.14) in Sect. 4.3.5.

WS	Weights			Goals				WS	Weights			Goals			
	W1	W2	W3	1	2	3	Total		W1	W2	W3	1	2	3	Total
1	0.33	0.33	0.33	0.23	0.53	0.31	0.44	18	0.4	0.2	0.4	0.23	0.71	0.18	0.41
2	1	0	0	0.00	2.00	2.00	0.05	19	0.4	0.4	0.2	0.22	0.37	0.56	0.45
3	0	1	0	1.00	0.00	2.00	0.00	20	0	0.5	0.5	0.26	0.56	0.25	0.41
4	0	0	1	1.00	2.00	0.00	0.00	21	0.5	0	0.5	0.07	2.00	0.00	0.10
5	0.8	0.1	0.1	0.16	0.73	0.62	0.43	22	0.5	0.5	0	0.10	0.02	2.00	0.13
6	0.1	0.8	0.1	0.24	0.10	1.60	0.29	23	0.5	0.33	0.17	0.21	0.38	0.58	0.45
7	0.1	0.1	0.8	0.31	1.06	0.02	0.19	24	0.17	0.5	0.33	0.24	0.43	0.42	0.44
8	0.6	0.2	0.2	0.20	0.56	0.37	0.45	25	0.33	0.17	0.5	0.24	0.82	0.11	0.37
9	0.2	0.6	0.2	0.23	0.27	0.77	0.42	26	0.67	0.33	0	0.09	0.05	2.00	0.17
10	0.2	0.2	0.6	0.26	0.83	0.09	0.33	27	0	0.67	0.33	0.24	0.37	0.53	0.42
11	0.5	0.25	0.25	0.21	0.54	0.34	0.45	28	0.33	0	0.67	0.08	2.00	0.00	0.06
12	0.25	0.5	0.25	0.23	0.37	0.54	0.44	29	0.56	0.33	0.11	0.20	0.30	0.86	0.44
13	0.25	0.25	0.5	0.25	0.72	0.15	0.39	30	0.11	0.56	0.33	0.24	0.41	0.45	0.44
14	0.4	0.3	0.3	0.22	0.53	0.31	0.45	31	0.33	0.11	0.56	0.24	0.98	0.07	0.32
15	0.3	0.4	0.3	0.23	0.46	0.38	0.45	32	0.22	0.33	0.44	0.24	0.61	0.22	0.41
16	0.3	0.3	0.4	0.23	0.60	0.24	0.43	33	0.44	0.22	0.33	0.22	0.64	0.24	0.43
17	0.2	0.4	0.4	0.24	0.53	0.29	0.43	34	0.33	0.44	0.22	0.22	0.37	0.55	0.44

Appendix 4.2: Results of the 22 Weight Scenarios (WSs) From the Improved Model (First Iteration)

In Appendix 4.2, we list the 22 weight scenarios (WSs) and the corresponding achievement of the goals. These are the results of the improved model in the first iteration. With these results, we get the *satisficing* area (Fig. 4.17) in Sect. 4.3.5.

WS	Weights			Goals				WS	Weights			Goals			
	W1	W2	W3	1	2	3	Total		W1	W2	W3	1	2	3	Total
1	0.33	0.33	0.33	0.39	0.43	0.30	0.37	12	0.25	0.5	0.25	0.38	0.26	0.54	0.36
2	1	0	0	0.01	2.00	2.00	0.01	13	0.25	0.25	0.5	0.44	0.62	0.13	0.33
3	0	1	0	2.00	0.01	2.00	0.01	14	0.4	0.3	0.3	0.38	0.42	0.31	0.37
4	0	0	1	5.91	2.00	0.00	0.00	15	0.3	0.4	0.3	0.39	0.35	0.38	0.37
5	0.8	0.1	0.1	0.28	0.53	0.63	0.34	16	0.3	0.3	0.4	0.41	0.50	0.21	0.36
6	0.1	0.8	0.1	0.37	0.06	1.33	0.22	17	0.2	0.4	0.4	0.43	0.44	0.26	0.36

(continued)

(continued)

WS	Weights			Goals				WS	Weights			Goals			
	W1	W2	W3	1	2	3	Total		W1	W2	W3	1	2	3	Total
7	0.1	0.1	0.8	0.54	0.91	0.01	0.16	18	0.4	0.2	0.4	0.39	0.62	0.16	0.34
8	0.6	0.2	0.2	0.34	0.43	0.38	0.37	19	0.4	0.4	0.2	0.36	0.26	0.58	0.36
9	0.2	0.6	0.2	0.37	0.18	0.75	0.33	20	0	0.5	0.5	0.54	0.48	0.20	0.34
10	0.2	0.2	0.6	0.47	0.72	0.07	0.28	21	0.5	0	0.5	0.14	2.00	0.00	0.07
11	0.5	0.25	0.25	0.36	0.42	0.34	0.37	22	0.5	0.5	0	0.12	0.02	2.00	0.07

Appendix 4.3: Mathematical Formulation of the CODP as a CDSP

Appendix 4.3 is referenced in Sect. 4.4.4. In this appendix we present the mathematical form of cDSP for the CODP design. We explore the design preferences and capacity in iterations using the algorithm in Table 4.13. After three iterations, the design is improved to overcome the original bottleneck, and the solutions are relatively insensitive to the migration of Product Life Cycle phases and changes in the market demand.

<u>**Given**</u>

ts	production time in supplier
tsm	transportation time from supplier to manufacturer
tm	production time in manufacturer
tmr	transportation time from manufacturer to retailer
tr	operation time in the retailer
trc	transportation time from retailer to customer
t	delivery time of this supply chain,
t	f_t(CODP, ts, tsm, tm, tmr, tr, trc)
VS	production volume of supplier
VS	f_{sm}(rls, rlsm, ts, tsm, SRM, SFG)
VM	production volume of manufacturer
VM	f_{mr}(rlm, rlmr, tm, tmr, MRM, MFG)
VR	preparation volume of retailer
VR	f_{rc}(rlr, rlrc, tr, trc, RRM, RFG)
CS	capacity limit of supplier
CM	capacity limit of the manufacturer
CR	capacity limit of retailer
tave	average delivery time of the industry
rlt	reliability of delivery time, $\text{rlt} = f_{rlt}(\text{t, tave})$
sl	service level, $\text{sl} = \text{rls} \cdot \text{rlsm} \cdot \text{rlm} \cdot \text{rlmr} \cdot \text{rlr} \cdot \text{rlrc} \cdot \text{rlt}$

slmin minimum service level of the industry
pmin minimum unit price of the industry
pmax maximum unit price of the industry
pave average unit price of the industry
p unit price of retailer's finished goods,
p f_p(sl, slave, slmin, pave)
D market demand for the retailer's finished goods,
D f_D(sl, slave, slmin, pave)

hcsrm/scsrm/hcsfg/scsfg/hcmrm/scmrm/hcmfg/scmfg/hcrrm/scrrm/hcrfg/scrfg

Unit holding/shortage cost in supplier's manufacturer's retailer's raw material / finished goods

pcs/pcm/pcr

Unit production/operation cost of supplier manufacturer retailer

tcsm/tcmr/tcrc

Unit transportation cost from supplier to manufacturer manufacturer to retailer retailer to customers

FAS/FAM/FAR

Forecast accuracy of supplier manufacturer retailer

FSFG/FMFG/FRFG

Forecast of the demand for finished goods at supplier manufacturer retailer (if it is MTS)

FSFGf_{SFG}(CODP, FAS, rls, rlsm, p)
FMFGf_{MFG}(CODP, FAM, rlm, rlmr, p)
FRFGf_{RFG}(CODP, FAR, rlr, rlrc, p)

Find

System Variables

Rls reliability of production of supplier
rlsm reliability of transportation from supplier to manufacturer
rlm reliability of production of the manufacturer
rlmr reliability of transportation from manufacturer to retailer
rlr reliability of operation of the retailer
rlrc reliability of transportation from retailer to customers
SRM CODP at the raw material of supplier
SFG CODP at finished goods of the supplier
MRM CODP at the raw material of the manufacturer
MFG CODP at finished goods of a manufacturer
RRM CODP at the raw material of retailer

RFG CODP at finished goods of retailer
SRM CODP at the raw material of supplier
SFG CODP at finished goods of the supplier
MRM CODP at the raw material of the manufacturer
MFG CODP at finished goods of a manufacturer
RRM CODP at the raw material of retailer
RFG CODP at finished goods of retailer

(SRM, SFG, MRM, MFG, RRM, and RFG are binary variables)

Deviation Variables

d_i^- Under-achievement of goal i
d_i^+ Over- achievement of goal i

Satisfy

System Constraints

$\mathbf{SRM + SFG + MRM + MFG + RRM + RFG = 1}$ Constraint 1

//There is one CODP in this supply chain

$\mathbf{sl \geq slmin}$ Constraint 2

//Service level is greater than or equal to the industry minimum service level

$\mathbf{PTOT \geq PTmin}$ Constraint 3

//Total profit is greater than or equal to the industry minimum profit

$\mathbf{t \leq tave}$ Constraint 4

//Delivery time of this supply chain is no more than the average of the industry

$\mathbf{p \leq pave}$ Constraint 5

//Unit price of the retailer's finished goods is no more than the average of the industry

System Goals

$\frac{\mathbf{PTOT}}{\mathbf{T1}} + \mathbf{d}_1^- - \mathbf{d}_1^+ = 1$ Goal 1

//To reach profit target.

$\frac{\mathbf{SL}}{\mathbf{T2}} + \mathbf{d}_2^- - \mathbf{d}_2^+ = 1$ Goal 2

//To reach service level target.

$\mathbf{V} + \mathbf{d}_3^- - \mathbf{d}_3^+ = 0$ Goal 3

//To reach the target of the variance of the reliability of the entities (three entities: supplier, manufacturer, and retailer)

Bounds

$\mathbf{SRM}, \mathbf{SFG}, \mathbf{MRM}, \mathbf{MFG}, \mathbf{RRM}, \mathbf{RFG} \in \{0, 1\}$

//The CODP variables are Boolean variables.

$0.997 \leq \mathbf{rls}, \mathbf{rlsm}, \mathbf{rlm}, \mathbf{rlmr}, \mathbf{rlr}, \mathbf{rlrc} \leq 0.999997$

//All reliabilities are between 3σ and 6σ

$0 \leq \mathbf{d_i^-}, \mathbf{d_i^+} \leq 1, \mathbf{i} = 1, 2, 3$

$\mathbf{d_i^-} \cdot \mathbf{d_i^+} = 0, \mathbf{i} = 1, 2, 3$

$\sum_{i=1}^{3} \mathrm{w_i} \cdot (\mathbf{d_i^-} + \mathbf{d_i^+}) = 1$

Minimize

$$\mathbf{Z} = \sum_{i=1}^{3} \mathrm{w_i} \cdot (\mathbf{d_i^-} + \mathbf{d_i^+})$$

//Minimize the weighted sum of the deviation variables.

References

Ahmadi, M., & Teimouri, E. (2008). Determining the order penetration point in auto export supply chain by the use of dynamic programming. *Journal of Applied Sciences, 8*(18), 3214–3220.

Ahmed, S., Goh, C.-H., Allen, J. K., Mistree, F., Zagade, P., & Gautham, B. (2014). Hot Forging of automobile steel gear blanks: An exploration of the solution space. In *ASME design automation conference*, Paper Number DETC2014-34197.

Aitken, J., Childerhouse, P., & Towill, D. (2003). The impact of product life cycle on supply chain strategy. *International Journal of Production Economics, 85*(2), 127–140.

Barros, M. T., Tsai, F. T., Yang, S.-L., Lopes, J. E., & Yeh, W. W. (2003). Optimization of large-scale hydropower system operations. *Journal of Water Resources Planning and Management, 129*(3), 178–188.

Bellman, R., & Kalaba, R. E. (1965). *Dynamic programming and modern control theory* (Vol. 81). Academic.

Box, G. E., & Draper, N. R. (1987). *Empirical model-building and response surfaces.* Wiley.

Byron, M. (1998). Satisficing and optimality. *Ethics, 109*(1), 67–93.

Changchit, C., & Terrell, M. (1993). A multiobjective reservoir operation model with stochatic inflows. *Computers and Industrial Engineering, 24*(2), 303–313.

Chen, W., Tsui, K.-L., Allen, J. K., & Mistree, F. (1995). Integration of the response surface methodology with the compromise decision support problem in developing a general robust design procedure. *Advances in Design Automation ASME DE, 82–2*, 485–492.

Chen, W., Allen, J. K., Tsui, K.-L., & Mistree, F. (1996). A procedure for robust design: Minimizing variations caused by noise factors and control factors. *ASME Journal of Mechanical Design, 118*(4), 478–485.

Choi, H.-J., Austin, R., Allen, J. K., McDowell, D. L., Mistree, F., & Benson, D. J. (2005). An approach for robust design of reactive powder metal mixtures based on non-deterministic micro-scale shock simulation. *Journal of Computer-Aided Materials Design, 12*(1), 57–85.

Choi, H.-J., McDowell, D. L., Allen, J. K., & Mistree, F. (2008a). An inductive design exploration method for hierarchical systems design under uncertainty. *Engineering Optimization, 40*(4), 287–307.

Choi, H., McDowell, D. L., Allen, J. K., Rosen, D., & Mistree, F. (2008b). An inductive design exploration method for robust multiscale materials design. *ASME Journal of Mechanical Design, 130*(3), Article 031402.

Choi, H.-J., Austin, R., Shepherd, J., Allen, J. K., McDowell, D., Mistree, F., & Benson, D. (2004). An approach for robust micro-scale materials design under unparameterizable variability. In *10th AIAA/ISSMO multidisciplinary analysis and optimization conference.*

Clayton, E. R., Weber, W. E., & Taylor, B. W., III. (1982). A goal programming approach to the optimization of multi response simulation models. *IIE Transactions, 14*(4), 282–287.

Crawley, P., & Dandy, G. (1993). A headworks optimisation model incorporating a graphical user interface. In *Watercomp 93: 2nd Australasian conference on computing for the water industry today and tomorrow* (pp. 319–324); Preprints of Papers, Institution of Engineers.

Daaboul, J., Da Cunha, C., Le Duigou, J., Novak, B., & Bernard, A. (2015). Differentiation and customer decoupling points: an integrated design approach for mass customization. *Concurrent Engineering, 23*(4), 284–295.

Daaboul, J., Laroche, F., & Bernard, A. (2010). Determining the CODP position by value network modeling and simulation. In *IEEE international technology management conference (ICE)* (pp 1–10).

Dahe, P., & Srivastava, D. (2002). Multireservoir multiyield model with allowable deficit in annual yield. *Journal of Water Resources Planning and Management, 128*(6), 406–414.

Dreyfus, S. E. (1957). Computational aspects of dynamic programming. *Operations Research, 5*(3), 409–415.

Eschenbach, E. A., Magee, T., Zagona, E., Goranflo, M., & Shane, R. (2001). Goal programming decision support system for multiobjective operation of reservoir systems. *Journal of Water Resources Planning and Management, 127*(2), 108–120.

Fok, K. L., & Chopra, A. K. (1986). Earthquake analysis of arch dams including dam–water interaction, reservoir boundary absorption and foundation flexibility. *Earthquake Engineering and Structural Dynamics, 14*(2), 155–184.

Ford, L., & Fulkerson, D. R. (1962). *Flows in networks.* Press.

Gaither, N. (1990) *Production and operation management: A problem–solving and decision–making approach.* Dryden Press.

Ghalehkhondabi, I., Sormaz, D., & Weckman, G. (2016). Multiple customer order decoupling points within a hybrid MTS/MTO manufacturing supply chain with uncertain demands in two consecutive echelons. *Opsearch, 53*(4), 976–997.

Ghalehkhondabi, I., Ardjmand, E., & Weckman, G. (2017). Integrated decision making model for pricing and locating the customer order decoupling point of A newsvendor supply chain. *Opsearch, 54*(2), 417–439.

Hajfathaliha, A., Teimoury, E., Khondabi, I. G., & Fathi, M. (2011). Using queuing approach for locating the order penetration point in a two-echelon supply chain with customer loss. *International Journal of Business and Management, 6*(1), 258.

Hsu, N.-S., & Cheng, K.-W. (2002). Network flow optimization model for basin-scale water supply planning. *Journal of Water Resources Planning and Management, 128*(2), 102–112.

Ignizio, J. P. (1982). *Linear programming in single and multi-objective systems.* Prentice Hall.

Iravani, S. M., Liu, T., & Simchi-Levi, D. (2012). Optimal production and admission policies in make-to-stock/make-to-order manufacturing systems. *Production and Operations Management, 21*(2), 224–235.

Isukapalli, S., Roy, A., & Georgopoulos, P. (1998). Stochastic Response Surface Methods (SRSMs) for uncertainty propagation: Application to environmental and biological systems. *Risk Analysis, 18*(3), 351–363.

Jacobs, F. R., & Chase, R. B. (2004). *Operations management for competitive advantage* (16th ed.). Mc-Graw Hill.

Jeong, I.-J. (2011). A dynamic model for the optimization of decoupling point and production planning in a supply chain. *International Journal of Production Economics, 131*(2), 561–567.

Kuczera, G., & Diment, G. (1988). General water supply system simulation model: WASP. *Journal of Water Resources Planning and Management, 114*(4), 365–382.

Lindo Online Help. (2024). Lindo Online Help. from https://www.lindo.com/index.php

Liu, W., Mo, Y., Yang, Y., & Ye, Z. (2015). Decision model of customer order decoupling point on multiple customer demands in logistics service supply chain. *Production Planning and Control, 26*(3), 178–202.

Liu, W., Wu, R., Liang, Z., & Zhu, D. (2018). Decision model for the customer order decoupling point considering order insertion scheduling with capacity and time constraints in logistics service supply chain. *Applied Mathematical Modelling, 54*, 112–135.

Loganathan, G., & Bhattacharya, D. (1990). Goal-programming techniques for optimal reservoir operations. *Journal of Water Resources Planning and Management, 116*(6), 820–838.

Loucks, D. P. (2000). Sustainable water resources management. *Water International, 25*(1), 3–10.

Loucks, D. P., Stedinger, J. R., & Haith, D. A. (1981). *Water resource systems planning and analysis: Solutions manual*. Prentice-Hall.

McPherson, R., et al. (2016). Impacts of climate change on flows in the red river Basin. Final Report to the South Central Climate Science Center University of Oklahoma.

Mistree, F., Hughes, O. F., & Phuoc, H. (1981). An optimization method for the design of large, highly constrained complex systems. *Engineering Optimization, 5*(3), 179–197.

Mistree, F., Hughes, O. F., & Bras, B. (1993). The compromise decision support problem and the adaptive linear programming algorithm. In M. P. Kamat (Ed.), *Structural optimization: Status and promise* (pp. 247–286). AIAA.

Mousavi, S. J., Moghaddam, K. S., & Seifi, A. (2004). Application of an interior-point algorithm for optimization of a large-scale reservoir system. *Water Resources Management, 18*(6), 519–540.

Nandalal, K., & Bogard, J. J. (2007). *Dynamic programming based operation of reservoirs: Applicability and limits*. Cambridge University Press.

Naylor, J. B., Naim, M. M., & Berry, D. (1999). Leagility: Integrating the lean and agile manufacturing paradigms in the total supply chain. *International Journal of Production Economics, 62*(1–2), 107–118.

Needham, J. T., Watkins, D. W., Jr., Lund, J. R., & Nanda, S. (2000). Linear programming for flood control in the iowa and Des Moines rivers. *Journal of Water Resources Planning and Management, 126*(3), 118–127.

Neely, W. P., Sellers, J., & North, R. M. (1980). Goal programming priority sensitivity analysis: An application in natural resource decision making processes. *Interfaces, 10*(5), 83–89.

Nellippallil, A. B., Song, K. N., Goh, C.-H., Zagade, P., Gautham, B., Allen, J. K., & Mistree, F. (2017). A goal-oriented, sequential, inverse design method for the horizontal integration of a multistage hot rod rolling system. *ASME Journal of Mechanical Design, 139*(3), Article 031403.

Nellippallil, A. B., Mohan, P., Allen, J. K., & Mistree, F. (2018). Robust concept exploration of materials, products and associated manufacturing processes. In *ASME design automation conference*. Paper Number: DETC2018-85913.

Parker, C. J. (2016). Human acceptance of 3D printing in fashion paradox: Is mass customization a bridge too far? *WIT Transactions on Engineering Sciences, 113*, 373–380.

Poff, N. L., Brown, C. M., Grantham, T. E., Matthews, J. H., Palmer, M. A., Spence, C. M., Wilby, R. L., Haasnoot, M., Mendoza, G. F., & Dominique, K. C. (2016). Sustainable water management under future uncertainty with eco-engineering decision scaling. *Nature Climate Change, 6*(1), 25.

Ponnambalam, K., Vannelli, A., & Unny, T. (1989). An application of Karmarkar's interior-point linear programming algorithm for multi-reservoir operations optimization. *Stochastic Hydrology and Hydraulics, 3*(1), 17–29.

Rani, D., & Moreira, M. M. (2010). Simulation–optimization modeling: A survey and potential application in reservoir systems operation. *Water Resources Management, 24*(6), 1107–1138.

Reddy, M. J., & Kumar, D. N. (2007). Multiobjective differential evolution with application to reservoir system optimization. *Journal of Computing in Civil Engineering, 21*(2), 136–146.
Rudberg, M., & Wikner, J. (2004). Mass customization in terms of the customer order decoupling point. *Production Planning and Control, 15*(4), 445–458.
Sabeghi, M., Smith, W., Allen, J. K. & Mistree, F. (2015, August). *Solution space exploration in model-based realization of engineered systems.* In ASME Design Automation Conference, Paper Number DETC2015-46457.
Sabeghi, M., Shukla, R., Allen, J. K., & Mistree, F. (2016). Solution space exploration of the process design for continuous casting of steel. In *ASME design automation conference.* Paper Number IDETC2016-59399.
Salvendy, G. (2001). *Handbook of industrial engineering: Technology and operations management.* Wiley.
Seada, H., & Deb, K. (2014). *U-NSGA-III: A unified evolutionary algorithm for single, multiple, and many-objective optimization.* COIN report 2014022.
Seifi, A., & Hipel, K. W. (2001). Interior-point method for reservoir operation with stochastic inflows. *Journal of Water Resources Planning and Management, 127*(1), 48–57.
Shidpour, H., Da Cunha, C., & Bernard, A. (2014). Analyzing single and multiple customer order decoupling point positioning based on customer value: A multi-objective approach. *Procedia CIRP, 17*, 669–674.
Simon, H. A. (1956). Rational choice and the structure of the environment. *Psychological Review, 63*(2), 129.
Simon, H. A. (1976). From substantive to procedural rationality. In T. J. Kastelein, S. K. Kulpers, W. A. Nijenhuis, & G. R. Wagenaar (Eds.), *25 years of economic theory* (pp. 65–86). Springer.
Simon, H. A. (1996). *The sciences of the artificial.* MIT Press.
Simon, H. A., & Kadane, J. B. (1975). Optimal problem-solving search: All-or-none solutions. *Artificial Intelligence, 6*(3), 235–247.
Smith, W. F., Milisavljevic, J., Sabeghi, M., Allen, J. K., & Mistree, F. (2015). The realization of engineered systems with considerations of complexity. In *ASME design automation conference.* Paper Number DETC2015-46211.
Sreenivasan, K., & Vedula, S. (1996). Reservoir operation for hydropower optimization: A chance-constrained approach. *Sadhana, 21*(4), 503–510.
Taguchi, G. (1993). *Taguchi on robust technology development: bringing quality engineering upstream.* ASME Press Series on International Advances in Design Productivity.
Taguchi, G., & Clausing, D. (1990). Robust quality. *Harvard Business Review, 68*(1), 65–75.
Talvitie, A. (1981). Inaccurate or incomplete data as a source of uncertainty in econometric or attitudinal models of travel behavior. In *New horizons in travel-behavior research. The fourth international conference on behavioral travel modeling.*
Teegavarapu, R. S., & Simonovic, S. P. (2000). Short-term operation model for coupled hydropower reservoirs. *Journal of Water Resources Planning and Management, 126*(2), 98–106.
Teimoury, E., & Fathi, M. (2013). An integrated operations-marketing perspective for making decisions about order penetration point in multi-product supply chain: A queuing approach. *International Journal of Production Research, 51*(18), 5576–5596.
Teimoury, E., Modarres, M., Khondabi, I., & Fathi, M. (2012). A queuing approach for making decisions about order penetration point in multiechelon supply chains. *The International Journal of Advanced Manufacturing Technology, 63*(1–4), 359–371.
Triantaphyllou, E., & Sánchez, A. (1997). A sensitivity analysis approach for some deterministic multi-criteria decision-making methods. *Decision Sciences, 28*(1), 151–194.
van Ackooij, W., Henrion, R., Möller, A., & Zorgati, R. (2014). Joint chance constrained programming for hydro reservoir management. *Optimization and Engineering, 15*(2), 509–531.
Vedula, S., Mujumdar, P., & Sekhar, G. C. (2005). Conjunctive use modeling for multicrop irrigation. *Agricultural Water Management, 73*(3), 193–221.
Velasquez, J., Khakifirooz, M., & Fathi, M. (2019). *Large scale optimization applied to supply chain and smart manufacturing: Theory and applications.* Springer Optimization and Its Applications.

Wikner, J., Naim, M. M., Spiegler, V. L., & Lin, J. (2017). IOBPCS based models and decoupling thinking. *International Journal of Production Economics, 194*, 153–166.
Wurbs, R. A. (1991). Optimization of multiple-purpose reservoir systems operations: A review of modeling and analysis approaches, Hydrologic Engineering Center, US Army Corps of Engineers.
Xu, X., & Liang, Z. (2011). CODP positioning based on extension superiority evaluation model. In *Proceeding of the IEEE 2011 international conference on electronic and mechanical engineering and information technology (EMEIT)* (pp. 4041–4047).
Xue, X., Zhang, K., Hong, Y., Gourley, J. J., Kellogg, W., McPherson, R. A., Wan, Z., & Austin, B. N. (2015). New multisite cascading calibration approach for hydrological models: Case study in the Red River basin using the VIC model. *Journal of Hydrologic Engineering, 21*(2), https://doi.org/10.1061/(ASCE)HE.1943-5584.0001282.
Yakowitz, S. (1982). Dynamic programming applications in water resources. *Water Resources Research, 18*(4), 673–696.
Zhou, H.-C., Zhang, G.-H., & Wang, G.-L. (2007). Multi-objective decision making approach based on entropy weights for reservoir flood control operation. *Journal of Hydraulic Engineering, 38*(1), 100–106.
Zhou, W., Huang, W., & Zhang, R. (2014). A two-stage queueing network on form postponement supply chain with correlated demands. *Applied Mathematical Modelling, 38*(11–12), 2734–2743.

Chapter 5
Type I, II, and III Robust Design Through Improving Approximation

Adaptive Linear Programing Algorithm with Parameter Learning (ALPPL)

In the world of engineering, challenges abound, each more complex than the last. Imagine grappling with nonlinear features, multiple objectives, and coupled decision-making, all while navigating the labyrinth of limited computational resources. It is a daunting task, but one that holds the promise of unlocking new realms of possibility.

Amidst the dynamic landscape of contemporary manufacturing processes, the hot rod-rolling process chain is increasingly used as a test problem by researchers. This multi-stage endeavor encompasses a series of intricately coordinated steps, commencing from the casting phase and culminating in the production of steel rods through a thermo-mechanical processing journey involving stages of casting, reheating, rolling, cooling, and rod-making.

Central to the hot rod process chain is the profound metamorphosis of steel microstructures. Columnar grains, characteristic of the initial cast material, undergo a transformative evolution, transitioning into equiaxed formations. The phase composition of the steel undergoes a significant alteration, with the predominant phase transformation occurring during the cooling phase subsequent to the rolling process. The resultant microstructure profoundly influences the mechanical properties of the end product. However, the endeavor of refining steel properties to meet stringent mechanical requirements poses a formidable challenge. Traditionally, the process necessitates extensive and costly plant trials, demanding both time and resources. Consequently, it is imperative to adopt model-based design methods to explore the solution space while concurrently minimizing operational overheads.

The adaptive linear programming (ALP) has been proven to be an effective solution algorithm to solve nonlinear, multi-goal, and coupled-decision models; see Mistree et al. (1993) and Sharma et al. (2022). However, the hyperparameters in the ALP may affect the convergence and accuracy of the solution space, which may severely influence the properties of the end product of a process chain. Therefore, in this chapter, we introduce an improved algorithm that incorporates parameter learning to realize model evolution by improving the determination of a critical hyperparameter in the solution algorithm—Adaptive Linear Programming Algorithm with

L. Guo et al., *Architecting a Satisficing Strategy for Designing Complex Systems*,
https://doi.org/10.1007/978-3-032-04577-5_5

Parameter Learning (ALPPL); see Guo et al. (2024a, 2024b). With ALPPL, we have a more comprehensive definition of linearization performance. The hyperparameter value that brings the most critical performance characteristics (defined as evaluation indices later in this chapter) into their desired range is returned as the best value, which ensures appropriate scenario of tradeoffs between the accuracy of the linearization and the robustness of the solutions; see Sect. 5.4.2. For our test problem, the hot-rod rolling process chain, the ALP returns the best hyperparameter value in twelve iterations considering only the deviation between the achieved value of the goals and their targets as the linearization performance index, whereas the ALPPL returns the best hyperparameter value in fourteen iterations considering multiple evaluation indices (EIs). The complexity of both the ALP and the ALPPL is $O(n^2)$.

5.1 Organization of This Chapter

The organization of this chapter is illustrated in Fig. 5.1. In Sect. 5.2, the reference to solution algorithms, specifically the ALP, is framed as a revisit and extension of Sect. 1.2.

In Sect. 5.3 the problem statement regarding the research gaps in approximation and solution algorithms within the satisficing strategy, based on Sects. 1.4.2 and Sect. 2.3 are discussed.

In Sect. 5.4 we introduce the Adaptive Linear Programming algorithm with Parameter Learning (ALPPL) in detail, based on the research gaps described in Sect. 1.5 and the Method 2 proposed in Sect. 3.3.2.

In Sect. 5.5, the ALPPL is applied to the hot rod process chain problem to identify the solution space that is relatively insensitive to parameter settings and model structure variations caused by problematic heuristics.

In Sect. 5.6, the role of this chapter in this monograph is summarized.

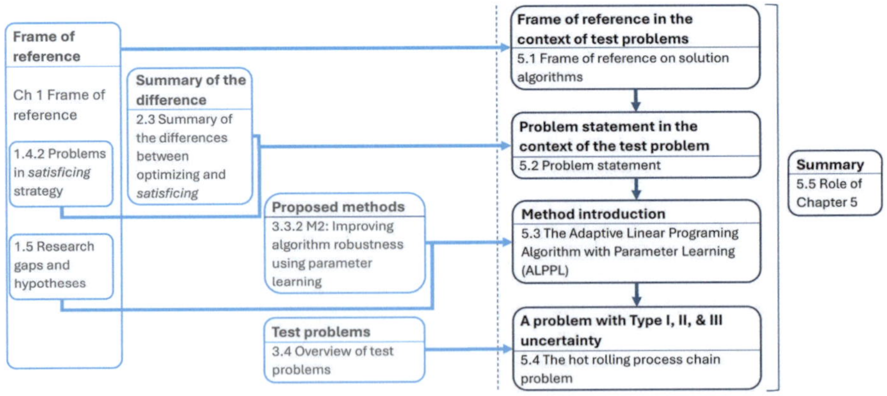

Fig. 5.1 Organization of this chapter

We interpret this chapter regarding the "what, how, and why" questions in Table 5.1.

The plan of specifying and answering Research Question 2 (RQ2 in Sect. 3.2.2) in the context of the test problems is shown in Table 5.2. In this chapter, the Proposed Method 2 (M2 in Sect. 3.3.2), Adaptive Linear Programming Algorithm with Parameter Learning (ALPPL), is empirically verified (EVe2) using Test Problem 2 (T2, problem statement is in Sect. 5.3), the cooling stage of the hot rod process chain. Research Question 2 (RQ2) is specified in the context of the test problem (SQT2) and answered (AQ2) by testifying M2. The empirical validation and theoretical validation are in Chaps. 8 and 9.

In this chapter, the Adaptive Linear Programming Algorithm with Parameter Learning (ALPPL) is proposed and tested by using an engineering-design problem—the hot rolling process chain. The Research Question 2 (RQ2) is answered.

Table 5.1 Summary interpretation of the problems investigated in this chapter

Elements	What?
Components	The approximation, exploration, and evaluation stages of the design evolution cycle
Connectors	The adaptive linear programming algorithm with parameter learning (ALPPL) enables to pass of information among approximation, exploration, and evaluation stages and learn knowledge to replace heuristics on the approximation (linearization) of a nonlinear model
Form	How?
Component roles	Provides the constructs for connecting the approximation, exploration, and exploration evaluation stages
Properties	Attributes or features of the approximation, exploration, and evaluation stages—in the approximation stage, designers apply heuristics (e.g., the value of the reduced move coefficient) to proceed with the model approximation in the early stage, whereas in the exploration and evaluation stages, it is feasible to acquire insight to modify or explore more ways of formulation
Relationship	The relations among the activities—heuristics application, learning the approximation performance associated with the heuristics, and use of association to develop the latter to replace the former
Rationale	Why?
Motivation	Provide support to assist designers in improving model approximation to realize Type I, II, & III Robust design through rule-based learning
Assumptions	The improvement in model approximation can be abstracted as learning tradeoff scenarios among the approximation accuracy, computational complexity, the optimality of the solution, and the robustness of the solution to uncertainties that lead to variation in the approximated model and apply the most appropriate scenario
Interpretation	Understanding how parameter tuning, performance indices development, and indices-based performance evaluation help in parameter learning and approximation improvement

Table 5.2 Plan of specifying Research Question 2 (RQ2) and empirically verifying the Adaptive Linear Algorithm with Parameter Learning (ALPPL) (M2)

Chapter	Ch 1	Ch 2	Ch 3	Ch 4–7			Ch 8	Ch 9
				Ch 4	Ch 5	Ch 6–7		
Actions	RG H	RD RQ SH	TVe M	EVe1 SQT1 AQ1	EVe2: use an engineering-design problem with uncertainties in parameters and metaheuristics to verify SH2 and demonstrate M2	EVe SQT AQ	CQ EVa	TE
					SQT2: What is the mathematics in the design method that allows approximation improvements in the realization of complex systems?			
					AQ2: by incorporating parameter learning in the approximation algorithm, designers can learn the association between parameter and approximation performance so they can guarantee the approximation performance through updating the heuristics in parameter settings			
Nomenclature	AQ—answer research questions CQ—closure the answers to research questions H—give hypotheses RD—tie to roust design RG—give research gaps RQ—pose research questions H—specify hypotheses TVe—theoretically verify hypotheses M—introduce methods Eve—empirically verify hypotheses SQT—specify research questions in the context of test problems Eva—empirically validate hypotheses TE—theoretically extend the research							

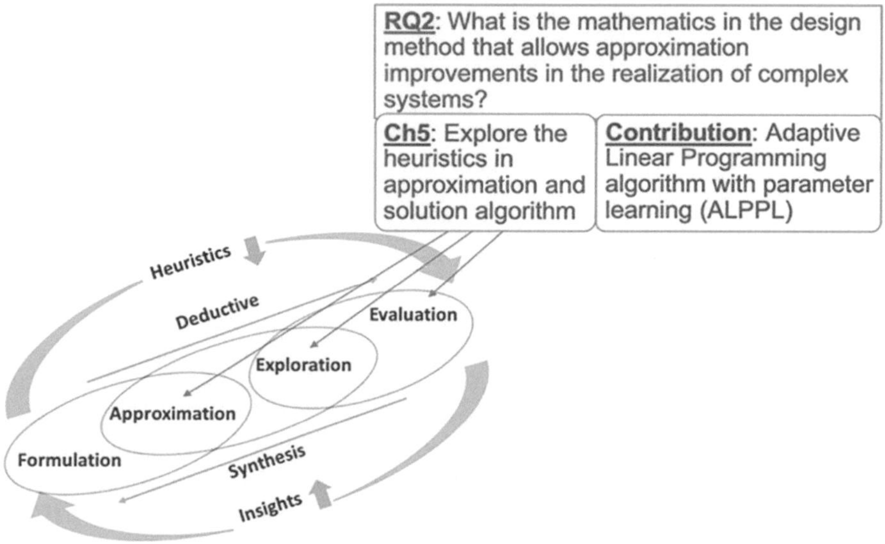

Fig. 5.2 Specified Research Question 2 and the relevant stages to be connected in design evolution cycle

RQ2: What is the method to evolve model to update metaheuristics?

In this chapter, the research question is specified into the context of the test problem—the cooling stage of the hot rod process chain.

SQT2: What is the mathematics in the design method that allows approximation improvements in the realization of complex systems?

To answer RQ2, interactions between the approximation, exploration, and evaluation of a design problem should be studied and the mechanisms of information sharing and intervention between the three stages are established; see Fig. 5.2.

In this chapter, an approximation and solution algorithm—the Adaptive Linear Programing algorithm with parameter learning (ALPPL) is applied to manage a test problem that is parameter-sensitive. Further, different issues are addressed in the approximation of complex-system design: (1) enabling scientific determination and updating of critical parameters to manage approximation efficiency, (2) meeting the requirement on problem fidelity, (3) meeting the requirement on computational complexity, and (4) managing uncertainty. The efficacy of the method is demonstrated using a hot rod rolling problem as Test Problem 3.

The establishment of the context is in Sect. 5.2. The answer to the Research Question 2 is the mechanism and process of improving the ALP by incorporating parameter learning, see Sects. 5.3 and 5.4. The empirical structural validity of the method is presented in Sect. 5.5.

5.2 Solution Algorithms for Designing Complex Engineering Problems

Solution algorithms for solving complex engineering-design problems fall into two categories (Table 5.3): (i) an optimizing strategy or (ii) a satisficing strategy. For typical engineering design problems with nonlinear, nonconvex properties, that cannot be solved straightforwardly using simplex, Lagrange multipliers, or other gradient-based solution algorithms, designers need to solve problems approximately when applying the optimizing strategy, for example, using pattern search methods (Rios and Sahinidis, 2013), particle swarm optimization (Vrahatis et al., 2020), memetic algorithms (Behmanesh and Pannek, 2021), penalty function methods (Viswanathan and Grossmann, 1990), or metaheuristic algorithms (Nagadurga et al., 2023). When methods in Category (i) are used, the solutions are desired to be near-Pareto solutions, often on or close to the boundary of the feasible space (bounded by constraints) but not necessarily at a vertex of the feasible space. Whereas when methods in Category (ii) are are used, such as sequential linear programming (Hajihashemi, 2023; Mistree et al., 1993) or FEAST (Teng and Lu, 2019), the solutions are always at one or multiple vertices of the linearized (approximated) solution space. Since the solution always includes a vertex, we can use information of the dual problem to explore the solution space (Mistree et al., 1993). Typically, in engineering design problems, the number of constraints in the primal formulation exceeds the number of variables (Rao and Mulkay, 2000). In these cases, solving the dual formulation is quicker (Mistree et al., 1993). For optimization strategies, there are state-of-the-art surveys summarizing and classifying linearization techniques thoroughly, such as in (Asghari et al., 2022). Linearization techniques are classified into two groups—nonlinear equations replaced by an exact equivalent LP formulation or linear approximations that find the equivalent of a nonlinear function with the least deviation around the point of interest or separate straight-line segments.

In this chapter, our focus is on a method in Category (ii), the satisficing strategy, namely, the compromise Decision Support Problem (cDSP) and adaptive linear programming (ALP) algorithm (Mistree et al., 1993), because it is more appropriate for engineering design problems by giving satisficing solutions and exploring solution space. Satisficing solutions are near-optimal and on the vertices of the approximated linear problems. The reasons are summarized in Sects. 5.2.1 and 5.2.2.

5.2.1 *Overcoming Assumptions and Challenges*

There are special requirements for engineering design methods. Engineering design involves multiple designers with different fields of expertise making decisions to meet a variety of requirements (Reich et al., 2010). Designers may define objective functions using different methods—utility theory (Fishburn, 1968), game theory (Nash Jr, 1950), analytic hierarchy process (AHP) (Saaty, 1988), Pareto-optimal methods

Table 5.3 The mechanisms supporting satisficing solutions convergence

Mechanisms	Advantage	Assumption removed
Using goals and minimizing deviation variables instead of objectives	At a solution point, only the necessary KKT condition is met, whereas the sufficient KKT condition does not have to be met Therefore, designers have a greater chance of finding a solution and a lower chance of losing a solution due to parameterizable and/or unparameterizable uncertainties	Assumption 1
Using second-order sequential linearization	Designers can have a balance between linearization accuracy and computational complexity	Assumption 2
Using accumulated linearization	Designers can manage nonconvex problems and deal with highly convex, nonlinear problems relatively more accurately	Assumption 2
Combining interior-point search and vertex search	Designers can avoid getting trapped in local optima to some extent and identify satisficing solutions which are relatively insensitive when the starting points change	Assumption 1
Allowing some violations of soft requirements, such as the bounds of deviation variables	Designers can manage rigid requirements and soft requirements in different ways to ensure feasibility As a result, goals and constraints with different scales can be managed	Assumptions 1 and 2

(Calpine and Golding, 1976), etc. For optimization problems with one or multiple objective function(s) to be minimized or maximized, one can define such problems based on the maximization of expected utility. Designers accept assumptions and simplifications when using utility theory to manage a problem, sometimes without realizing it. Such assumptions include the abstraction of mathematical relationships between variables to be 100% accurate and unchanging, the levels of fidelity of different sub-models or segments of a model being the same, the design preferences among multiple objectives are fully captured (Guo et al., 2023). For many problems, these assumptions are incorrect, and this may cause problems. For example, when implementing an optimal solution, a system may not give optimal output because of the inaccurate equations used in the model, or because uncertainty breaks the equilibrium of any Karush–Kuhn–Tucker (KKT) conditions, thus destroying the optimality of the solution (Speakman and Francois, 2021).

In addition, when the problem is nonconvex, specifically, when any nonzero linear combination of the constraints is more convex than the objective function (Souza et al., 2016), Category (i) methods cannot return a feasible solution because the sufficient KKT condition cannot be satisfied (Guo et al., 2024a, 2024b). In this case, formulating the problem using the compromise Decision Support Problem (cDSP) and solving it using the adaptive linear programming (ALP) algorithm is a way to identify satisficing solutions that meet the necessary KKT condition (Guo, 2021).

5.2.2 *Mechanisms to Ensure That the cDSP and ALP Find Satisficing Solutions*

In Chap. 2, we introduced five mechanisms in the cDSP and ALP that enable designers to identify satisficing solutions to nonlinear, nonconvex engineering design problems, especially when optimization methods fail to converge. Here in this chapter, we summarize those mechanisms again in Table 5.3.

Two assumptions in standard optimization methods make optimal solutions unavailable for some nonlinear, nonconvex problems:

Assumption 1 Mathematical models are assumed to be 100% complete and accurate abstractions of physical models, so the optimal solution to the mathematical problem is considered optimal for the physical problem.

Assumption 2 The convexity degree of at least one nonzero linear combination of all constraints is assumed to be higher than the convexity degree of the objective function.

However, for Assumption 1, designers need boundary information to deal with uncertainties, but such information is normally unavailable (Su et al., 2023). Using the cDSP and ALP, designers can identify satisficing solutions by removing the two assumptions.

In Chap. 2, we have a demonstration of the comparison of five examples. The details are also in this video: https://www.youtube.com/watch?v=7apDZO-9A74, and a tutorial of using DSIDES to formulate and solve a problem in this video: https://www.youtube.com/watch?v=tUpVC97Y1L8.

5.3 Problem Statement—Limitations of the ALP Regarding Parameter Determination

Section 5.3 is an extension of Sect. 1.4.2, with more detailed description of the mechanisms and limitations of the ALP algorithm.

5.3.1 *Adaptive Linear Programing (ALP) Algorithm*

The Adaptive Linear Programming (ALP) algorithm can be used to approximate and solve nonlinear problems. The procedures of the ALP algorithm are shown in Fig. 5.3. The ALP is implemented in DSIDES (Hajihashemi, 2023), a decision support system. In Fig. 5.3, the processes within the dotted rounded rectangle A are executed in a synthesis cycle. During the synthesis cycle, continuous improvement (or repeated modification) occurs, involving two parts: approximating the nonlinear

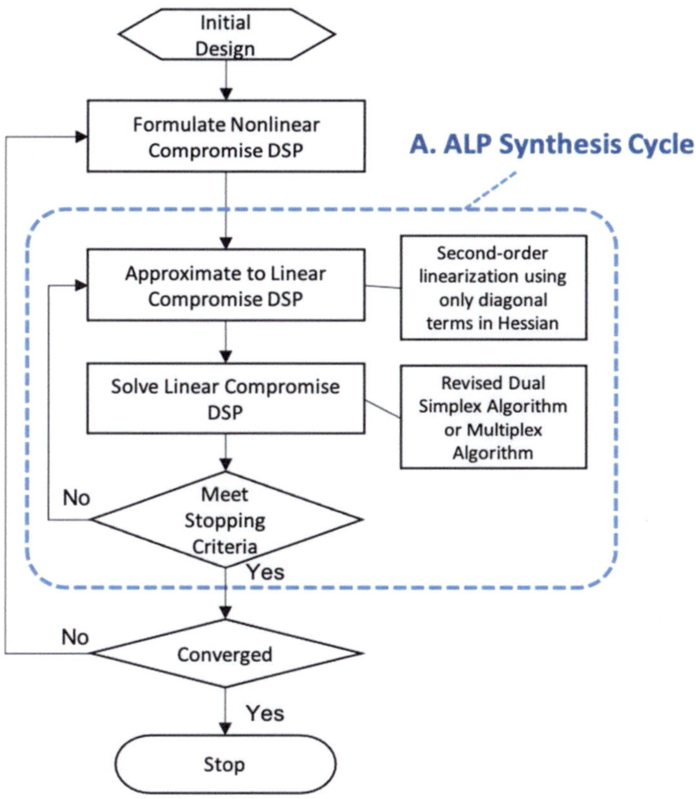

Fig. 5.3 The ALP algorithm

problem and solving the approximated linear problem using the revised dual simplex algorithm. If the stopping criteria[1] are met, the solution with the best value of the deviation function Z is returned as the *satisficing* solution.

The ALP incorporates a local approximation algorithm (Barthelemy and Haftka, 1993), where a secant plane of the paraboloid (using the second-order derivatives at a point as coefficients) replaces the original high-order nonlinear function.

In Fig. 5.4, the projection of an n-dimension problem onto a two-dimension plane and how the design problem is approximated and solved in two iterations using the ALP are illustrated. In the first iteration, the starting point, X_0^0 (user-defined), may not be in the feasible region. The Hook-Jeeves pattern search algorithm is used to bring X_0^0 into the feasible region, as X_1^0, *allowing it to enter the first iteration, where*

[1] The stopping criteria include maximum number of synthesis cycle (NITER), desired stationarity of deviation function (EPSZ), desired stationarity of system variables (EPSX), and desired stationarity of the RMC (threshold value ϵ) (Hajihashemi, 2023).

The program is stopped if any one of the four stopping criteria is met.

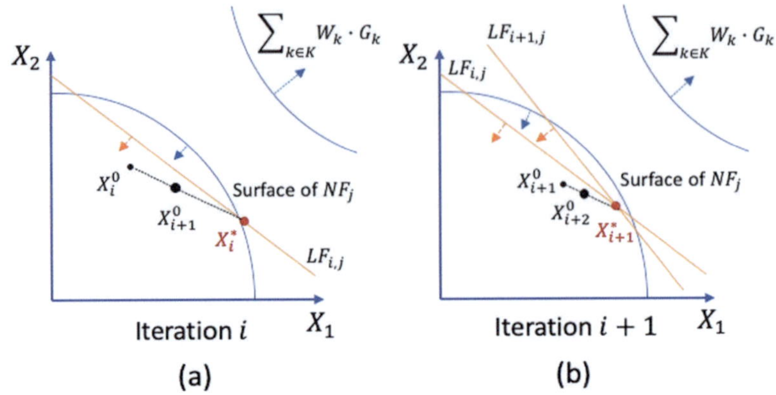

Fig. 5.4 The approximation and obtained solution using the ALP in two iterations

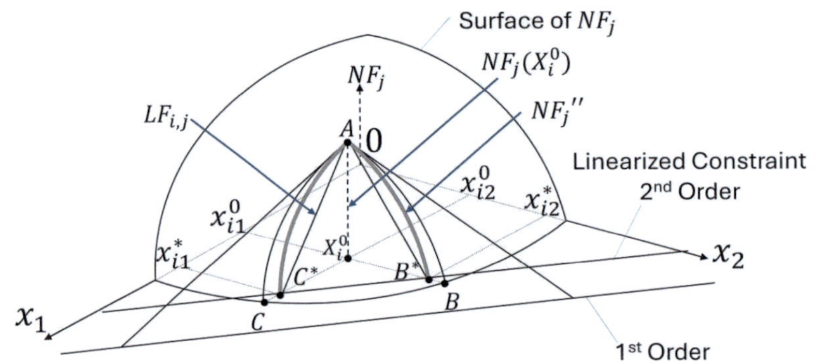

Fig. 5.5 The original nonlinear constraint, the second-order paraboloid, and the secant plane [4]

the problem is linearized at X_1^0. In Iteration i, Fig. 5.4a, a nonlinear constraint,[2] NF_j, is approximated at X_i^0, so an approximated constraint $LF_{i,j}$ is obtained. Using the revised simplex dual algorithm, a solution X_i^* is returned. With the reduced move coefficient (RMC), we obtain the point X_{i+1}^0, a point between X_i^0 and X_i^*, as the starting point of Iteration $i+1$. In Iteration $i+1$, Fig. 5.5b, the approximated linear constraints of both iterations, $LF_{i,j}$ and $LF_{i+1,j}$ are accumulated, and a solution X_{i+1}^* is returned as the starting point of Iteration $i+2$. The approximation continues until one of the stopping criteria is met: either the solution points or the fulfillment of the goals remain in a stationary range, or the total iterations reach an upper limit.

To illustrate the approximation, we use a three-dimensional graph—Fig. 5.5. The approximation takes place in two steps. First, NF_j (Paraboloid ABC in Fig. 5.5) is approximated to a paraboloid $NF_{i,j}''$ (Paraboloid AB^*C^* in Fig. 5.5) by using the diagonal terms of its Hessian matrix at X_i^0 as the coefficient. Then, NF_j'' is approximated

[2] If NF_j is an inequality constraint, $NF_j \geq 0$, then the curve denoted as NF_j in Fig. 5.2 is the surface of NF_j; if NF_j is an equality constraint, $NF_j = 0$, then the curve denoted as NF_j in Fig. 5.2 is NF_j.

to a secant Plane $LF_{i,j}$ (Plane AB^*C^* in Fig. 5.5). The calculations of NF_j'' and $LF_{i,j}$ are given as follows.

$NF_{i,j}''$ is obtained by using only the second-order full derivatives at X_i^0, shown in Eq. (5.1), because the second-order partial derivatives are proved to have very limited impact on the gradient (Mistree et al., 1981).

$$NF_{i,j}'' = NF_j(X_i^0) + \sum_{p=1}^{n}\left(x_{ip} - x_{ip}^0\right)\left(\frac{\partial NF_j}{\partial x_{ip}}\right)_0 + \frac{1}{2}\sum_{p=1}^{n}\left(x_{ip} - x_{ip}^0\right)^2\left(\frac{\partial^2 NF_j}{\partial x_{ip}^2}\right)_0 \tag{5.1}$$

Then, from Eq. (5.1), for the pth dimension, the quadratic to be solved to determine $\left(x_{ip} - x_{ip}^0\right)$ is Eq. (5.2).

$$NF_j(X_i^0) + \left(x_{ip} - x_{ip}^0\right)\left(\frac{\partial NF_j}{\partial x_{ip}}\right)_0 + \frac{1}{2}\left(x_{ip} - x_{ip}^0\right)^2\left(\frac{\partial^2 NF_j}{\partial x_{ip}^2}\right)_0 = 0 \tag{5.2}$$

If Eq. (5.2) has real roots, solving Eq. (5.2) and selecting the root between Eqs. (5.3) and (5.4) with the smaller absolute value for each dimension, we obtain the intersection that is closer to the paraboloid in each dimension, such as B^* and C^*.

$$\left(\frac{\partial NF_j}{\partial x_{ip}}\right)_0^* = \frac{-NF_j(X_i^0)\left(\frac{\partial^2 NF_j}{\partial x_{ip}^2}\right)_0}{-\left(\frac{\partial NF_j}{\partial x_{ip}}\right)_0 - \sqrt{\left(\frac{\partial NF_j}{\partial x_{ip}}\right)_0^2 - 2NF_j(X_i^0)\left(\frac{\partial^2 NF_j}{\partial x_{ip}^2}\right)_0}} \tag{5.3}$$

$$\left(\frac{\partial NF_j}{\partial x_{ip}}\right)_0^* = \frac{-NF_j(X_i^0)\left(\frac{\partial^2 NF_j}{\partial x_{ip}^2}\right)_0}{-\left(\frac{\partial NF_j}{\partial x_{ip}}\right)_0 + \sqrt{\left(\frac{\partial NF_j}{\partial x_{ip}}\right)_0^2 - 2NF_j(X_i^0)\left(\frac{\partial^2 NF_j}{\partial x_{ip}^2}\right)_0}} \tag{5.4}$$

If Eq. (5.2) has no real roots, for example, as the case shown in Fig. 5.6, where NF_j'' does not have intersect with Plane x_1x_2, then the first-order derivative function at X_i^0 is used, so $\left(\frac{\partial NF_j}{\partial x_{ip}}\right)_0^*$ is estimated by the tangent at x_{ip}^0, as Eq. (5.5).

$$\left(\frac{\partial NF_j}{\partial x_{ip}}\right)_0^* = \left(\frac{\partial NF_j}{\partial x_{ip}}\right)_0 \tag{5.5}$$

Using the intersections in each dimension, such as B^* and C^*, we can calculate $LF_{i,j}$ using Eq. (5.6).

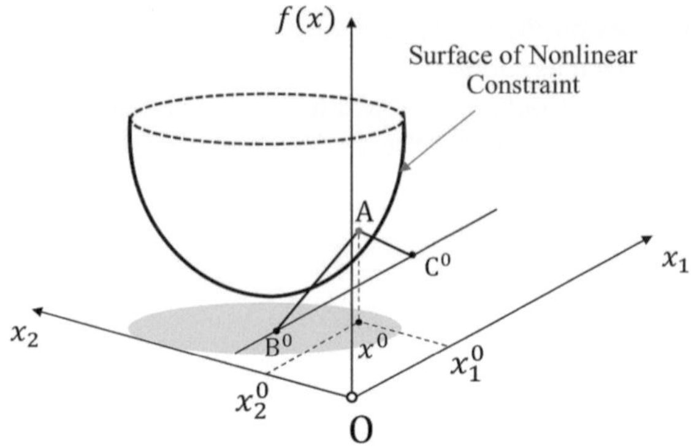

Fig. 5.6 When the second-order paraboloid has no intersection with plane x_1x_2, The first-order tangent is used to approximate NF_j

$$LF_{i,j} = \sum_{p=1}^{n} x_{ip}\left(\frac{\partial NF_j}{\partial x_{ip}}\right)_0^* - \left(\sum_{p=1}^{n} x_{ip}^0\left(\frac{\partial NF_j}{\partial x_{ip}}\right)_0^* - NF_j\left(X_i^0\right)\right) \quad (5.6)$$

If the the convexity of NF_j is relatively large—greater than 0.015 at any starting point in and before the ith iteration, we use the accumulated constraints $\bigcup_{q\le i} LF_{q,j}$ to replace NF_j as shown in Eq. (5.7). If the convexity of NF_j at all starting point in and before the ith iteration is less than or equal to 0.015, we use the single linear constraint in the ith iteration to replace NF_j. as shown in Eq. (5.8).

Constraint Accumulation Algorithm

In the ith iteration,

for every j in J.

if $max\left\{\left|\left[conv\left(NF_j\right)\right]_{X_q^0}\right|\right\}_{q\le i} > 0.015.$

$$P_i^L = \left(P^0 \backslash \{NF_j\}\right) \cup \left\{\bigcup_{q\le i} LF_{q,j}\right\} \quad (5.7)$$

Else

$$P_i^L = \left(P^0 \backslash \{NF_j\}\right) \cup \left\{LF_{i,j}\right\} \quad (5.8)$$

Then, the revised simplex dual algorithm is applied to solve the linear problem $\mathrm{P}_\mathrm{i}^\mathrm{L}$, yielding a solution X_i^*. X_{i+1}^0, a point between the starting point X_i^0 and the solution X_i^*, becomes the starting point of the next iteration. The Reduced Move Coefficient (RMC) is used to determine X_{i+1}^0 as shown in Eq. (5.9).

$$X_{i+1}^0 = X_i^0 + RMC \cdot \left(X_i^* - X_i^0\right) \quad (5.9)$$

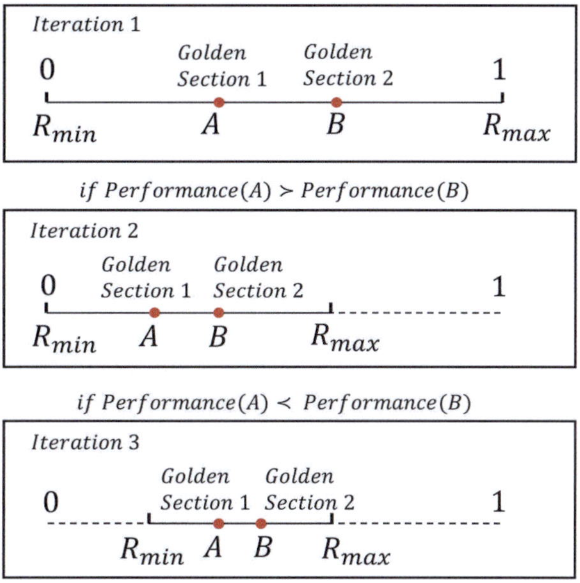

Fig. 5.7 7 The golden section search for the RMC in the ALP

5.3.2 Reduced Move Coefficient (RMC)

The value[3] of RMC ranges 0 to 1. In the ALP, the RMC is either set by the designer as a fixed value or updated using the golden section search algorithm. In (Mistree et al., 1981), the authors recommend setting 0.5 to the RMC as a fixed value through all iterations based on the observations from around 1100 experiments conducted by them in the seventies. However, 0.5 may not be the best value for every design problem. Therefore, the golden section search algorithm was added to improve the RMC determination, in the sense of using metaheuristics to explore the solution space. Users can choose whether to set the RMC as a fixed value or to trigger the golden section search algorithm to allow the RMC to adapt to new situations.

The use of golden section search reduces the range of RMC by cutting off the region with undesired deviation function value or bigger violation of the constraints, till the range is smaller than a threshold ε, ultimately returning the best solution. The range updating mechanism is illustrated in Fig. 5.7.

Two factors, the deviation function value and the violation value of constraints, are considered to evaluate the performance. In the $(i-1)$th iteration, two golden section points, A and B, are used to acquire two starting points for the ith iteration,

[3] RMC does not have to be in [0, 1]. We set its range as [0, 1] because we want to explore the region between X_i^0 and X_i^* in case the approximation fidelity in this region is poor as we assume this is a relatively important region. However, it is fine to have RMC < 0 or RMC > 1 so that designers may explore the region out of the $[X_i^0, X_i^*]$ to avoid being stuck in local *satisficing*. Yet as in this section we deal with nonlinear problems, there is no hassle of local *satisficing*, so we bound the RMC in [0, 1] to stay focused on this relatively important region.

$X_i^0(A)$ and $X_i^0(B)$. In the ith iteration, P^0 is approximated at $X_i^0(A)$, and $X_i^0(B)$ and two solutions $X_i^*(A)$ and $X_i^*(B)$ are obtained by solving the linear problems using revised dual simplex algorithm. We then compute (i) the violation values of constraints $V(X_i^*(A))$ and $V(X_i^*(B))$, by adding all the violation values of the constraints, and (ii) the values of the deviation functions, $Z(X_i^*(A))$ and $Z(X_i^*(B))$, by plugging in the solutions. Then, the decisions on how to update the RMC range are made based on these four values. The performance evaluation rules and the RMC updating rules follow:

RMC Range Updating using Golden Section Search

if $V(X_i^*(A)) == 0$ and $V(X_i^*(B)) == 0$

  if $Z(X_i^*(A)) \leq Z(X_i^*(B))$

    $Performance(A) \succcurlyeq Performance(B)$

  else

    $Performance(A) \prec Performance(B)$

else

  if $V(X_i^*(A)) \leq V(X_i^*(B))$

    $Performance(A) \succcurlyeq Performance(B)$

  else

    $Performance(A) \prec Performance(B)$

if $Performance(A) \succcurlyeq Performance(B)$

  $R_{max} = B$

else

  $R_{min} = A$

In this way, the RMC range is reduced through iteration, until the range is smaller than a threshold, $R_{\max} - R_{\min} < \varepsilon$, at which point we end the RMC search and return the solution X^* with the best performance.

5.3.3 Limitations of the ALP Regarding the Determination of the RMC

Using golden section search, the desired sub-range of RMC may be missed because it is based on two assumptions.

First, the approximation is a continuous function of the RMC and has no more than one inflection point within the RMC range. In other words, the performance function of the RMC is either convex or concave, as shown in Fig. 5.8a. However, for many design problems, the performance function is neither convex nor concave;

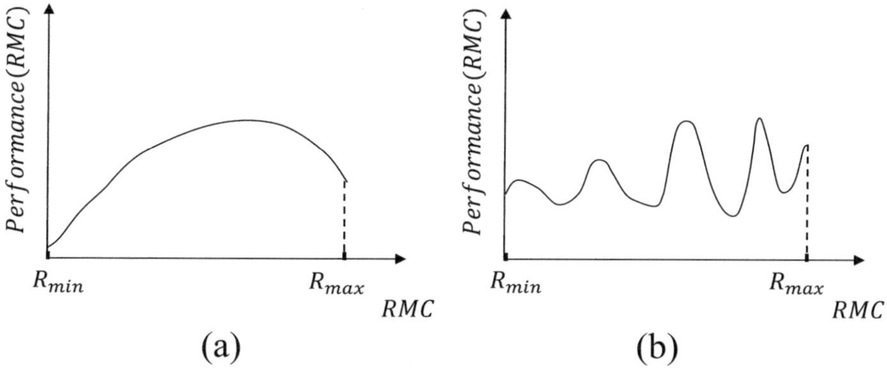

Fig. 5.8 Possible patterns of the performance of the RMC in a sub-range

it fluctuates without any pattern, as shown in Fig. 5.8b. In such cases, the golden section search may miss a sub-range of the RMC with the desired performance.

Second, the criterion for evaluating approximation performance is oversimplified—only the values of the deviation function and constraint violation are considered. Other critical aspects relevant to engineering design problems, such as the robustness of the results to uncertainties and approximation accuracy, are not considered.

Given these two limitations, we propose using parameter learning to improve the ALP algorithm by learning the relationship between the RMC value and the quality of the solutions. This approach will help determine the RMC value more effectively, thereby enhancing the robustness of the approximation algorithm.

5.3.4 Hypothesis for Improving the ALP Algorithm

We hypothesize that by incorporating parameter learning into the ALP algorithm, we can identify the range of RMC that yields more desired performance, meaning more stable results in terms of goal achievement and insensitivity to uncertainties. Therefore, we need to define desired performance, capture the relationship between the RMC and the approximate performance, and use this relationship to tune the appropriate RMC value. There are three steps to verify the hypothesis.

Step 1. *Identify the criteria to evaluate the approximation performance.*
Step 2. *Develop evaluation indices (EIs) based on the identified criteria to quantify the approximation performance regarding RMC values.*
Step 3. *Learn the desired range of each EI (DEI) and tune the RMC to produce results within the DEI.*

In Sect. 5.3, we propose an improved algorithm—the ALP with parameter learning (ALPPL)—to realize the three steps and verify the hypothesis.

5.4 The Adaptive Linear Programing Algorithm with Parameter Learning (ALPPL)

In this section, we implement the three steps proposed in Sect. 5.3.4 to improve the ALP algorithm.

Given the drawbacks of the golden section search, we propose the adaptive linear programming algorithm with parameter learning (ALPPL). See Fig. 5.9. We embed the ALP algorithm into a loop, incorporating the parameter updating, results extraction, performance evaluation, and feedback on parameter updating. The loop is ALPPL. In this chapter, the parameter is specifically referred to the RMC, but the parameter learning algorithm can be used to learn any parameter of an algorithm.

5.4.1 *Step 1—Identify the Criteria for Evaluating the Quality of a Solution*

In the RMC search method currently used in the ALP, the golden section search, two criteria are considered—the fulfillment of the goals and the feasibility of the solutions. In ALPPL, we build upon them.

Criterion 1—fulfillment of the goals.

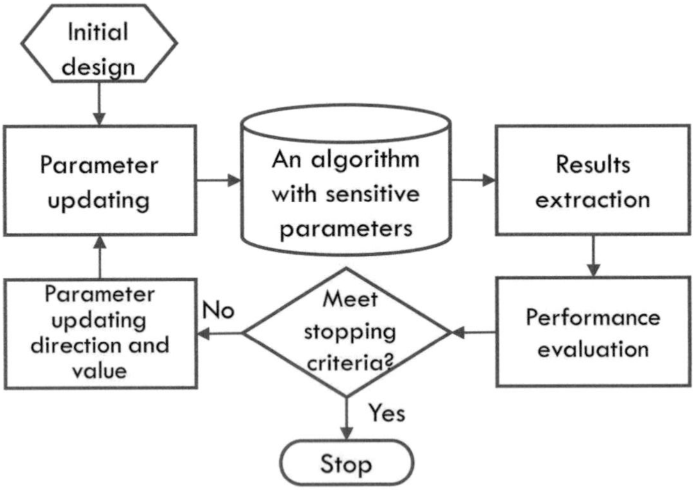

Fig. 5.9 The concept of Adaptive Linear Programming Algorithm with Parameter Learning (ALPPL) (Guo et al., 2020)

Fulfillment of the goals. In engineering design problems, the fulfillment of the goals is a basic criterion to assess whether the system's potential has been sufficiently explored (Jiang et al., 2012). Therefore, we retain this criterion in ALPPL.

Criterion 2—the robustness of the solution.

When we discuss the feasibility of the solutions, we ensure that the solution is feasible in a deterministic situation but cannot guarantee its feasibility under uncertainties. As all models are approximations of reality (Box and Draper, 1987), we cannot perfectly capture all real-world information in a model. We need to manage unexpected uncertainty by making solutions relatively insensitive to model errors and incompleteness. So, we extend the notion of "feasibility" to "robustness," which means the solutions should be feasible under multiple design scenarios and robust to certain model errors and uncertainties, such as wrong parameter values, inaccurate equations, or unstable functional relationships between parameters and variables. In this chapter, we define the robustness of a solution as its insensitivity to model errors and uncertainties. We assume all errors and uncertainties result in changes to the boundary, specifically the change in the right-hand-side value of each constraint. Therefore, the robustness of the solution can be measured by its distance from the boundary. The robustness of the solution and the fulfillment of the goals often contradict each other. An interior solution (one that is away from the boundary) often fulfills the goals less than a boundary solution does. Therefore, we consider both criteria and make decisions based on their trade-offs. In Fig. 5.10, we show two solutions with different levels of robustness. We take the X_i^* in Fig. 5.10a as a robust solution (relatively insensitive solution) and take the X_i^* in Fig. 5.10b as a sensitive solution. The solution in Fig. 5.10a is not on the boundary of the feasible space, making it relatively insensitive (or more robust) to errors of the mmodel or variations. NF_j Fig. 5.10a is not an active constraint. On the contrary, the solution in Fig. 5.10b is on the boundary of the feasible space bounded by NF_j, so it is sensitive (or less robust) to errors or variations of the constraint. NF_j Fig. 5.10b is an active constraint. The approximation does have an impact on the robustness of the solution, and the RMC affects the approximation result. We want the approximation to bring more robust solutions by adjusting the RMC.[4]

Criterion 3—Approximation accuracy.

In general, an increase in accumulated constraints leads to an improvement in approximation accuracy, but not necessarily the more the better; see Fig. 5.11. Therefore, we want sufficient and useful accumulated constraints to bring desired solutions rather than having as many accumulated constraints as possible. Details on developing evaluation indices (EIs) for approximation accuracy are given in Sect. 5.4.2.

Table 5.4, we summarize the criteria for the evaluation of the approximation performance. Based on these criteria, we develop the Evaluation Indices (EIs) in Sect. 5.4.2.

[4] In this chapter, the starting point of the test problem is a user-defined value and unchangeable, so the RMC is the only factor that affects the approximation quality.

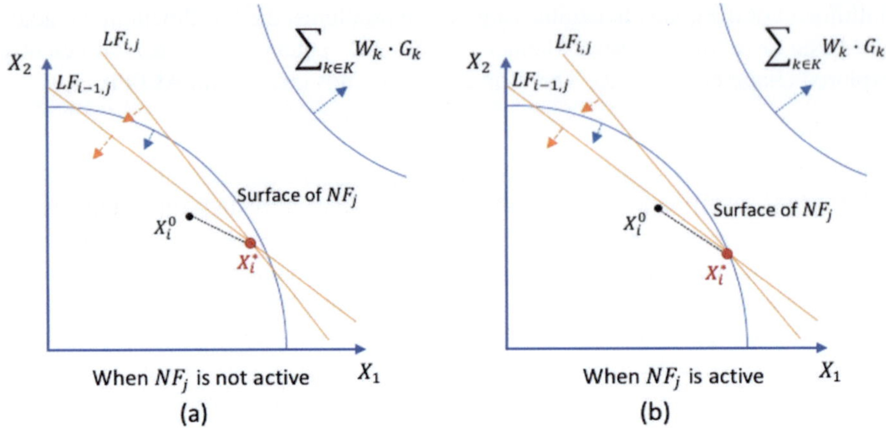

Fig. 5.10 A relatively sensitive solution and a robust solution (relatively insensitive to uncertainties)

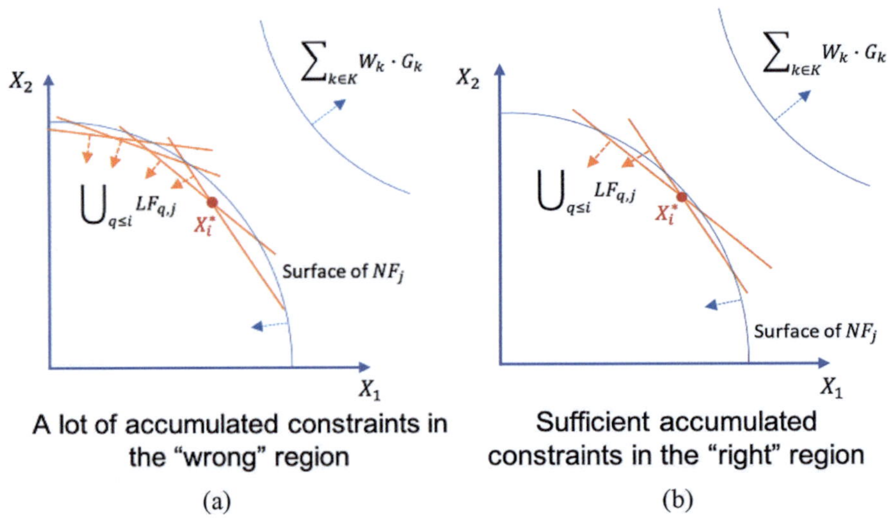

Fig. 5.11 Unnecessary accumulated constraints versus necessary accumulated constraints

Table 5.4 Criteria for the evaluation of approximation performance

Criteria	Meaning and representation
Goals fulfillment	Distance between achieved value and the target value of each goal
Robustness	Whether a solution is away from the physical boundary of the system
Approximation accuracy	Whether the nonlinear constraints are approximated well in the sub-region that contains more desired solutions

5.4.2 Step 2—Developing the Evaluation Indices (EIs)

To manage different preferences for multi-goal design problems, we obtain several solutions using multiple design scenarios.[5] Since the design scenarios we use are discrete and do not cover all situations, we obtain limited, discrete solutions and use them to predict the *satisficing* solution space.

By running the Synthesis cycle (Rectangle A in Fig. 5.3) using multiple design scenarios, we acquire information such as the fulfillment of the goals, the activeness of the constraints, and the accumulated constraints. Based on this information and the criteria in Table 5.4, the evaluation indices (EIs) are developed (Table 5.5). The descriptions of the EIs are given as follows.

The development of the EIs is based on the index for the robust concept exploration method, EMI (error margin index) (Chen et al., 1996) and DCI (design capacity index) (Choi et al., 2005). This is based on the observation that the results of each criterion follow a Gaussian distribution, and the mean (μ) and standard deviation (σ) can represent their characteristics. We assume that Gaussian distribution applies in general engineering design problems with continuous solution space. Therefore, we use the mean (μ) and the standard deviation (σ) as the EIs and tune the RMC by minimizing the mean and the variance of each EI.

Index for evaluating the fulfillment of the goals—μ_Z and σ_Z. We obtain a set of solutions satisficing a set of design scenarios (Sabeghi et al., 2016). We minimize the mean and standard deviation of Z, μ_Z and σ_Z, to maximize the fulfillment of the goals and minimize the dispersion of goal fulfillment.

Index for evaluating robustness—μ_{Nab}, σ_{Nab}, μ_{Naoc} and σ_{Naoc}. To develop the indices based on the robustness criterion, we have the following definitions and assumptions.

In this chapter, we assume that all errors and incompleteness of a model can be reflected as changes in the slack or surplus of an inequality constraint or a variable bound (as another type of inequality constraint).

If an inequality constraint has zero slack or surplus when plugged into a feasible solution, we define it as an active constraint. If the active constraint belongs to the original nonlinear problem P^0, as NF_j in Fig. 5.10a, then the solution is sensitive to the physical boundary because an error or variation that takes place to the active constraints may make the solutions infeasible. We define an original constraint with zero slack as an active original constraint (AOC). We want to minimize the number of AOCs.

If a solution makes a variable reach its upper or lower bound, the solution is also relatively sensitive to physical boundary. We define such bound as an active bound, noted as AB. We want to minimize the number of ABs as well.

[5] In this chapter, design scenarios include different weight vectors of the goals, different scenarios of constraint capacities (may result from different parameter values, different scenarios of quantitative relationship between variables, etc.), and any other factors that lead to model variations.

Table 5.5 Develop the Evaluation Indices (EIs) from the information obtained from ALP running

Criteria	Information	EIs	Meaning
Goals fulfillment	Weighted sum of the deviation variables: $Z = \sum_{k \in K} W_k \cdot \left(d_k^- + d_k^+\right)$	μ_Z	The average goals fulfillment in multiple design scenarios
		σ_Z	The standard deviation (Zarfl et al., 2015) of fulfillment of the goals (stability of performance in fulfillment of the goals) in multiple design scenarios
Robustness	The number of active bounds: *Nab*	μ_{Nab}	The average sensitivity to variable bounds in multiple design scenarios
		σ_{Nab}	The SD of sensitivity to variable bounds in multiple design scenarios
	The number of active original constraints: *Naoc*	μ_{Naoc}	The average sensitivity to original constraints in multiple design scenarios
		σ_{Naoc}	The SD of sensitivity to original constraints in multiple design scenarios
Approximation accuracy	The number of accumulated constraints:*Nacc*	μ_{Nacc}	The average complexity of the approximated problem in multiple design scenarios
		σ_{Nacc}	The SD of the complexity of the approximated problem in multiple design scenarios
	The number of iterations: Nit	μ_{Nit}	The average convergence speed in multiple design scenarios
		σ_{Nit}	The SD of the convergence speed in multiple design scenarios

Any solution returned using simplex (or any solution algorithms that are derived from simplex, such as dual simplex, revised dual simplex, etc.) is a vertex solution to the linear problem. Therefore, there should be at least one active constraint or active bound in the approximated linear problem. However, we prefer fewer active original constraints (AOCs) and active bounds (ABs) to make the solution more robust (or relatively insensitive) to potential errors or variations.

In summary, we simplify the robustness of a solution by examining the number of active bounds *Nab* and its number of active original constraints *Naoc*. When the design scenarios change, we maximize the robustness of the design and minimize the variation of the robustness of the design by minimizing the mean and standard deviation of *Nab* and *Naoc*. Therefore, the EIs for robustness are μ_{Nab}, σ_{Nab}, μ_{Naoc}, and σ_{Naoc}.

***Index for evaluating computational complexity*—μ_{Nacc}, σ_{Nacc}, μ_{Nit} *and* σ_{Nit}.** To evaluate how accurate the approximation is, we use the number of accumulated constraints *Nacc* and the number of iterations *Nit*. Since the approximation accuracy does not always get improved with the increase of accumulated constraints or the number of approximation iterations (Fig. 5.10), we need to learn the range of *Nacc* and *Nit* that associated with good goal fulfillment and robustness. We desire the *Nacc* and *Nit* to be acceptable under all design scenarios and not too sensitive to the design change, so we measure their mean and standard deviation, μ_{Nacc}, σ_{Nacc}, μ_{Nit}, and σ_{Nit}.

In summary, we tune the RMC by satisfying the EIs—μ_Z, σ_Z, μ_{Nab}, σ_{Nab}, μ_{Naoc}, σ_{Naoc}, μ_{Nacc}, σ_{Nacc}, μ_{Nit} and σ_{Nit} to fall within their desired range (DEI). The formulation of the RMC tuning process is given as follows. Based on this formulation, we can learn the DEI and tune the RMC.

Given

$$\begin{aligned}&\text{Design scenarios}\\&\text{EIs}: \mu_Z, \sigma_Z, \mu_{Nab}, \sigma_{Nab}, \mu_{Naoc}, \sigma_{Naoc}, \mu_{Nacc}, \sigma_{Nacc}, \mu_{Nit}, \sigma_{Nit}\\&\text{Iteration}: \ i\\&\text{Number of EIs}: \ n\end{aligned}$$

Find

$$\begin{aligned}&RMC\\&DEI\\&\text{Value of EIs of Iteration } i: \mu_{Z_i}, \sigma_{Z_i}, \mu_{Nab_i}, \sigma_{Nab_i}, \mu_{Naoc_i}, \sigma_{Naoc_i}, \mu_{Nacc_i},\\&\sigma_{Nacc_i}, \mu_{Nit_i}, \sigma_{Nit_i}\\&\text{Rank of the EIs in the } i\text{th iteration among EIs of all iterations}:\\&Rank(\mu_{Z_i}), Rank(\sigma_{Z_i}), Rank(\mu_{Nab_i}), Rank(\sigma_{Nab_i}),\\&Rank(\mu_{Naoc_i}), Rank(\sigma_{Naoc_i})\\&Sort[Rank_i]: \text{Sort the Rank of the EIs in the } i\text{th iteration among}\\&\text{EIs of all iterations}\end{aligned}$$

Satisfy

$$\{\mu_{Z_i}, \sigma_{Z_i}, \mu_{Nab_i}, \sigma_{Nab_i}, \mu_{Naoc_i}, \sigma_{Naoc_i}, \mu_{Nacc_i}, \sigma_{Nacc_i}, \mu_{Nit_i}, \sigma_{Nit_i}\} \in DEI$$

Minimize

$$\text{First } \kappa \text{ items of } Sort[Rank_i]$$

5.4.3 Step 3—Learning the DEI and Tuning the RMC

We learn the desired range of the EIs (DEI), and the connections between the RMC value and the result of the EIs so that we can find the RMC that outputs *satisficing* solutions. As iterations progress, the DEI is expected to better facilitate the evaluation of approximation performance.

To make the learning process efficient, we first adopt an offline learning process using a sample of RMC values to initialize the parameters and then adopt an online learning process to tune the RMC. In Fig. 5.12, we illustrate the two processes in the dotted Rectangles B and C and show their relationship with the synthesis cycle A in Fig. 5.3 (the main process of the ALP).

During parameter initialization (Rectangle B in Fig. 5.12), a sample of RMC values is generated (B1). By using each RMC value to run the synthesis cycle (A), the corresponding EIs are obtained (B2). Based on the results, we perform evaluations (B2). We pick the RMC value that gives the best EIs as the starting RMC value for tuning (B3-1), and as the initial best RMC to be updated during the tuning process (B3-2). If any EI is insensitive to RMC changes, has no correlation with other EIs, or is not important for the current problem, we remove it to reduce computational complexity (B3-3). We initialize the DEI based on the sample results (B3-4), to allow a certain percentage of (e.g., 75%) RMC values to fall into the DEI. These evaluation results are aggregated (B3-5) and used as the input for the RMC tuning cycle (C).

With the new RMC value, we run the synthesis cycle (A) and obtain the results (C2). By evaluating the new results using the DEI and comparing them with previous cycles (C3), we determine the next RMC value (C3-1), evaluate if the current DEI needs to be updated to either restrict or relax the satisficing solution space based on the trade-offs between EIs (C3-2), and evaluate if the best RMC needs to be updated (C3-3). These evaluation results are aggregated (C3-4). After judging whether the iteration should stop (C4), the program either goes to the next iteration of RMC tuning with the aggregated results (C3-4) as input or stops with the best RMC as the returned value. The stopping criteria include the total number of iterations and the number of iterations without updating the best RMC. The RMC tuning process is summarized in Table 5.6.[6] In Sect. 5.5, a test problem is used to verify the efficacy of the ALPPL.

[6] The details of the determination algorithms are given in Appendix 5.1.

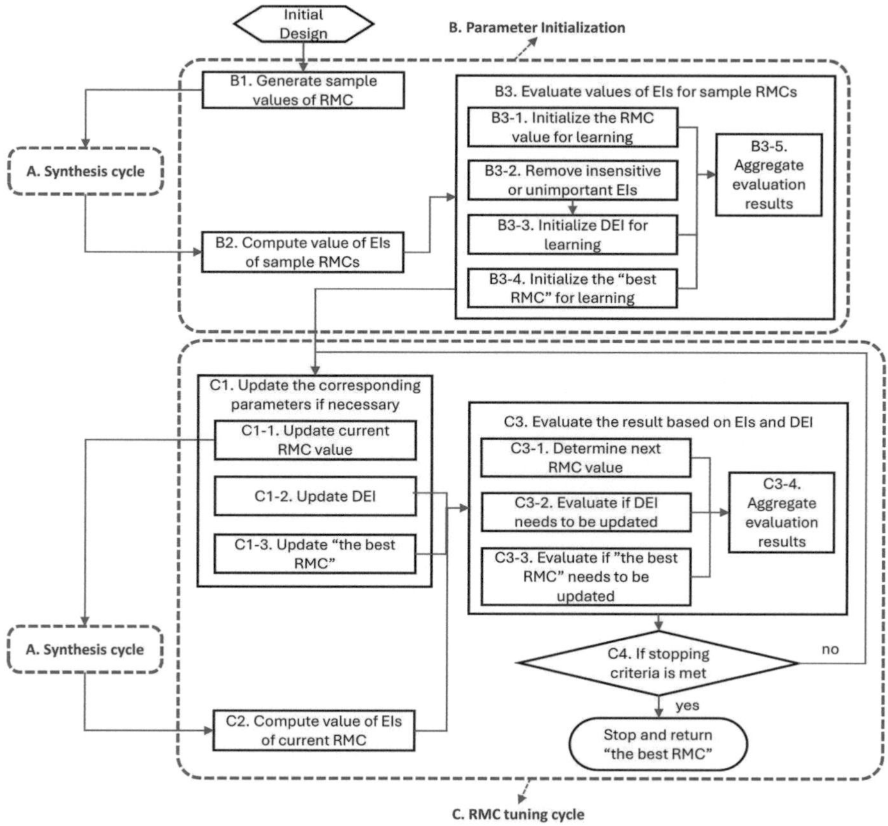

Fig. 5.12 ALPPL includes parameter initialization and the RMC tuning

5.5 The Hot Rolling Process Chain Problem

– ***Test Problem 3: Applying the ALPPL algorithm to engineering-design***

In this section, we demonstrate the effectiveness of the ALPPL using an industry-inspired test problem—the integrated design of a hot rolling process chain for producing a steel rod (Nellippallil et al., 2018). We choose this problem because it is a nonlinear problem formulation, and the RMC value significantly impacts the result.

5.5.1 Statement of Test Problem 2

Hot rolling is a multi-stage manufacturing process in which a reheated billet, slab, or bloom produced after the casting process is further thermo-mechanically processed

Table 5.6 The parameter learning process—for RMC tuning

<u>*Given*</u>: *DEI*, the best RMC sample value, *DEI* updating rules
<u>*Initialize*</u>: t <- 0, *best* <- the best RMC sample value, RMC_0 <- the best RMC sample value, the maximum iteration number T, stopping criterion 2 <- {*best* has not been updated in n iterations}
While t ≤ T do // Define stopping criterion 1
    RMC_t<- Next_RMC
    Run synthesis cycle
    Calculate $EI[RMC_t]$
    if $EIs[RMC_t] > EIs[best]$
        best <- RMC_t // Update the best RMC
    if *DEI* should be updated
        update *DEI*
    if stopping criterion 2 is met
        break
    else
        Next_RMC <- Determine_next$\{EIs[RMC_t], EIs[RMC_{t-1}], EIs[RMC_{t-1}], best\}$
// Use EIs of the tth, (t-1)^th^, (t-2)^th^, and *best* to determine the RMC of the next iteration[6]
        t <- t+1
return *best*

by passing through a series of rollers, see Nellippallil et al. (2017, 2018). During the thermo-mechanical processing, there is an evolution of microstructure of the material. The columnar grains in the material are broken down into equiaxed grains. Along with the evolution of grain size, the phase transformation of the steel occurs. Phase transformation is predominant during the cooling stage that follows the hot rolling process chain. The transformation of the austenite phase of steel to other phases like ferrite, pearlite, martensite, etc., occurs during this stage. The final microstructure of the material after the rolling and cooling process defines the mechanical properties of the product. Producing a new steel grade with improved properties and performance requires many plant trials, which are usually expensive and time-consuming. Therefore, there is a need to address the problem from a simulation-based design perspective to explore solutions that *satisfice* multiple conflicting property/performance goals. In this problem, the requirement is to produce steel rods with improved mechanical properties like yield strength (*YS*), tensile strength (*TS*), and hardness (*HV*). These mechanical properties are defined by the microstructure after cooling, which includes the phase fractions (only ferrite and pearlite phases are considered in this problem), pearlite interlamellar spacing, ferrite grain size, and chemical compositions, see Nellippallil et al. (2018) for the mechanical property-microstructure relationships identified. Using a goal-oriented inverse design method, Nellippallil et al. (2018) identify the microstructural requirements after the cooling stage to meet the mechanical properties of the rod. The microstructural requirements are to achieve a high ferrite fraction value, low pearlite interlamellar spacing, and low ferrite grain size value within the defined ranges. The goal is to carry out the integrated design of the material and the process by managing the cooling rate (cooling process variable), final austenite grain size after rolling (rolling microstructure variable), and the chemical compositions of the material. Therefore, our interest in this problem is to explore the solution space of the defined variables using ALPPL to meet the target values identified for the microstructure after the cooling stage, ensuring that the mechanical property requirements of the steel rod are met. Our focus in this chapter is to use this example problem to test the solution algorithm rather than to detail the design of the material and the manufacturing process.

The initial design formulation of the problem using the *Given*, *Find*, *Satisfy*, and *Minimize* keywords is shown as follows.

Given

(1) Target values for microstructure after cooling.

$$\text{Ferrite Grain Size Target, } D_{\alpha_{Target}} = 8\ \mu\text{m}$$

$$\text{Ferrite Fraction Target, } X_{f_{Target}} = 0.9$$

$$\text{Pearlite Interlamellar Spacing Target, } S_{o_{Target}} = 0.15$$

(2) Well established empirical and theoretical correlations, response surface models, and complete information flow from the end of rolling to the end product mechanical properties (Details provided in Nellippallil et al. (2018)).
(3) System variables and their ranges.

Find

System Variables

$$X_1,\ \text{Cooling Rate } (CR)$$
$$X_2,\ \text{Austenite Grain Size } (D)$$
$$X_3,\ \text{the carbon concentration } ([C])$$
$$X_4,\ \text{the manganese concentration after rolling } ([Mn])$$

Deviation Variables

$$d_i^-,\ d_i^+,\ \text{i} = 1, 2, 3$$

Satisfy

System Constraints

$$\text{Minimum ferrite grain size constraint} \quad D_\alpha \geq 8\,\mu\text{m} \tag{5.10}$$

$$\text{Maximum ferrite grain size constraint} \quad D_\alpha \leq 20\,\mu\text{m} \tag{5.11}$$

$$\text{Minimum pearlite interlamellar spacing constraint} \quad S_o \geq 0.15\,\mu\text{m} \tag{5.12}$$

$$\text{Maximum pearlite interlamellar spacing constraint} \quad S_o \leq 0.25\,\mu\text{m} \tag{5.13}$$

$$\text{Minimum ferrite phase fraction constraint (manage banding)} \quad X_f \geq 0.5 \tag{5.14}$$

$$\text{Maximum ferrite phase fraction constraint (manage banding)} \quad X_f \leq 0.9 \tag{5.15}$$

$$\text{Maximum carbon equivalent constraint} \quad C_{eq} = (C + Mn)/6;\, C_{eq} \leq 0.35 \tag{5.16}$$

Mechanical Property Constraints

$$\text{Minimum yield strength constraint} \quad YS \geq 250\,\text{MPa} \tag{5.17}$$

$$\text{Maximum yield strength constraint} \quad YS \leq 330\,\text{MPa} \tag{5.18}$$

$$\text{Minimum tensile strength constraint} \quad TS \geq 480\,\text{MPa} \tag{5.19}$$

$$\text{Maximum tensile strength constraint} \quad TS \leq 625\,\text{MPa} \tag{5.20}$$

$$\text{Minimum hardness constraint} \quad HV \geq 130 \tag{5.21}$$

$$\text{Maximum hardness constraint} \quad HV \leq 150 \tag{5.22}$$

System Goals

The target values for system goals are identified in Nellippallil et al. (2018) and are listed in the Given keyword above.

$$\text{Goal 1: Achieve Ferrite Grain Size Target} \quad \frac{D_{\alpha Target}}{D_\alpha(X_i)} + d_1^+ - d_1^- = 1 \tag{5.23}$$

$$\text{Goal 2 : Achieve Ferrite Fraction Target} \quad \frac{X_f(X_i)}{X_{fTarget}} + d_2^- - d_2^+ = 1 \tag{5.24}$$

$$\text{Goal 2 : Achieve Pearlite Interlamellar Spacing Target} \quad \frac{S_{oTarget}}{S_o(X_i)} + d_3^+ - d_3^- = 1 \tag{5.25}$$

Variable Bounds

$$11 \leq X_1 \leq 100\ (\text{K/min})$$
$$30 \leq X_2 \leq 100\ (\mu m)$$
$$0.18 \leq X_3 \leq 0.3\ (\%)$$
$$0.7 \leq X_4 \leq 1.5\ (\%)$$

Bounds on deviation variables

$$d_i^-, d_i^+ \geq 0 \text{ and } d_i^- * d_i^+ = 0, \quad \text{i} = 1, 2, 3 \tag{5.26}$$

Minimize

Minimize the deviation function in the initial design

$$Z = \sum_{i=1}^{3} W_i\left(d_i^- + d_i^+\right); \ \sum_{i=1}^{3} W_i = 1 \tag{5.27}$$

There are three goals in the problem discussed: (i) Minimize Ferrite Grain Size (D_α), (ii) Maximize Ferrite Fraction $\left(X_f\right)$, and (iii) Minimize Interlamellar Spacing (S_o). The target values and acceptable values of the three goals are determined and defined. All three goals require nonlinear goal formulation. There are four design/system variables—(i) cooling rate (CR), (ii) final austenite grain size after rolling (D), (iii) carbon concentration ($[C]$), and (iv) manganese concentration after rolling ($[Mn]$). We obtain: (i) the range of the system variables that *satisfice* the goals for the different design preferences assigned, and (ii) the *satisficing* weight set, $W^s = \bigcap_{k \in K} W_k^s$, the set of weight scenarios that *satisfices* all three goals for the different design preferences assigned.

In Nellippallil et al. (2018), the authors formulate and execute the initial design compromise Decision Support Problem for the hot rolling process chain problem and carry out weight sensitivity analysis to identify the *Satisficing* Weight Set of the three goals. Ternary plots are generated to visualize and explore the weight set. In each ternary plot, the three axes represent the weights assigned to the three goals, respectively, and the color contours indicate the achievement of each goal, $\frac{G_k}{T_k}$, $k = 1, 2, 3$. Since the goals conflict with each other and the total resources for the goals are limited, compromise solutions are desired. Weight sensitivity analysis is a way to mediate compromise or innovate around the conflicts between goals. W_1^s, W_2^s, W_3^s, and W^s are the satisficing weight regions identified in the ternary plots, see areas identified using arrows in Fig. 5.11a–d, respectively. To discuss and compare the achievements of multiple goals, $\frac{G_k}{T_k}$ are normalized within the range [0, 1]. An acceptable value of each goal is identified and plotted as a dashed line in each ternary plot. The area between the corner with the best $\frac{G_k}{T_k}$ value and the dashed line (the acceptable value) is the *satisficing* weight area of one goal, W_k^s, and the superimposed area is the *satisficing* weight area of all three goals, W^s.

In Figs. 5.13, 5.14 and 5.15, we illustrate how different RMC values affect W_k^s and W^s. When RMC is 0.1, as shown in Fig. 5.13, W^s is large, whereas when RMC is 0.8, as Fig. 5.15 shows, W^s is small. However, we do not have rules to evaluate which RMC value results in a relatively accurate approximation and thereby gives us the most robust W^s. Therefore, the ALPPL is applied to fill this gap.

We implement the parameter initialization (Process B in Fig. 5.12) and the RMC tuning (Process C in Fig. 5.12) using Python and incorporate executing the ALP (Synthesis cycle A in Fig. 5.12) to obtain the results.

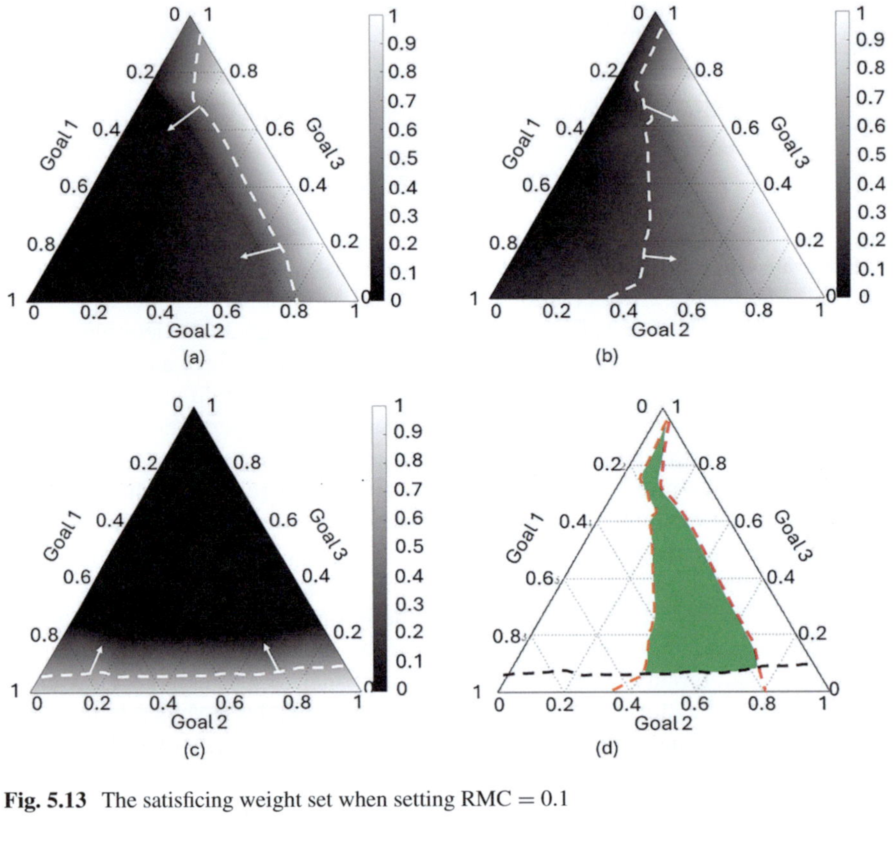

Fig. 5.13 The satisficing weight set when setting RMC = 0.1

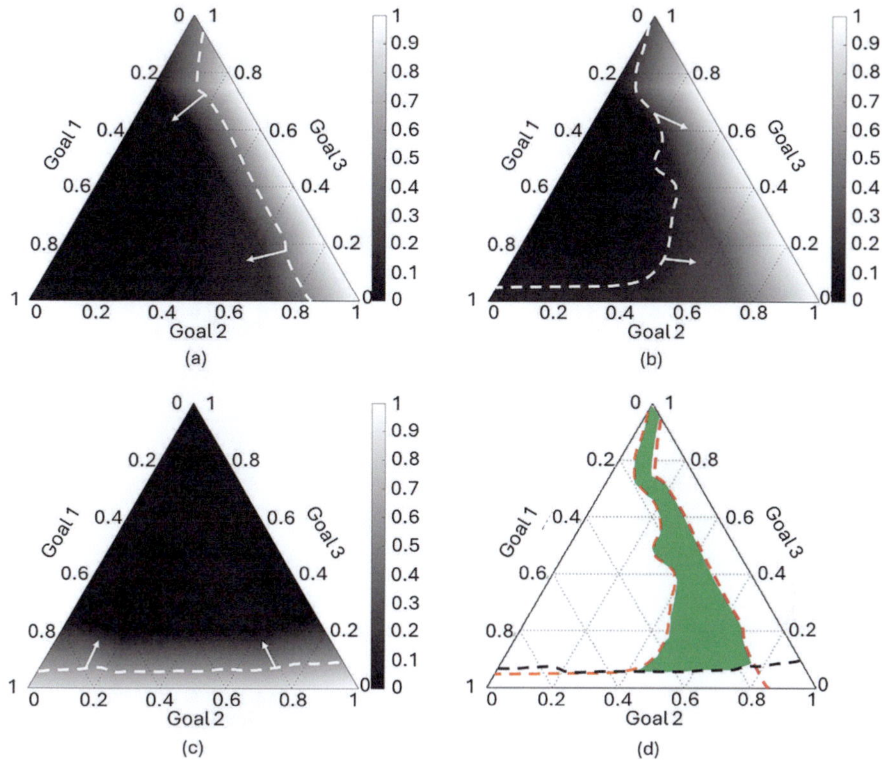

Fig. 5.14 **The satisficing weight set** when setting RMC = 0.5

5.5.2 *Applying ALPPL*

Parameter Initialization

The design scenarios used in Nellippallil et al. (2018) are nineteen weight vectors (weight scenarios) of the goals (Table 5.7), that represent a variety of design preferences. Then we go through Processes B1, B2, and B3 in Fig. 5.12 as follows.

B1 We randomly choose some values—0.1, 0.2, 0.5 and 0.8 as the RMC sample.
B2 By using the four RMC values and the nineteen weight vectors, we solve the problem for $4 \times 19 = 76$ times and obtain 76 results of each EI. The results of the EIs of the four RMC values are in Table 5.8.
B3 Based on the EIs of the sample RMC (Table 5.5), we proceed as follows.
B3-1 Initialize the RMC value for learning. Set $\text{RMC}_0 \leftarrow$ the best RMC sample value, hence, $\text{RMC}_0 = 0.5$.
B3-2 Remove insensitive or unimportant EIs—μ_{Nacc}, σ_{Nacc}, μ_{Nit}, and σ_{Nit}.

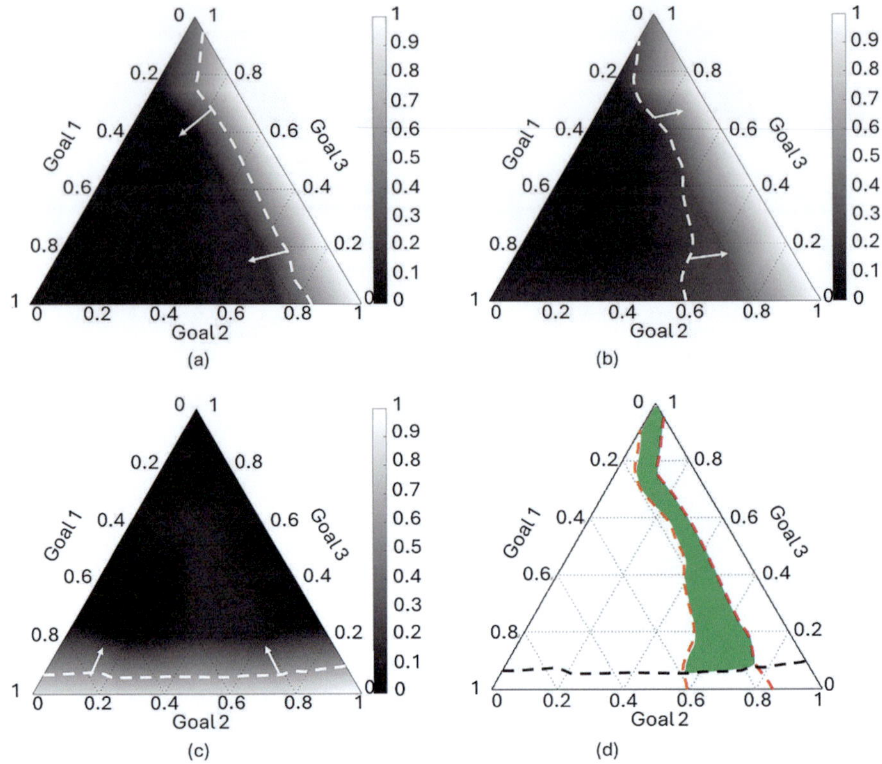

Fig. 5.15 The satisficing weight set when setting RMC = 0.8

Table 5.7 Weight vectors used in Nellippallil et al. (2018) as different design scenarios

W	W1	W2	W3	*W*	W1	W2	W3
1	1	0	0	11	0	0.75	0.25
2	0	1	0	12	0	0.25	0.75
3	0	0	1	13	0.33	0.33	0.33
4	0.5	0.5	0	14	0.2	0.2	0.6
5	0.5	0	0.5	15	0.4	0.2	0.4
6	0	0.5	0.5	16	0.2	0.4	0.4
7	0.25	0.75	0	17	0.6	0.2	0.2
8	0.25	0	0.75	18	0.4	0.4	0.2
9	0.75	0	0.25	19	0.2	0.6	0.2
10	0.75	0.25	0				

Table 5.8 Results of EIs using sample RMC values with nineteen design scenarios

RMC	Statistics	Z	*Nit*	*Nacc*	*Nab*	*Naoc*
0.1	μ	0.1480	46.58	18.74	1.79	0.84
	σ	0.0679	5.95	0.87	0.63	0.50
0.2	μ	0.1474	34.16	19.00	1.84	0.84
	σ	0.0682	7.88	0.82	0.60	0.60
0.5	μ	0.1467	20.42	19.47	2.05	0.79
	σ	0.0675	9.47	0.84	0.62	0.42
0.8	μ	0.1480	8.32	14.16	2.21	0.79
	σ	0.0675	5.56	6.94	0.63	0.71

B3-3 Initialize DEI. For each EI, we define the range from the median of the four sample results to the ideal value as the initial DEI. For example, the ideal value of μ_Z is zero because ideally, we want the deviation function Z to be zero so that the goals can be fully achieved; the median of μ_Z for the four RMC sample values is 0.1477; therefore, the initial DEI of μ_Z is [0, 0.1477]. We visualize the results of the EIs for the four sample RMC values in Fig. 5.16. For each EI, there is a graph showing its mean and standard deviation, as in Fig. 5.16a–c. In each graph, the left vertical axis represents the mean value, the right vertical axis represents the standard deviation value, and the horizontal axis represents RMC value. The bars are means and the lines are standard deviations. The left arrows and the right arrows show the desired range of the mean and standard deviation, respectively. The initial DEI are summarized in Table 5.9.

B3-4 Initialize the best RMC. Set best $\leftarrow$ the best RMC sample value, hence, best = 0.5.

B3-5 We aggregate the results of the parameter initialization—RMC_0, EIs, DEI, and best—as the input of RMC tuning.

RMC Tuning

In RMC tuning, we establish rules to proceed with each procedure based on heuristics. The heuristics are generalized from parameter learning and can be adjusted through the search process.

In the first iteration of RMC tuning, Process C1 is based on the output of Process C3 and Process C2 is the same as Process B2, so we start making rules for Process C3 and return to C1.

C3 Evaluate the result of the current RMC based on EIs and DEI.

Table 5.9 The initial DEI

DEI of μ_Z	DEI of σ_Z	DEI of μ_{Naoc}	DEI of σ_{Naoc}	DEI of μ_{Nab}	DEI of σ_{Nab}
[0, 0.1477]	[0, 0.0677]	[0, 0.82]	[0, 0.55]	[0, 1.95]	[0, 0.63]

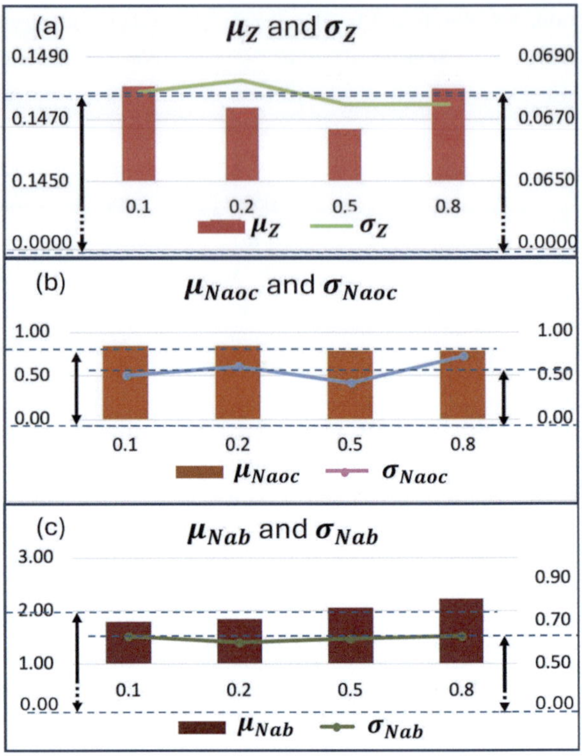

Fig. 5.16 EIs and DEI of the sample RMC values

C3-1 Determine the next RMC value.

Rule 1: Compare the performance of multiple EIs and define the comparison rules (Table 5.6, Line 14). Lines 22–30 in Appendix 5.1 are an expansion of this rule. We define RMC A as better than RMC B if "no less than κ of the EI(A) are better than EI(B), whereas other EI(A) do not exceed γ of the upper and lower bound of DEI. This rule can be applied differently to other problems. In this problem, we set $\kappa = 1/2$ and $\gamma = 30\%$.

Rule 2: Determine when and how the RMC should be updated. The code for this rule is shown in Lines 6–13 in Appendix 5.1. We use a hill-climbing approach to update the RMC (Appendix 5.1, Line 5–8). If the updating in the previous RMC tuning cycle improve performance, then the previous updating is in the "hill-climbing direction", as a result, we keep updating the RMC in this direction with a step size α; otherwise, we need the best RMC to help us "get back on track" with a portion β, hence we update the RMC as the linear combination of the best RMC (elite) and the RMC from two cycles ago (parent). In this problem, we set α as a random value that is uniformly distributed in [0, 1] and set β as a random value that is uniformly distributed in [0.5, 1]. In this way, we incorporate greediness (hill-climbing), elitism use the best RMC), and rrandomness in evolution (the randomness of α and β).

C3-2 Evaluate if DEI needs to be updated.

Rule 3: Determine when and how the DEI should be updated. See Appendix 5.1, Lines 18–38. If in an RMC-tuning cycle, the RMC leads us to get more than κ EIs better than the EIs of the previous cycle, and more than ι EIs are in the desired range (DEI) whereas only $(1 - \iota)$ EIs have minor violations (Appendix 5.1, Line 27), we consider the current RMC as performing better than the previous RMC and update the DEI. In this problem, we set $\kappa = \iota = 2/3$ because the number of EIs is small. For problems with many EIs, κ and ι can be tuned using the performance improvement rate or the proportion of acceptable results among all the results as the tuning goal. In this way, we prevent insensible DEI stopping us from walking to a better range, while ensuring gradual and relatively conservative updating of DEI.

C3-3 Evaluate if "the best RMC" needs to be updated.

Rule 4: Determine when and how the current best RMC is updated. See Appendix 5.1, Lines 34–36. We use a variable ("best") to store the current best RMC. If more than κ EIs of current RMC are better than the EIs of the best, we set the current RMC as the new best.

C3-4 We aggregate the results of the RMC tuning—RMC_{t+1}, DEI, and the best, as the input of the next tuning iteration.

C4 Determine if the iterating should stop.

Rule 5: Establish the stopping criteria. To stop the RMC tuning at the appropriate time avoid overwhelming computation, we use two stopping criteria—the maximum number of RMC tuning iterations and maximum number of RMC-tuning iterations without updating the best RMC. See Appendix 5.1, Lines 41–42.

5.5.3 Parameter Learning Results and Discussion

After running the RMC tuning for fourteen iterations, the tuning stops. We identify 0.55 as the best RMC for the cooling problem. Compared to the initial "best RMC" 0.5, the finalized best RMC 0.55 brings improvement in σ_Z, σ_{Naoc}, and σ_{Nab}. The RMC values in the fourteen iterations and their EIs are given in Table 5.10. During the fourteen iterations, the DEI is updated four times, and the best RMC is updated three times. The final best RMC is identified in the ninth iteration. In the first seven cycles, the RMC value varies considerably. In the last seven cycles, the RMC value varies little, indication that the desired sub-range of RMC is sufficiently explored and the appropriate value can be identified.

To verify the efficacy of ALPPL, we evaluate the adequacy and necessity of the algorithm in obtaining robust solutions.

Adequacy—The best RMC (0.55) ensures the solutions fall in a relatively insensitive range.

Table 5.10 **The** record of the EIs, DEI, RMC, Best RMC of the fourteen iterations of RMC tuning

Iteration	RMC	Deviation function		Number of active original constraints		Number of active bounds		Better than cycle (t − 1)	Better than best RMC	Update DEI
		μ_Z	σ_Z	μ_{Naoc}	σ_{Naoc}	μ_{Nab}	σ_{Nab}			
1	0.5	0.147	0.068	0.79	0.42	2.05	0.62	–	–	μ_{Nab} ~[1, 1.95] → μ_{Nab} ~[1, 2.05]
2	1.0	0.152	0.071	0.95	0.78	2.26	0.45	N	N	–
3	0.8	0.148	0.068	0.79	0.71	2.21	0.63	N	N	–
4	0.6	0.147	0.067	0.95	0.52	2.00	0.58	Y	Y	$\mu_{Naoc} \sim$ [0, 0.82]→$\mu_{Naoc} \sim$ [0, 0.95]
5	0.4	0.147	0.068	0.68	0.48	2.00	0.58	Y	Y	–
6	0.2	0.147	0.068	0.84	0.60	1.84	0.60	N	N	$\sigma_{Naoc} \sim$ [0, 0.55]→ $\sigma_{Naoc} \sim$ [0, 0.6]
7	0.3	0.154	0.069	0.95	0.52	1.79	0.71	N	N	–
8	0.45	0.147	0.068	0.89	0.66	2.00	0.58	Y	N	–
9	0.55	0.147	0.067	0.84	0.37	2.05	0.52	Y	Y	–
10	0.65	0.147	0.067	1.00	0.58	2.05	0.62	N	N	$\mu_{Naoc} \sim$ [0, 0.95] → $\mu_{Naoc} \sim$ [0, 1.00]
11	0.48	0.147	0.068	0.89	0.66	2.05	0.62	N	N	–
12	0.53	0.147	0.067	0.84	0.69	2.00	0.58	Y	N	–
13	0.57	0.147	0.067	0.84	0.37	2.11	0.57	N	N	–
14	0.43	0.145	0.069	0.84	0.60	2.00	0.58	N	N	–

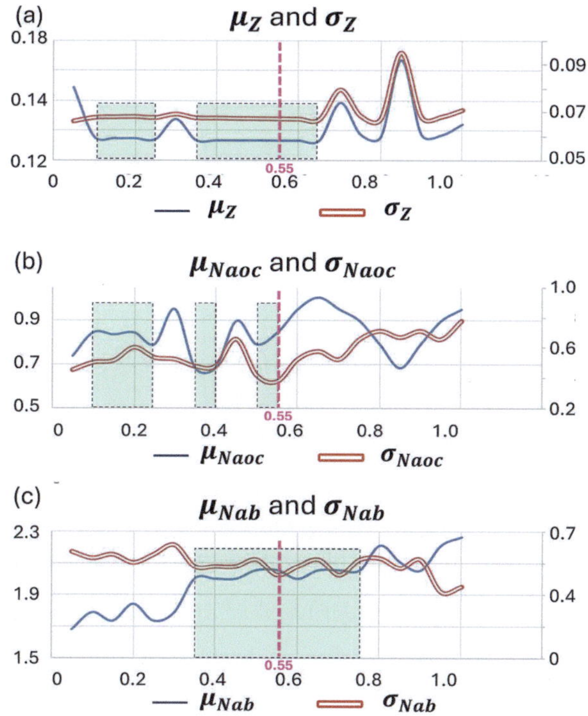

Fig. 5.17 Identifying the insensitive range of RMC value using twenty RMC values

Necessity—The insensitive range is sufficiently explored during the RMC tuning.

First, we identify the insensitive range of RMC value. We test 20 RMC values uniformly distributed in [0, 1] and obtain their EIs (Fig. 5.17). In each graph of Fig. 5.17, the range in the dotted rectangle is the insensitive range[7] of RMC for each EI based on the twenty results.

Verification of the adequacy: We observe from Fig. 5.17 that 0.55 is in the insensitive RMC range for all EIs, that when RMC is 0.55, it gives a relatively robust performance.

Verification of the necessity: From Fig. 5.17, we observe that the ranges [0.35, 0.4] and [0.5, 0.55] are insensitive ranges in all three EIs, so within the RMC values in these two ranges, the solutions are relatively insensitive. In Fig. 5.18, we illustrate all the fourteen RMC values used during RMC tuning. The horizontal axis represents the iteration number, and the vertical axis represents the RMC value. Four out of the fourteen RMC values falling in the two insensitive ranges, so 28.5% of the RMC values we test during the tuning fall in the insensitive ranges, whereas the insensitive

[7] We identify the insensitive range by using quantitative methods such as searching for $[x_1, x_2]$ where $\frac{\max[y(x)]-\max[y(x)]}{x_2-x_1} \le \gamma$, $x \in [x_1, x_2]$, and determine the value of γ as a rate of $(\min[y(x)] - \min[y(x)])$, $x \in [0, 1]$, for each EI.

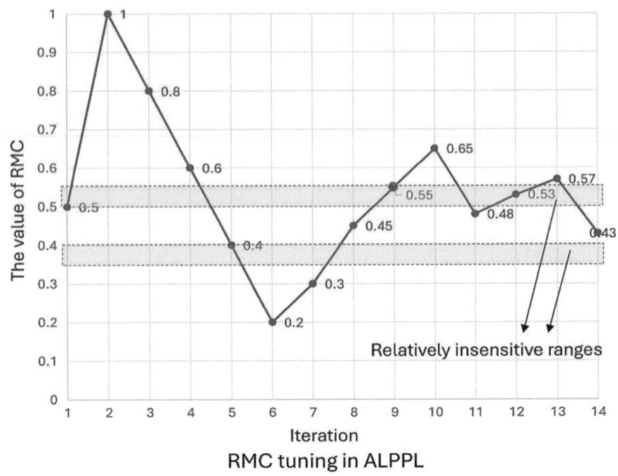

Fig. 5.18 The fourteen RMC values in the RMC tuning

ranges only occupy 10% of the whole RMC range. Hence, we conclude that our rules in the RMC tuning enable a relatively sufficient exploration of the insensitive ranges.

Verification of the improvement of ALPPL over ALP: By visualizing the RMC values tested using ALPPL versus the RMC values tested using golden section search, in Fig. 5.19, we observe that the best RMC identified using golden section search is 0.65, which is not in the insensitive range; in addition, the RMC values tested in the golden section search are concentrated in [0.57, 0.77], which misses the insensitive ranges [0.35, 0.4] and [0.5, 0.55]. Other improvements of ALPPL over ALP are summarized in Table 5.11.

5.6 Role of This Chapter in This Monograph

5.6.1 Summarizing How We Finish Task 2: Connecting Approximation, Exploration, and Evaluation

For the model of a complex system that contains nonlinear and possibly non-convex equations, when approximating the problem to a linear and convex problem for solving, designers apply heuristics or metaheuristics, which simplify the problem and guarantee feasible solutions but may lose information. Even those metaheuristics are not the best ones to help acquire the approximations and solutions with an acceptable quality—where quality may include, but is not limited to, approximation accuracy, solution robustness to multiple types of uncertainty, and the computational complexity of the approximation and solution searching—there is no mechanism to evaluate such quality and learn the association between the heuristics and the quality.

Table 5.11 ALPPL with RMC tuning versus ALP with golden section search

		ALPPL	ALP
General comparison	Search method	Rule-based parameter learning	Golden section search
	Criteria used for evaluation of the RMC	Deviation (fulfillment of the goals), the robustness of the solution	Fulfillment of the goals
	If the approximation is sensitive to the scenario changing	Considering different scenarios, the most appropriate RMC is identified. The approximation is relatively insensitive to scenario changing	In each scenario, the best RMC is identified, and it may vary as scenario changing. The approximation is relatively sensitive to scenario changing
	Stopping criteria	The best RMC has not been updated for n iterations, or the total iteration number reaches a threshold	The distance between two golden section points are less than a threshold ε
Comparison of the cooling problem results	Number of search iterations	14	12
	If the identified best RMC is in the insensitive range	Yes	No
	Number of tested RMC values falling into insensitive range	4	2
	Is the insensitive range explored sufficiently	Relatively sufficiently	Insufficiently

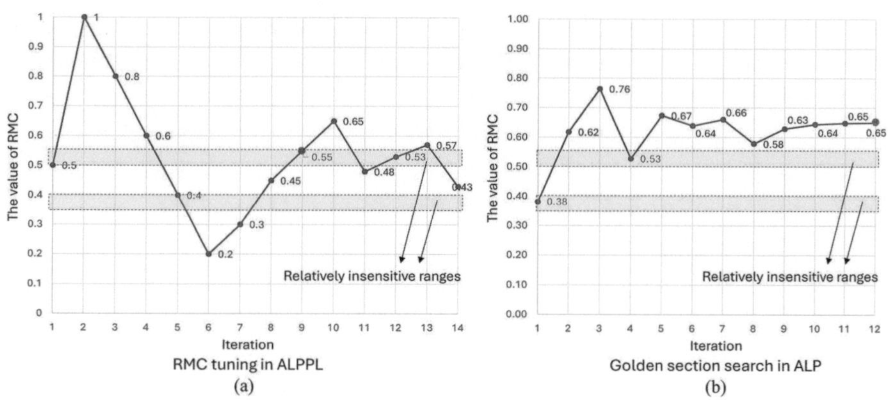

Fig. 5.19 The comparison of ALPPL and ALP regarding the RMC updating

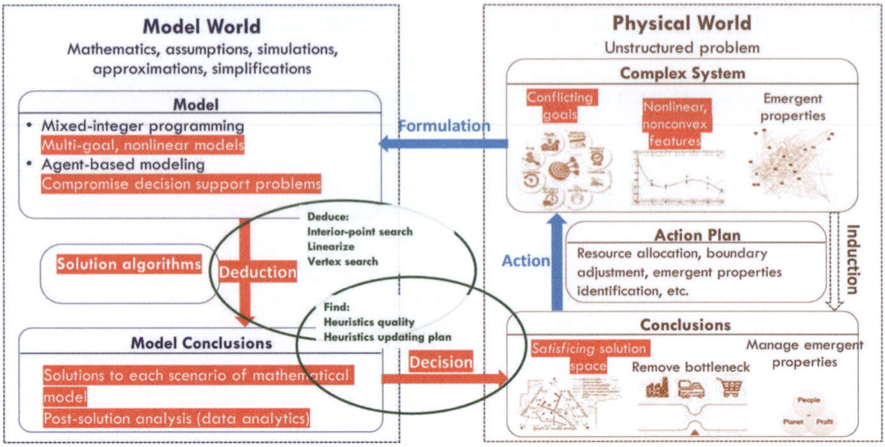

Fig. 5.20 The Procedures involved in approximation-exploration-evaluation—establish the information exchange, knowledge awareness, and instructions sharing between deduction and decision

The essence of Hypothesis 2 of this monograph (proposed in Sect. 3.2.2), "*learn, evaluate, and update metaheuristics to improve model performance*," is to improve the model approximation, evaluate and update the heuristics, and obtain knowledge on tradeoffs between exploration and exploitation of the problem approximation and post-solution analysis. Designers need to explore different heuristics and establish the connection among approximation, exploration, and evaluation.

In this chapter, we summarize the process of exploring, learning, and updating heuristics that need to be customized for each problem into a more generally useful algorithm, the parameter learning process in Table 5.6. The customization of the algorithm for the test problem, the hot rod rolling process chain, is in Appendix 5.1.

In other words, we strengthen the connections between deduction and decision, as shown in Fig. 5.20. In this monograph, we establish the information exchange, knowledge awareness, and instructions sharing between the three processes and standardize it; see Table 5.6 Appendix 5.1.

5.6.2 *Summarizing How We Realize Type I, II, and III Robust Design*

For the test problem (T2), the cooling stage of the hot rod process chain, Type I uncertainty is identified as the various values of the reduced move coefficient (RMC) of the adaptive linear programming (ALP) algorithm (Table 5.8, Fig. 5.16). Type II uncertainty is implemented as the design scenarios that represent the different priorities of the goals (Table 5.7). Type III uncertainty results from the incorporation of Type I uncertainty—when using different RMC values, the model structure changes as the linearized problem varies with the RMC. A representation of the influence of Type III uncertainty on the result is shown in Figs. 5.13, 5.14 and 5.15. Type I, II, and III uncertainties are managed through the parameter learning algorithm in Table 5.6.

By implementing the parameter learning algorithm and customizing it in the rod rolling process chain problem as described in Appendix 5.1, we can identify the solution space that is relatively insensitive to the Type I, II, and III uncertainties that we aim to manage. In this way, we realize Type I, II and III robust design; see the summary in Table 5.12 as the closing remarks of Table 3.2 regarding the robust design realization and uncertainty management for Test Problem 2.

Table 5.12 Summary of test problem 2 regarding Type I, II, and III uncertainty management

RD Type	**RDI-II**				
			RDIII		
				RDIV	
Method	M1: Formulation-Exploration Framework		**M2: Adaptive Linear Programming with Parameter Learning (ALPPL)**	M3: Adaptive Leveling-Weighting-Clustering Algorithm (ALWC)	M4: Scenario Planning in Agent-Based Modeling
Chapter	Ch 4		**Ch 5**	Ch 6	Ch 7
Uncertainty \ Test Problem	T1: Dam network	T2: Supply chain	**T3: Hot rolling process chain**	T4: Thermal system	T5: Promoting second season farming
Type I	*Uncertainty in timing and amount of inflow – Table 4.8*	*Uncertainty in demand side – Figure 4.25*	***Uncertainty in hyper parameter setting – Table 5.8, Figure 5.16***	*Uncertainty in parameter setting in solution algorithm (Starting point of searching)*	Uncertainty in price (Price of agriculture products)
Type II	*Uncertainty in outflow (water release target) – Table 4.6*	*Uncertainty in supply side - Table 4.15*	***Uncertainty in user preferences – Table 5.7***		*Promotion effort and timing*
Type III			***Uncertainty in model approximation due to heuristics in approximation – Table 5.6***	*Uncertainty in model approximation (ways of combining multiple goals)*	
Type IV				*Uncertainty in using domain knowledge to simplify the model (fixing decision variables and selecting design scenarios)*	*Interventions that change the mathematical relation among promotion and result (developing local market)*

RD – robust design
M – method
EVe – empirical verification of the method
T – test problem

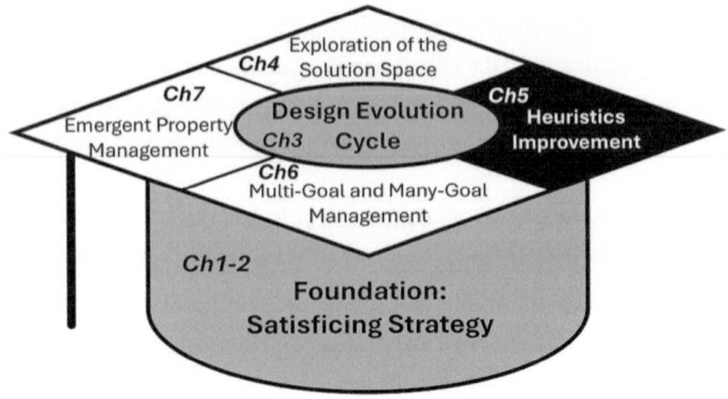

Fig. 5.21 The role of this chapter in the knowledge foundation

5.6.3 *Role of This Chapter*

In this chapter, we use parameter learning to improve the adaptive linear programming (ALP) algorithm. In ALP, one critical parameter, the reduced move coefficient (RMC), which has a severe impact on the approximation performance, is determined using golden section search, and no mechanism of obtaining insight to improve the approximation during the search, which may result in missing the sub-range of RMC value with good approximation performance; the best RMC value is sensitive to design scenario changes; no criteria other than the fulfillment of the goals are taken into account when evaluating the approximation performance.

To improve the ALP, we hypothesize that by incorporating parameter learning in the ALP, as ALPPL, we can improve the approximation performance, especially for multi-goal engineering-design problems. To verify the hypothesis, we implement the parameter learning in three steps—identifying the criteria for approximation performance evaluation, developing evaluation indices (EIs), and using them to tune the RMC.

With an application of the hot rolling process chain problem, we depict the procedure of applying ALPPL and demonstrate the improvements of ALPPL over ALP. With ALPPL, using different design scenarios, we initialize the parameters and use them to tune the RMC and update them during the tuning. We validate that the ALPPL helps find the RMC value facilitating the identification of more robust solutions, and the insensitive range of RMC gets explored more sufficiently.

The proposed algorithm ALPPL can be applied to multi-goal engineering-design problems, especially when goals conflict with one another, the priority of the goals evolves with the environment changing, the model is a medium part of a multi-stage process chain so its input and output may affect or be affected by other stages, and the outputs of the model need to be insensitive to model errors and variations. The "Heuristics Improvement" piece of the "Design Evolution Cycle" is finished (see Fig. 5.21).

The proposed rule-based parameter learning can be used to improve other algorithms, especially when there are no customizable criteria for evaluating the algorithm performance, or the algorithm performance is highly sensitive to some critical parameters that are determined with heuristics or human intuition while the critical parameters do not get updated based on the algorithm performance during the design iterations.

Glossary

ALP	Adaptive Linear Programming
ALPPL	Adaptive Linear Programming with Parameter Learning
DEI	Desired Range of Evaluation Index
DSP	Decision Support Problem
EI/EIs	Evaluation Index/Indices
RMC	Reduced Move Coefficient
Completion of a Goal (or Fulfillment of a Goal)	Designers define this as how closely a goal meets its target.
RMC Performance	Designers assess this as how well a solution meets specified criteria for a given RMC value.
Quality of a Solution	Designers determine this as the overall performance level of a solution based on criteria such as goal fulfillment, the number of active constraints, and the number of accumulated constraints.
Active Constraints	Designers label a constraint as active when, after they plug a solution's decision variables into it, the left-hand-side value equals the right-hand-side value nt.
Accumulated constraints	Designers acquire multiple linear constraints when they linearize a nonlinear constraint. Then they accumulate those linear constraints to replace the original nonlinear constraint.
X	$X = \{x_1, x_2, \ldots, x_n\}$. System variables (decision variables).

G_k	The kth goal of a K-goal compromise formulation.
T_k	The target value of the kth goal of a K-goal compromise formulation.
d_k^-, d_k^+	Negative deviation variables and positive deviation variables of Goal k. d_k^- and d_k^+ are the under-achievement and over-achievement of Goal k. The formulation of Goal k is: $\frac{G_k}{T_k} + d_k^- - d_k^+ - 1 = 0$. $d_k^- \cdot d_k^+ \equiv 0$.
$\frac{G_k}{T_k}$	Fulfillment of Goal k—measures how well Goal k achieves its target.
W	Weight Scenarios—The different weight vectors used to combine multiple goals to represent various design preferences linearly.
W_k^s	Satisficing Weight set for Goal k—The range of weight values that satisfice Goal k.
W^s	$W^s = \bigcap_{k \in K} W_k^s$. Satisficing Weight Set—The set of weight scenarios that satisfice all K goals.
Z	$Z = \sum_{k \in K} W_k \cdot \left(d_k^- + d_k^+\right)$. Deviation function Z is the weighted sum of deviation variables of all goals. By minimizing Z, goals are achieved in a compromise way. Z measures how much the goals do not achieve their targets.
P^0	The original problem—may be nonlinear.
P_i^L	The approximate linear problem in the ith iteration.
X_i^*	$X_i^* = \left\{x_{i1}^*, x_{i2}^*, \ldots, x_{in}^*\right\}$. The solution of the ith iteration.

X_i^0	$X_i^0 = \{x_{i1}^0, x_{i2}^0, \ldots, x_{in}^0\}$. The starting point of the *i*th iteration. The original problem P^0 is linearized at X_i^0 in the *i*th iteration. See Fig. 5.5 and Eqs. (5.1)–(5.3). X_i^0 is determined using Eq. (5.4).
AOC	Active Original Constraints—The inequality constraints with zero slack or the surplus when plugging in a solution. See Fig. 5.11a.
AB	Active Bounds—For a solution, if the value of a variable is on its upper or lower bound, then the bound is an active bound.
J	The set of nonlinear constraints of P^0.
NF_j	The *j*th nonlinear constraint of P^0, $j \in J$. NF_j maintains the same in every iteration. See Figs. 5.5 and 5.6.
$NF'_{i,j}$	The approximated second-order parabolic of NF_j using the second-order derivatives at X_i^0. See Fig. 5.6 and Eqs. (5.1) and (5.2).
$LF_{i,j}$	The approximated linear constraint of NF_j in the *i*th iteration. $LF_{i,j}$ is a part of P_i^L. See Figs. 5.4 and 5.5, and Eqs. (5.1)–(5.6).
ACC	Accumulated Constraints. When the convexity of a nonlinear constraint NF_j is relatively large—greater than 0.015, approximated linear constraints from multiple iterations are accumulated to replace NF_j. See Fig. 5.4 and Eq. (5.7).

Appendix 5.1: The RMC Tuning Algorithm Customized for the Hot Rolling Process Chain Problem

Appendix 5.1 is the RMC tuning algorithm (Table 5.6) customized for the Cooling Procedure of the Hot Rolling Problem. Appendix 5.1 is referenced in Sect. 5.5.2. This algorithm is an extension of the algorithm in Table 5.6. More auxiliary parameters are defined to assist RMC tuning—T, α, β, γ, θ, ι, κ, M. For the parameters that are relatively more important (the results are more sensitive to their values), e.g. θ, we tune their values. For the parameters that are relatively less important, e.g. ι, κ, M, we set values to them with heuristics.[8]

---

t <- 0 // Initiate the number of synthesis cycles

b <- 0 // Initiate the time of updating the best RMC

RMC_t <- a random value // Initiate RMC with a random value (here we set RMC_0<- 0.5)

$best_b$ <- a random value // Initiate the "best RMC" with a random value (here we set $best_0$<- 0.5)

While $t \leq T$ **Do** // Search for best RMC for T synthesis cycles (the first stop criterion, here we set T <- 20)

  if $EI[RMC_{t-1}] \succ EI[RMC_{t-2}]$ // If RMC_{t-1} performs better than RMC_{t-2}

    RMC_t <- $RMC_{t-1} + \alpha \cdot (RMC_{t-1} - RMC_{t-2})$ // Update RMC_t in the improving direction ($\alpha \in [0,1]$, and here we set α as random values that uniformly distributed in [0, 1])

  else if $best_b \neq RMC_{t-2}$

    RMC_t <- $\beta \cdot best_b + (1-\beta) \cdot PEI^{-1}[better\{PEI[RMC_{t-2}], PEI[best_{b-1}]\}]$ // Update RMC_t as the linear

    combination of best RMC and the last best RMC ($best_{b-1}$) iff $best_{b-1}$ performs better than RMC_{t-2}; otherwise

    update RMC_t as the linear combination of best RMC and RMC_{t-2} ($\beta \in [0,1]$, and here we set β as random

    values that uniformly distributed in [0.5, 1])

  else if $best_{b-1}$

      RMC_t <- $\beta \cdot best_b + (1-\beta) \cdot best_{b-1}$ // Update RMC_t as the linear combination of the current best

      RMC and the last best RMC

    else

      RMC_t <- $\beta \cdot RMC_{t-1} + (1-\beta) \cdot RMC_{t-2}$

  if $RMC_t > 1$ // If RMC_t is larger than its upper bound

    RMC_t <- 1 // Pull RMC_t back to range

  if $RMC_t < 0$ // if RMC_t is lower than its lower bound

---

[8] $EI^i[RMC_t] \succ EI^i[\text{best}]$ means for the ith PEI, comparing with the EI of current best, the EI of RMC_t is closer to the ideal value of the PEI.

$RMC_t = 0$ // Pull RMC_t back to range
I <- 0 // Initiate the number of EIs of RMC_t in DEI
J <- 0 // Initiate the number of EIs of RMC_t that are better than the EIs of best RMC
K <- 0 // Initiate the number of EIs of RMC_t that violate DEI within θ. We tune θ maximizing the $L_2 - norm$ distance between EIs of two adjacent iterations and get 10%
L <- 0 // Initiate the number of EIs of RMC_t that are better than the EIs of RMC_{t-1}
for i in 1 to n // For all the EIs (n is the number of EIs)
if$PEI[RMC_t]_i \in DEI_i$ // If the ith EI of RMC_t is in the desired range of the ith EI
I <- $I + 1$ // Update the number of EIs of RMC_t in DEI
if $EI[RMC_t]_i \geqslant EI[\text{best}]_i$[8] // If for the ith EI RMC_t performs better than or equal to best RMC
J <- $J + 1$ // Update the number of EIs of RMC_t that are better than best RMC
if $EI[RMC_t]_i \notin DEI_i$ and $EI[RMC_t]_i \in [DEI_i]_{-\gamma_i}^{+\gamma_i}$ // If the ith EI of RMC_t violates the desired range of the ith EI within γ_i (in this problem, γ_i=30%)
K <- $K + 1$ // Update the number of EIs of RMC_t that violate DEI within γ_i
if $EI[RMC_t]_i \geqslant EI[RMC_{t-1}]_i$ // If for the ith EI, RMC_t performs better than or equal to RMC_{t-1}
L <- $L + 1$ // Update the number of EIs that RMC_t improves versus RMC_{t-1}
if $I \geq \iota \cdot n$ **and** $I + K = n$ // If for at least ι (we set it as 2/3) EIs are in DEI, and the violation rate are all within γ_i
if $L \geq \kappa \cdot n$ // If at least κ (we set it as 2/3) EIs are better than previous synthesis cycle
$EI[RMC_t] \succ EI[RMC_{t-1}]$ // We define that RMC_t overall performs better than RMC_{t-1}
if $J \geq \kappa \cdot n$ // If at least κ EIs is better than best RMC
$best$ <- RMC_t // Update best RMC
nupdt <- -1 // Reset no updating pointer "nupdt" as "-1"
if $K \geq 1$ // If at least one violation EI
DEI_i <- $EI[RMC_t]_i$ which $EI[RMC_t]_i \notin DEI_i$ and $EI[RMC_t]_i \in [DEI_i]_{-\theta}^{+\theta}$ // Update DEI_i
nupdt <- nupdt + 1 // Increase no updating pointer "nupdt" by 1
else
nupdt <- nupdt + 1 // Increase no updating pointer "nupdt" by 1
if nupdt $\geq M$ // If no updating in M synthesis cycles in a row (the second stop criterion, and here we set M<-5)
Break
t <- t+1 // Move on to the next synthesis cycle
Return best // Return the final best RMC as the appropriate RMC

Steps in the RMC-tuning algorithm:

Determine the evaluation indices (EIs) based on multiple criteria to classify good results from the bad ones;

Initialize the desired range of each EI (DEI) of the test problem;

Identify auxiliary parameters to assist RMC tuning;

Bring the EIs into DEI by tuning the auxiliary parameters;

Update DEI to ensure a proportion of good results out of all results;

Tradeoff between elitism and randomness to ensure a diversity while getting fast convergence.

References

Asghari, M., Fathollahi-Fard, A. M., Mirzapour Al-E-Hashem, S., & Dulebenets, M. A. (2022). Transformation and linearization techniques in optimization: A state-of-the-art survey. *Mathematics, 10*(2), 283.

Barthelemy, J.-F., & Haftka, R. T. (1993). Approximation concepts for optimum structural design—A review. *Structural Optimization, 5*(3), 129–144.

Behmanesh, E., & Pannek, J. (2021). A comparison between memetic algorithm and genetic algorithm for an integrated logistics network with flexible delivery path. *Operations Research Forum, 2*, 47.

Box, G. E., & Draper, N. R. (1987). *Empirical model-building and response surfaces.* Wiley.

Calpine, H., & Golding, A. (1976). Some properties of Pareto-optimal choices in decision problems. *Omega, 4*(2), 141–147.

Chen, W., Allen, J. K., Tsui, K.-L., & Mistree, F. (1996). A procedure for robust design: Minimizing variations caused by noise factors and control factors. *ASME Journal of Mechanical Design, 118*(4), 478–485.

Choi, H.-J., Austin, R., Allen, J. K., McDowell, D. L., Mistree, F., & Benson, D. J. (2005). An approach for robust design of reactive powder metal mixtures based on non-deterministic micro-scale shock simulation. *Journal of Computer-Aided Materials Design, 12*(1), 57–85.

Fishburn, P. C. (1968). Utility theory. *Management Science, 14*(5), 335–378.

Guo, L. (2021). Model evolution for the realization of complex systems. Ph.D. Dissertation, The School of Industrial and Systems Engineering, University of Oklahoma, Norman, OK.

Guo, L., Balu Nellippallil, A., Smith, W. F., Allen, J. K., & Mistree, F. (2020). Adaptive linear programming algorithm with parameter learning for managing engineering-design problems. In *ASME Design Automation Conference*. Paper Number DETC2020-22602.

Guo, L., Milisavljevic-Syed, J., Wang, R., Huang, Y., Allen, J. K., & Mistree, F. (2023). Managing multi-goal design problems using adaptive leveling-weighting-clustering algorithm. *Research in Engineering Design, 34*(1), 39–60.

Guo, L., Allen, J. K., & Mistree, F. (2024a). Optimize or satisfice in engineering design? *Research in Engineering Design, 35*(3), 239–267.

Guo, L., Nellippallil, A. B., Smith, W. F., Allen, J. K., & Mistree, F. (2024b). An adaptive linear programming algorithm with parameter learning. *Algorithms, 17*(2), 88.

Hajihashemi, S. (2023). A user-friendly wrapper for DSIDES (Decision support in the design of engineering systems). MS Thesis, The School of Industrial and Systems Engineering, University of Oklahoma.

Jiang, R., Wang, J., & Guan, Y. (2012). Robust unit commitment with wind power and pumped storage hydro. *IEEE Transactions on Power Systems, 27*(2), 800.

Mistree, F., Hughes, O., & Phuoc, H. (1981). An optimization method for the design of large, highly constrained complex systems. *Engineering Optimization, 5*(3), 179–197.

Mistree, F., Hughes, O. F., & Bras, B. (1993). The compromise decision support problem and the adaptive linear programming algorithm. In M. P. Kamat (Ed.), *Structural optimization: Status and promise* (pp. 247–286). AIAA.

Nagadurga, T., Devarapalli, R., & Knypiński, Ł. (2023). Comparison of meta-heuristic optimization algorithms for global maximum power point tracking of partially shaded solar photovoltaic systems. *Algorithms, 16*(8), 376.

Nash Jr, J. F. (1950). The bargaining problem. *Econometrica: Journal of the Econometric Society, 18*, 155–162.

Nellippallil, A. B., Song, K. N., Goh, C.-H., Zagade, P., Gautham, B., Allen, J. K., & Mistree, F. (2017). A goal-oriented, sequential, inverse design method for the horizontal integration of a multistage hot rod rolling system. *ASME Journal of Mechanical Design, 139*(3), Article 031403.

Nellippallil, A. B., Rangaraj, V., Gautham, B., Singh, A. K., Allen, J. K., & Mistree, F. (2018). An inverse, decision-based design method for integrated design exploration of materials, products, and manufacturing processes. *ASME Journal of Mechanical Design, 140*(11), Article 111403.

Rao, S., & Mulkay, E. (2000). Engineering design optimization using interior-point algorithms. *AIAA Journal, 38*(11), 2127–2132.

Reich, D., Green, R. E., Kircher, M., Krause, J., Patterson, N., Durand, E. Y., Viola, B., Briggs, A. W., Stenzel, U., & Johnson, P. L. (2010). Genetic history of an Archaic Hominin Group from Denisova Cave in Siberia. *Nature, 468*(7327), 1053–1060.

Rios, L. M., & Sahinidis, N. V. (2013). Derivative-free optimization: A review of algorithms and comparison of software implementations. *Journal of Global Optimization, 56*(3), 1247–1293.

Saaty, T. L. (1988). *What is the analytic hierarchy process?* Springer.

Sabeghi, M., Shukla, R., Allen, J. K., & Mistree, F. (2016). Solution space exploration of the process design for continuous casting of steel. In *ASME Design Automation Conference*. Paper Number IDETC2016-59399.

Sharma, G., Allen, J. K., & Mistree, F. (2022). Designing concurrently and hierarchically coupled engineered systems. *Engineering Optimization, 55*(9), 1556–1576.

Souza, J. C. O., Oliveira, P. R., & Soubeyran, A. (2016). Global convergence of a proximal linearized algorithm for difference of convex functions. *Optimization Letters, 10*(7), 1529–1539.

Speakman, J., & Francois, G. (2021). Robust modifier adaptation via worst-case and probabilistic approaches. *Industrial and Engineering Chemistry Research, 61*(1), 515–529.

Su, X., Yang, X., & Xu, Y. (2023). Adaptive parameter learning and neural network control for uncertain permanent magnet linear synchronous motors. *Journal of the Franklin Institute, 360*(16), 11665–11682.

Teng, Z., & Lu, L. (2019). A FEAST algorithm for the linear response eigenvalue problem. *Algorithms, 12*(9), 181.

Viswanathan, J., & Grossmann, I. E. (1990). A combined penalty function and outer-approximation method for MINLP optimization. *Computers and Chemical Engineering, 14*(7), 769–782.

Vrahatis, M. N., Kontogiorgos, P., & Papavassilopoulos, G. P. (2020). Particle swarm optimization for computing Nash and Stackelberg equilibria in energy markets. *Operations Research Forum, 1*, 20, https://doi.org/10.1007/s43069-020-00021-4.

Zarfl, C., Lumsdon, A. E., Berlekamp, J., Tydecks, L., & Tockner, K. (2015). A global boom in hydropower dam construction. *Aquatic Sciences, 77*(1), 161–170.

Chapter 6
Type I, III, and IV Robust Design Through Unsupervised Learning

Adaptive Leveling-Weighting-Clustering Algorithm (ALWC)

Imagine you are a social entrepreneur tasked with designing a mini-power plant that provides electricity and support water purification through reverse osmosis. One solution is a thermal system based on the Rankine cycle. The challenge is to design this system robustly, balancing multiple conflicting goals such as maximizing efficiency, minimizing operational costs, and ensuring sustainable resource usage.

The Rankine cycle involves thermodynamic processes, including adiabatic pumping, isobaric heating, expansion, and condensation. These processes must be finely tuned to achieve the desired performance, which requires managing many parameters, such as the maximum and minimum pressure and temperature in the Rankine cycle.

Designing a robust Rankine cycle system is complicated when multiple goals must be met. Traditional methods for managing multi-goal problems often rely on prior knowledge of how the goals interact but fall short when such knowledge is lacking or when the goals compete in non-straightforward ways.

In this chapter, we describe an application of the Adaptive-Leveling-Weighting-Clustering algorithm (ALWC) in a real-world scenario. We demonstrate how this algorithm can be used to design a Rankine cycle-based system that achieves multiple goals, such as improving efficiency, enhancing power output, and achieving zero moisture, all without increasing computational complexity.

6.1 Organization of this Chapter

The layout of this chapter is shown in Fig. 6.1. In Sect. 6.2, we extend the discussion from Sect. 1.3 by providing an overview of modeling constructs and solution algorithms for multi-goal or multi-objective problems. In Sect. 6.3, we describe the Rankine cycle design problem. In Sect. 6.4, based on the research gaps identified in

L. Guo et al., *Architecting a Satisficing Strategy for Designing Complex Systems*,
https://doi.org/10.1007/978-3-032-04577-5_6

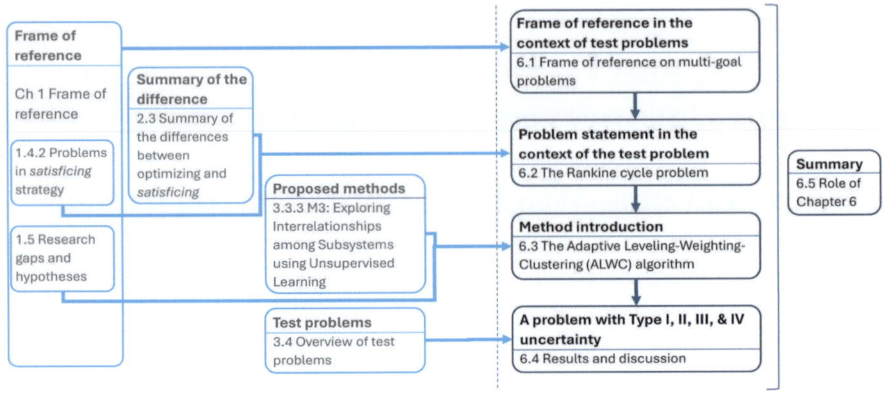

Fig. 6.1 Organization of this chapter

Sect. 1.6 and Method 3 proposed in Sect. 3.4.3, we introduce the Adaptive Leveling-Weighting-Clustering (ALWC) algorithm. In Sect. 6.5, we explain the application of the ALWC algorithm to the Rankine cycle design problem. Finally, in Sect. 6.6, we summarize the role of this chapter and interpret it regarding the "what, how, and why" questions in Table 6.1.

The plan for answering Research Question 3 (Sect. 3.3.3) within the context of the test problems is shown in Table 6.2. In this chapter, the Proposed Method 3 (M3 Sect. 3.4.3), the Adaptive Leveling-Weighting-Clustering (ALWC) algorithm, is empirically verified (EVe3) using a test problem, designing a Rankine cycle thermal system (T3). Research Question 3 (RQ3) is formulated in the context of the test problem (SQT3) and answered (AQ3) by testifying M3. The empirical validation and theoretical validation are in Chaps. 8 and 9.

In this chapter, we propose and test the Adaptive Leveling-Weighting-Clustering (ALWC) algorithm using a concurrent engineering design problem focusing on the thermal system around a Rankine cycle. Research Question 3 (RQ3), "**How can learning the system's behavior be accelerated?**" is framed around the Rankine cycle design problem (T3) as follows (SQT3) and then further specified into two sub-questions indicating the tasks and answers.

SQT3: How can we manage multiple conflicting goals and explore tradeoffs of the performance of multiple subsystems?

SQT3.1: How can we facilitate the exploration of the structure of the goals if there are more than three goals in a system?

Table 6.1 Summary interpretation of the Problems Investigated in this chapter

Elements	*What?*
Components	The formulation, exploration, and evaluation stages of the design evolution cycle
Connectors	The adaptive leveling-weighting-clustering algorithm (ALWC) enables the passing of information among formulation, exploration, and evaluation stages and learning knowledge to replace heuristics on the formulation (e.g., the formulation of the multiple, conflicting goals of a model)
Form	*How?*
Component roles	Provides the constructs for connecting the formulation, exploration, and exploration evaluation stages
Properties	Attributes or features of the formulation, exploration, and evaluation stages—In the formulation stage, designers apply heuristics (e.g., ways of combining multiple goals) in the early stages, whereas in the exploration and evaluation stages, it is feasible to acquire insight to modify or explore other ways of formulation
Relationship	The relations among the activities—heuristics application, learning the model performance associated with the heuristics, and use of association to evolve ways to improve the formulation
Rationale	*Why?*
Motivation	Provide support to assist designers in improving model formulation to realize Type I, III, and IV robust design through rule-based learning
Assumptions	The management of multiple conflicting goals of a model can be realized by clustering the goals based on their interrelationships and alternating the priority of each cluster with obtaining tradeoff scenarios among the achievement of the goals
Interpretation	Understanding how cluster analysis, iteration of lexicographic strategy, weight generation, and identifying mixed strategy to help in exploring combination scenarios of multiple goals

SQT3.2: How can we use the knowledge in the post-solution analysis be leveraged to support design improvement if domain knowledge is missing?

The interactions among the formulation, exploration, and evaluation of a design problem should be studied, and the mechanisms of information sharing and intervention between the three procedures should be established. See Fig. 6.2.

In this chapter, we address the challenge of solving many-goal decision problems, **specifically those involving more than three goals**. There are **limitations in managing such problems, including the need for domain expertise to integrate the goals**. In this chapter, we propose the domain-independent Adaptive Leveling-Weighting-Clustering (ALWC) algorithm, to manage the process of exploring the design scenarios of many goals and concurrent design problems. The ALWC is designed to allow designers to explore goal combinations based on their interrelationships iteratively. Design scenarios with better achievement rates of the goals are obtained through iteration without increasing computational complexity.

Table 6.2 Plan of specifying research question 3 (RQ3) and empirically verifying the adaptive leveling-weighting-clustering (ALWC) algorithm (M3)

Chapters	Ch 1	Ch 2	Ch 3	Ch 4–7			Ch 8	Ch 9
				Ch 4–5	Ch 6	Ch 7		
Actions	RG H	RD RQ SH	TVe M	EVe 1–2 SQT 1–2 AQ 1–2	*EVe3: Use a concurrent, multi-goal problem with uncertainties and unknown features in the interrelationships among the subsystems to verify SH3 and demonstrate M3*	EVe SQT AQ	CQ EVa	TE
					SQT3: What method facilitates managing multiple conflicting goals and exploring the tradeoffs of the performance of multiple subsystems? *SQT3.1: What methods facilitate the exploration of the structure of the goals if there are more than three goals in a system?* *SQT3.2: How can the knowledge in the post-solution analysis be obtained and used to support design improvement if the domain knowledge is missing?*			

(continued)

Table 6.2 (continued)

Chapters	Ch 1	Ch 2	Ch 3	Ch 4–7			Ch 8	Ch 9
				Ch 4–5	Ch 6	Ch 7		
					AQ3: Using the ALWC algorithm, with increasing weight vectors, the interrelationship among goals based on their deviations or achievement rates evolves and converges. Based on their interrelationships, goals are grouped into clusters to represent different subsystems. The combinations of the goals are explored iteratively, using either the Pre-emptive or Archimedean strategy. This facilitates assigning each cluster a different level (leveling) and combining the goals in each level using weight vectors (weighting). Through iteration, more design scenarios are identified, and the corresponding solutions are obtained so that designers can choose the appropriate design scenario and improve the design. As a tool to acquire insight when domain knowledge is lacking, the combination of the goals is explored so that better solutions regarding the average deviations, standard deviations, worst case, and Euclidean distance to the Utopia point are identified, while the computational complexity is reduced			

(continued)

Table 6.2 (continued)

Chapters	Ch 1	Ch 2	Ch 3	Ch 4–7			Ch 8	Ch 9
				Ch 4–5	Ch 6	Ch 7		
Nomenclature	AQ—answer research questions CQ—closure–answers to research questions Eva—empirically validate hypotheses Eve—empirically verify hypotheses RG—give research gaps H—give hypotheses M—introduce methods RD—tie to roust design RQ—pose research questions SH—specify hypotheses SQT—specify research questions in the context of test problems TE—theoretically extend the research TVe—theoretically verify hypotheses							

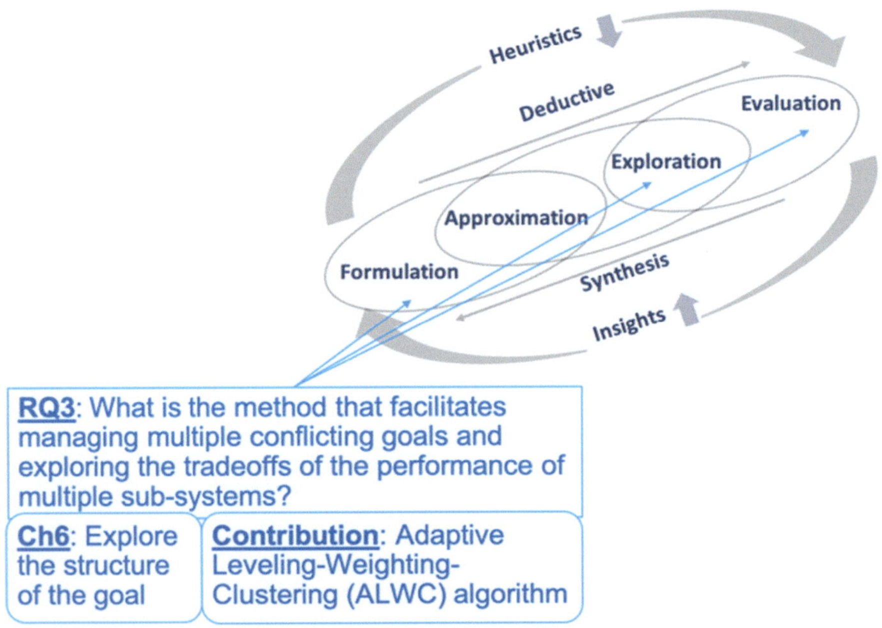

Fig. 6.2 Specified research question 3 and the relevant stages to be connected in the design evolution cycle

Additionally, knowledge of the relationships among subsystems is acquired, which is beneficial for concurrent design. The ALWC algorithm is illustrated using a thermal-system design problem as a test problem. In this chapter we **emphasize the method itself rather than the specific results.**

As manufacturing systems become increasingly complex, designers face challenges in addressing many-goal concurrent design problems—**problems with more than three goals, which are associated with the diversity, robustness, and computational efficiency of the solution space**. In this chapter, we address the issue of solving many-goal decision problems. There are limitations in managing many goal problems documented in the published literature, such as the need for domain expertise to combine the goals. Moreover, **designers may gain insight into the interrelationships among subsystems by learning the correlations or orthogonality between the goals**. Given this knowledge, designers can reorganize subsystems based on their interrelationships to enhance system performance.

As information within a solution space can improve design, it must be systematically passed through post-solution analysis to design modifications in an iterative manner Therefore, a method that facilitates information exchange between solution space and design space is needed. In this chapter, we show that many goal problems can be managed by exploring the goal combinations. The possible structures can be Preemptive (leveling), Archimedean (weighting), or a combination of the two in various forms.

To address the specified research questions, we propose the Adaptive Leveling-Weighting-Clustering (ALWC) algorithm. Using ALWC, the structure of the goals is explored by running three loops—leveling, weighting, and clustering. With ALWC, the information from the post-solution analysis can be used to improve the design and make corresponding decisions, **all without relying on domain-specific knowledge**.

6.2 Frame of Reference: Multi-goal Problems

6.2.1 Features of Concurrent Engineering Problems

In concurrent design, a designer must consider the interactions and coupling effects among subsystems and tradeoffs among conflicting requirements concurrently (Wang, 1994). Additionally, a designer must meet multidisciplinary constraints, such as cost, physical properties, manufacturing capacity, etc. This results in goals with different units, making it challenging to evaluate trade-offs directly. Hence, the need for *satisficing* solutions which meet multiple design preferences and are relatively insensitive to certain uncertainties (Wang et al., 2018).

In concurrent design, knowledge of downstream activities must be infused into the design process to iteratively improve model formulation and approximation. The goal is to explore a satisficing solution space and improve the design in terms of goal achievement rates, robustness, and computational complexity. In this chapter, design improvement is realized by exploring of goal combinations.

6.2.2 Two Categories of Studies on Multi-goal Problems

In this chapter, as an extension of Sect. 1.3, we divide multi-goal problems into two categories, namely, the performance of the solution algorithms and the indication of the design improvement. In papers on the solution algorithms, the main focus is on sorting near-Pareto front (Deb et al., 2002, Seada and Deb, 2014). Algorithm performance is assessed using criteria such as the optimality of the solutions, the diversity of the solutions, and the algorithm's computational complexity (Soltani et al., 2002). In contrast, in papers on design improvement, the authors focus on improving the design or acquiring more knowledge about the problem (Tang et al., 2010). Some typical methods from both categories are shown in Table 6.3. In this chapter, we address both categories.

The limitations of the literature focusing on the solution algorithms are summarized as follows:

Authors predominantly focus on identifying the near Pareto front. Over the past decades, advancements in solution algorithms have largely centered on improving the distribution of solution points along the near Pareto front and enhancing the speed of identification. However, there is a notable lack of discussion on how solutions from various regions of the near Pareto front can be applied in practice.

The limitations of the literature focusing on design improvement are:

The reliance on domain-specific knowledge or case-by-case analyses to explore goal combinations or decompose problems limits the generality and reusability of these methods. Considering these limitations in both categories of the literature, we propose leveraging data analysis to derive insights and offer decision support for goal integration, particularly in situations where domain knowledge is limited.. We hypothesize that exploring goal combinations through their interrelationships can enhance the achievement rate of each goal, uncover more diverse and practical design scenarios, and make the processes of knowledge discovery and reuse independent of domain constraints. In this chapter, objectives are defined as goals with target values as right-hand sides, and the design problem is framed as minimizing the deviation between the left-hand side and the target of each goal.

6.2.3 Differences Between a Goal and an Objective

There are differences between a goal and an objective. A goal is defined as an objective with a target value specified as its right-hand side (Parra et al., 2001) as the target value to be achieved (Parra et al., 2001). In this chapter, we discuss goals instead of objectives. In nonlinear problems, maximizing an objective without a specified right-hand side yields an optimal solution, x^*, which satisfies both necessary and sufficient Kuhn-Tucker conditions. Conversely, minimizing the distance between the left-hand side and right-hand side of the goal, the solution x^s is *satisficing*, which only satisfies the necessary Kuhn-Tucker condition.

Table 6.3 The features and limitations of some classic multi-objective (multi-goal) solution algorithms and methods

Emphases	Algorithm or method	Description	Searching for interior solutions	Having adaptable ways of producing more solutions (offspring)	Giving information on tradeoffs between multiple characteristics (such as solution accuracy vs convergence speed)	Use the knowledge of tradeoffs between goals (objectives) to improve model formulation	Explore multiple forms of goals (objectives) structure, such as decomposing the problem, reducing goals, etc
Solution algorithms	VEGA (Schaffer, 1985)	Using vector-valued feedback with adaptive procedures for searching high-order multi-objective problems		*			*
	SPEA2 (Zitzler et al., 2001)	Using fitness assignment, archiving, and truncating (the near-Pareto front) to evolve solutions to approach the Pareto-optimal set	*	*	*		
	MOEA/D (Zhang and Li, 2007)	Decomposing a problem into scalar optimization subproblems and optimizing them simultaneously	*	*	*		*
	NSGA-II/III (Deb et al., 2002, Seada and Deb, 2014)	Using the nondominated sorting evolutionary algorithm to update reference points to approach the Pareto front adaptively	*	*			*

(continued)

Table 6.3 (continued)

Emphases	Algorithm or method	Description	Searching for interior solutions	Having adaptable ways of producing more solutions (offspring)	Giving information on tradeoffs between multiple characteristics (such as solution accuracy vs convergence speed)	Use the knowledge of tradeoffs between goals (objectives) to improve model formulation	Explore multiple forms of goals (objectives) structure, such as decomposing the problem, reducing goals, etc
	REDGA (Jaimes et al., 2009)	Reducing the number of objectives by removing redundant (to some degree) objectives			*	*	*
	HypE (Bader and Zitzler, 2011)	Using Monte Carlo simulation to approximate the exact hypervolume values and seeking ranking of solutions	*	*	*		
	Multi-Level Decisions (Mistree et al., 1994)	Using two design strategies and multi-level decisions to foster discussion on multi-objective problems			*	*	
Design improvement	RCEM (Chen et al., 1997; Choi et al., 2005)	Improving the robustness of the design using indices based on the results of exploring the solution space				*	*

(continued)

Table 6.3 (continued)

Emphases	Algorithm or method	Description	Searching for interior solutions	Having adaptable ways of producing more solutions (offspring)	Giving information on tradeoffs between multiple characteristics (such as solution accuracy vs convergence speed)	Use the knowledge of tradeoffs between goals (objectives) to improve model formulation	Explore multiple forms of goals (objectives) structure, such as decomposing the problem, reducing goals, etc
	Interval analysis (Hao and Merlet, 2005)	Using parallel robots based on interval analysis to determine geometries to satisfy all compulsory requirements				*	*
	Level diagrams (Reynoso-Meza et al., 2013)	Comparing multiple Pareto fronts based on different design concepts using level diagrams, to support decision-making on design concept selection	*	*			*
	XPLORE in DSIDES (Sabeghi et al., 2016; Smith et al., 2015)	Exploring the design space by exploring different goal structures using a compromise Decision Support Problem			*	*	*
	CORTHOG (Warwick, 2019)	Removing poor measurement degrees-of-freedom iteratively until the pseudo-orthogonality check was optimized	*	*		*	*

The essence of the necessary conditions is that, at a solution point x^s, where both the primal and the dual are feasible, the gradient vector of the objective (the left-hand side of the goal) can be expressed as a linear combination of the gradient matrix of all equality constraints and the active inequality constraints.[1] The essence of the sufficient conditions is that at a solution point x^*, the degree of convexity of the objective should not exceed the convexity degree of the constraints combined by Lagrange multipliers.

To prevent the absence of solutions due to the strong convexity of the objective or the loss of solutions caused by uncertainties breaking the Kuhn-Tucker conditions, satisficing solutions are obtained by using goals instead of objectives. A target value, t, is assigned to the right-hand side of the objective. The objective $\boldsymbol{f}(\boldsymbol{x})$ becomes a goal $\boldsymbol{G}(\boldsymbol{x})$, whose position is fixed in the solution space. In $\mathcal{F}$, the feasible space bounded by all active constraints, the point, or several points, or an area, that is/are on the goal $\boldsymbol{G}(\boldsymbol{x})$ or closest to it (using Euclidean distance in this chapter) are the *satisficing* solution(s) $\boldsymbol{x}^s$, see (Guo et al., 2024). Deviation variables, $\boldsymbol{d} = (\boldsymbol{d}^-, \boldsymbol{d}^+)$, are introduced to quantify the under-achievement and over-achievement of a goal relative to its target.. *By minimizing these variables, the goal* $\mathbf{G}(\boldsymbol{x})$ is converted into an equality constraint to be satisfied. The deviation variables $\boldsymbol{d}$ become the decision variables forming the new objective $\boldsymbol{z}(\boldsymbol{d})$, while the original decision variables $\boldsymbol{x}$ become auxiliary variables excluded from the objective. The combination form of merit function $\boldsymbol{z}(\boldsymbol{d})$ is discussed in Sect. 6.2.4.

$$G(x) : f(x) + d^- - d^+ = t, \text{ where } 0 \le d^-, d^+ \le 1, d^- \cdot d^+ = 0 \tag{6.1}$$

$$\boldsymbol{Minimize}\, z(\boldsymbol{d}) \tag{6.2}$$

Such a construct is the compromise Decision Support Problem (cDSP) (Mistree et al., 1993).

Given: $\mathcal{P}, t$

Find: x, d

Satisfy:

$$x \in \mathcal{F} : \{g(x) \ge 0, h(x) = 0, lowerbound \le x \le upperbound\},$$

$$G(x), 0 \le d^-, d^+ \le 1, d^- \cdot d^+ = 0$$

Minimize: $z(d)$

For an n-dimension, K-goal problem, by adding deviation variables, we increase the dimensionality of a design problem, from $[\boldsymbol{x}_1, \boldsymbol{x}_2, \ldots \boldsymbol{x}_n]^T$ to

[1] An active inequality constraint has a left-hand side that is equal to its right-hand side at the solution point.

$[\boldsymbol{x}_1, \boldsymbol{x}_2, \ldots \boldsymbol{x}_n, \boldsymbol{d}_1^-, \boldsymbol{d}_1^+, \boldsymbol{d}_2^-, \boldsymbol{d}_2^+, \ldots \boldsymbol{d}_K^-, \boldsymbol{d}_K^+]^T$, and thus make it possible to absorb the risk of uncertainty that breaks the second-order sufficient condition. This results in a robust solution that is relatively insensitive to uncertainties. We decrease the dimensionality by returning solutions consisting only of the original decision variables, $\boldsymbol{x} = [\boldsymbol{x}_1, \boldsymbol{x}_2, \ldots \boldsymbol{x}_n]^T$. Such "dimension expansion and reduction" ensures a solution that is relatively insensitive to the uncertainties embodied in the modeling of an optimization problem.

In the cDSP construct, the nonlinear problem is linearized using the Adaptive Linear Programming (ALP) algorithm (Hajihashemi, 2023; Mistree et al., 1993). This linearization enables solving the problem with the Simplex algorithm, yielding a vertex solution. The advantages of a vertex solution over an interior point solution are:

1. A vertex solution ensures the closest distance to the target while being "good enough."
2. It eliminates the need for additional computing power to explore interior point solutions.

6.2.4 *Common Methods of Combining the Goals* $z(\boldsymbol{d})$

To address nonlinear multi-goal problems, a scalarizing function is often used (Bandaru et al., 2017). This approach, also applicable to many-goal problems, combines all goals into a single function, providing a compromise solution that balances their achievements or deviations. A widely adopted scalarization technique is the Archimedean strategy, or weighted sum method (Eq. 6.3). In some cases, multiple weight vectors are applied, generating various solutions for designers to choose from based on their preferences (Fig. 6.3).

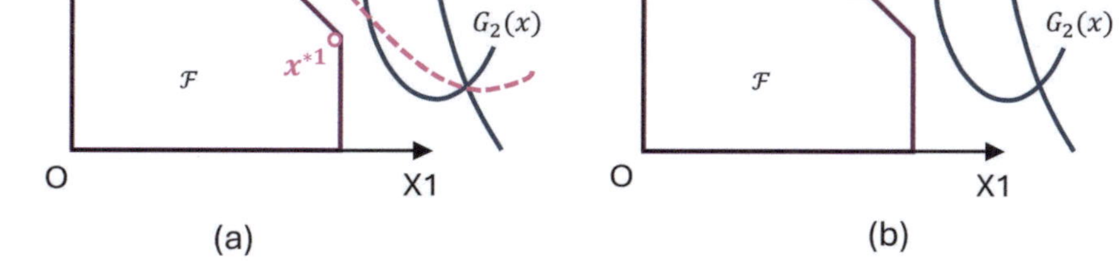

Fig. 6.3 Archimedean strategy (weighted sum)

$$\boldsymbol{Minimize} : \sum\nolimits_{k=1}^{K} W_k \cdot ({d_k}^- + {d_k}^+) \tag{6.3}$$

However, an even distribution of solutions in the solution space is not guaranteed using an even distribution of weight vectors (Messac and Mattson, 2002). Determining appropriate weight vectors requires prior knowledge of goal priorities and expected tradeoffs, and a weight sensitivity analysis can be done (Vevea and Woods, 2005). However, the weight sensitivity analysis is computationally expensive, and the metaheuristics applied in weight generation and selection are not generic as the interdependence among the goals may vary from problem to problem (Seada and Deb, 2014). A generic priori technique that integrates scalarization and domain knowledge is necessary.

Another priori approach for managing multi-goal or many-goal problems is the Pre-emptive strategy, also known as Lexicographic ordering, which converts all goals but one to constraints by iterating (Rae, 1972); see Fig. 6.4. The goals are placed at multiple levels of priority. The first-level goal function will be satisfied as far as possible, and then while holding it within a tolerance, the second-level goal function will be addressed, and so on, in an attempt to address all the goals across all levels (Ignizio, 1976). This process iterates across all priority levels. Like the Archimedean strategy, some domain knowledge on the priority among the goals is needed when using the Pre-emptive strategy. Otherwise, the computational cost is high—for a K-goal problem, if each goal is set to one level, the number of design scenarios for prioritization is $K!$. Designers must carefully evaluate the interrelationships between goals to minimize unnecessary computations and explore only meaningful design scenarios.

A strategy that combines the Archimedean and Pre-emptive approaches is also an option; see Fig. 6.5. At each level, there may be one or multiple goals. The goals at the same level aggregated into a single goal using weight vectors. The levels are switched to explore the solution space thoroughly based on various prioritizing

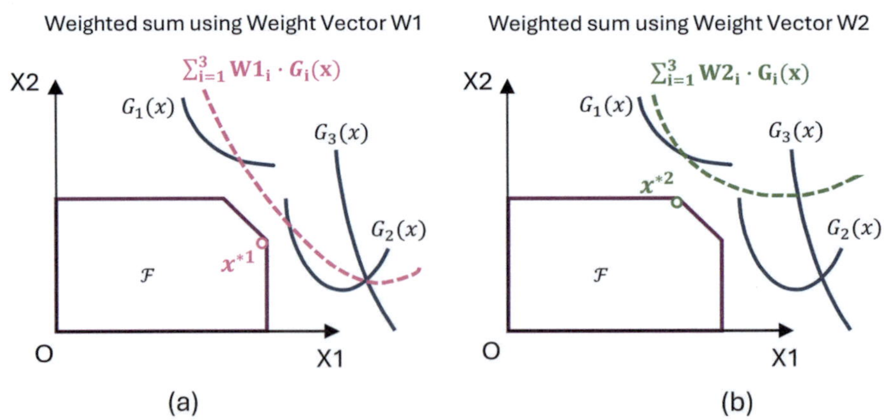

Fig. 6.4 Pre-Emptive strategy (lexicographic ordering)

scenarios. The drawback of the ensemble strategy is that insight into the problem is also required to explore the design scenarios effectively. Otherwise, the permutations and combinations of levelling and weighting increase exponentially as the number of goals increases.

In this chapter, we leverage data analysis to bridge gaps in domain knowledge and minimize the overuse of heuristics without adequate evaluation or refinement. We investigate the following key question in knowledge management for multi-goal problems:

What is a domain-independent method to capture and reuse the knowledge of a many-goal, concurrent-engineering-design problem to facilitate the exploration and selection of design scenarios?

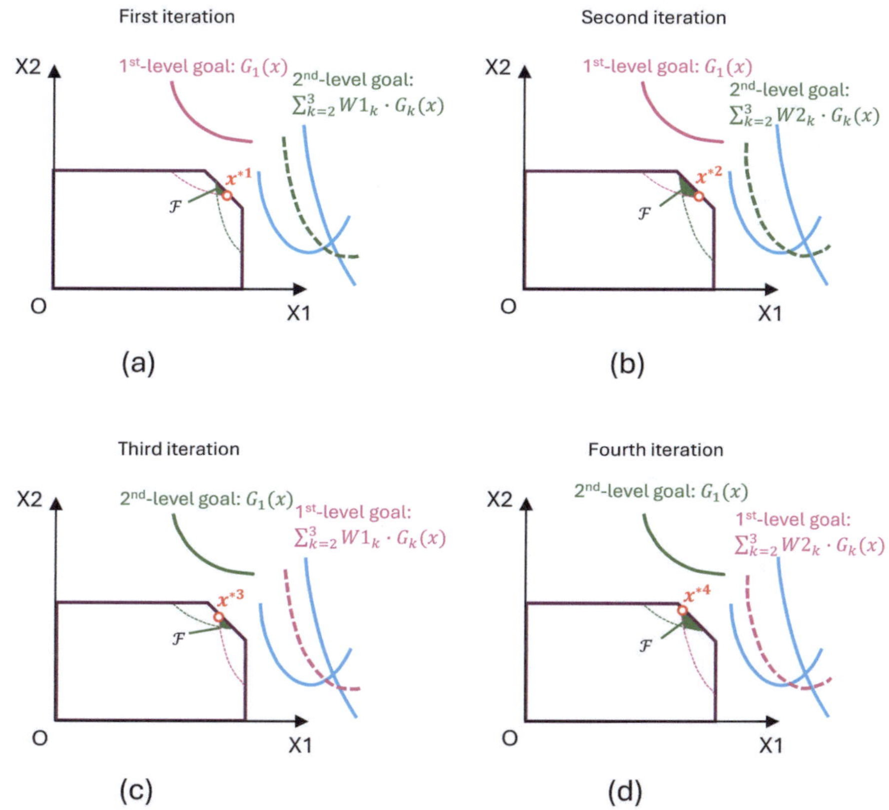

Fig. 6.5 An ensemble strategy using a mixture of archimedean and pre-emptive strategy

6.3 Problem Statement—Test Problem 3: The Rankine Cycle Problem

–Test Problem 3: applying ALWC to a concurrent, multi-goal problem

We illustrate a concurrent, multi-goal problem using a thermal system design example originally published by Smith et al. (2015).

6.3.1 Problem Description

There are various applications for small scale "power" plant systems that run small generators to produce electricity or that make direct mechanical use of the power produced. For example, these systems can be used for providing power to equipment in irrigation systems, driving reverse osmosis systems to produce fresh water for underdeveloped areas, and generating electricity for general use in small collectives.

Building a system around the Rankine cycle is a common approach when a heat source is available. The Rankine cycle is a mathematical representation of a heat engine that converts heat into mechanical work while undergoing phase change (Macquorn Rankine, 1853; Wikipedia, 2019). A schematic representation of the Rankine cycle, where the major components of the system are a power producing turbine, a pump to pressurize the flow to the turbine and two heat exchangers, a condenser, and a heater are shown in Fig. 6.6.

From a design and decision-support perspective, such thermal systems pose complex challenges that must be carefully managed and resolved. Expansion within the laboratory will deal with heat source issues (left side of Fig. 6.6) and power use

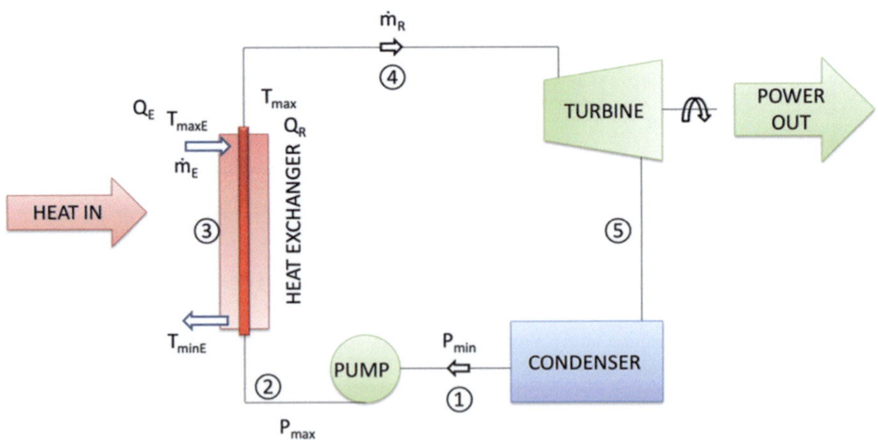

Fig. 6.6 The thermal system

issues (right side of Fig. 6.6) and the choice of working fluids. A common working fluid in a Rankine cycle is water.

The foundational example model is defined by the cycle's maximum and minimum pressures and maximum temperature (PMAX, PMIN and TMAX). Energy is transferred to the closed loop Rankine cycle through a heat exchanger. The heat exchanger is assumed to be of a counter flow design where the key characteristic is the maximum temperature of the heating flow (TMAXE).

6.3.2 Model Formulation

The ideal Rankine cycle involves four processes, as shown graphically in the Temperature (T) versus Entropy (S) plot in Fig. 6.7. There are two adiabatic isentropic processes (constant entropy) and two isobaric processes (constant pressure).

The Rankine cycle consist of 4 cycles, referring to Fig. 6.7,

①–② adiabatic pumping of the saturated liquid from PMIN to PMAX.

②–④ isobaric heat addition in heat exchanger to TMAX,

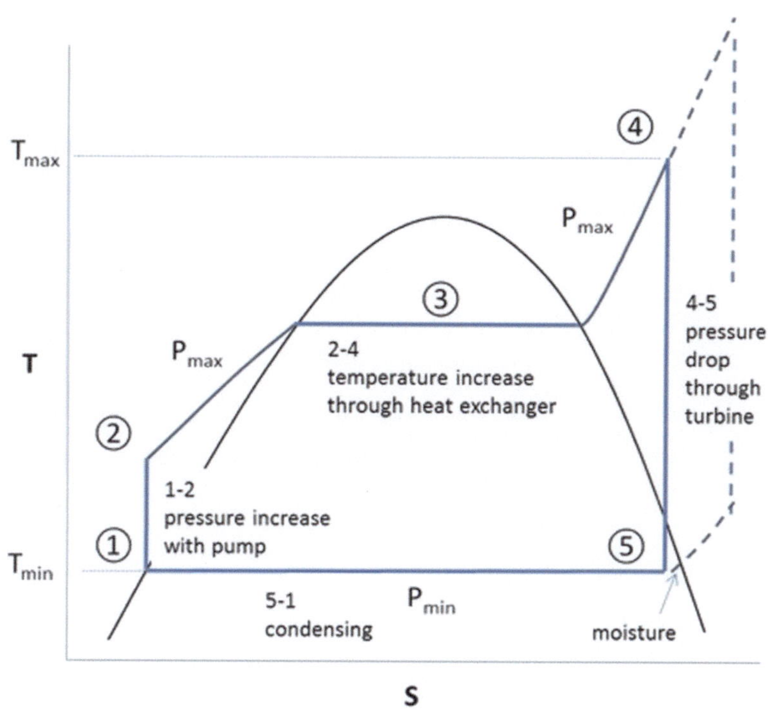

Fig. 6.7 Rankine cycle (temperature and entropy)

④–⑤ adiabatic expansion in the turbine from PMAX to PMIN producing power with the possibility of wet steam exiting the turbine, and

⑤–① isobaric heat loss in the condenser.

The isothermal segments represent moving from saturated liquid to saturated vapor in the case of ③ in the heater and the reverse in the condenser between ⑤–①. The key thermodynamic properties of the working fluid(s) are determined using REFPROP (Lemmon and Huber 2013). The basic features of the problem: there are four decision variables, one linear constraint, nine nonlinear inequality constraints, and six nonlinear goals. The cDSP is given as follows.

Given

Parameters including dependent system variables P1–P51 (units in abbreviated SI[2])

CARNOT	Carnot cycle efficiency (%)	Parameter 1 (P1)
CPEE	Specific heat value for input line in exchanger	P2
CPRE	Specific heat value for Rankine (output) feedline in exchanger (J/(kg · K))	P3
DBTMNR/DBTMNE = 273.16	Lower temperature limit (freezing point) in the Rankine cycle (*DBTMNR*) and in the heat exchanger (*DBTMNE*) for every fluid (K)	P4/P5
DBTMXR/DBTMXE = 2000.0	Upper temperature limit in the Rankine cycle (*DBTMXR*) or in the heat exchanger (*DBTMXE*) for every fluid (K)	P6/P7
DELTLM	Logarithmic main temperature difference (K)	P8
DENSi	$i = 1, 2, ...5$, Density at ①-⑤ (kg/m3)	P9, P10, P11, P12, P13
EDIA/ELEN	Diameter/length of heat exchanger (m)	P14/P15
ENTHi	$i = 1, 2, ...5$, Specific enthalpy at ①-⑤ (J/kg)	P16, P17, P18, P19, P20
ENTHMX/ENTHMN	Enthalpy at TMINE/TMAXE in exchanger	P21/P22
FLOWR/FLOWE	Mass flow rate in the Rankine cycle / exchanger (kg/s)	P23
FRMXR	The upper limit of Rankine cycle mass flow rate (kg/s)	P24
HTEFF	Heat transfer effectiveness (%)	P25
PPUMP/PTURB	Power of the pump/turbine (W)	P26
PRESi	$i = 1, 2, ...5$, Pressure at ①-⑤ (kPa)	P27, P28, P29, P30, P31
QINR	Heat transfer in the heat exchanger (W)	P32
QOUTE	Exchanger heat transfer (W)	P33
QUALi	$i = 1, 2, ...5$, Quality of stream at ①-⑤ (%)	P34, P35, P36, P37, P38
RCEFF	Rankine cycle efficiency (%)	P39
RCMIT	Rankine cycle moisture in turbine (%)	P40
REQPOW	Required power at the Rankine cycle (kW)	P41
RFEEDL	Calculated Rankine cycle length required given diameter (m)	P42
SAREAE	Surface area of the heat exchanger (m^2)	P43
SYSEF1	System efficiency 1 (%)	P44
STSEF2	System efficiency 2 (%)	P45
TDELE=10	Requirement of minimum temperature change in the heat exchanger (K)	P46
TDELC	Minimum temperature gap between the minimum temperature in the heat exchanger and the temperature at ②	P47
TEFFEX	Temperature exchanger efficiency (%)	P48
TEMPi	$i = 1, 2, ...5$, Temperature at ①-⑤ (K)	P49
TMINE	Minimum temperature in exchanger (K)	P50
UHTC	Overall heat transfer coefficient ($W/(m^2 \cdot K)$)	P51

[2] International System of Units.

Functional relationship between parameters and system variables F1–F14

$$CARNOT = 1.0 - \frac{TEMP1}{TEMP4}$$ ````Function 1 (F1)

$$CPEE = \frac{ENTHMX - ENTHMN}{TMAXE - TMINE}$$ F2

$$CPRE = \frac{ENTH4 - ENTH2}{TEMP4 - TEMP2}$$ F3

$$DELTLM = \frac{(TMINE - TEMP2) - (TMAXE - TEMP4)}{ln\left(\frac{TMINE - TEMP2}{TMAXE - TEMP4}\right)}$$ F4

$$FLOWE \cdot (ENTHMX - ENTHMN) = FLOWR \cdot (ENTH4 - ENTH2)$$ F5

$$FLOWR = \begin{cases} FRMXR & if\ EMP4 - TEMP5 = TEMP2 - TEMP1 \\ \frac{REQPOW}{(TEMP4 - TEMP5) - (TEMP2 - TEMP1)} & otherwise \end{cases}$$ F6

$$HTEFF = 1.0 - e^{-\frac{UHTC \cdot SAREAE}{FLOWR \cdot CPRE}}$$ F7

$$PPUMP = (ENTH2 - ENTH1) \cdot FLOWR$$ F8

$$PPUMB = (ENTH4 - ENTH5) \cdot FLOWR$$ F9

$$QINR = FLOWR \cdot CPRE \cdot (TEMP4 - TEMP2)$$ F10

$$QOUTE = FLOWE \cdot CPEE(TMAXE - TMINE)$$ F11

$$RCEFF = \frac{PTURB - PPUMP}{QINR}$$ F12

$$RCMIT = \begin{cases} 0 & if\ QUAL5 > 1 \\ 1 & if\ QUAL5 < 0 \\ 1 - QUAL5 & if\ 0 \leq QUAL5 \leq 1 \end{cases}$$ F13

$$RFEEDL = \frac{SAREAE}{\pi \cdot I}$$ F14

$$SAREAE = \pi \cdot EDIA \cdot ELEN$$ F15

$$SYSEF1 = \frac{PTURB - PPUMP}{QOUTE}$$ F16

$$SYSEF2 = RCEFF \cdot TEFFEX$$ F17

$$TEFFEX = \frac{TMAXE - TMINE}{TMAXE - TEMP2}$$ F18

$$UHTC = \frac{QINR}{SAREAE \cdot DELTLM}$$ F19

Find

X, the decision variables (system variables) ×1– ×4

PMAX	Maximum pressure in the Rankine cycle	Variable 1 (x1)
PMIN	Minimum pressure in the Rankine cycle	x2
TMAX	Maximum temperature in the Rankine cycle	x3
TMAXE	Maximum temperature of the heating fluid in the exchanger	x4

Deviation variables d

$\mathbf{d_k^-}, \mathbf{d_k^+}$	$k = 1, 2, ...6$, Under-achievement and over-achievement of Goal k	*d*

Satisfy

The system constraints

Linear constraints C1

TMAXE – TMAX ⩾ *DELTLM*

Temperature delta (10 K) for maximums in exchanger — Constraint 1 (C1)

Nonlinear constraints C2-C10

RCMIT ⩽ *TMXL*	Moisture in turbine (RCMIT) less than upper limit (TMXL)	C2
FLOWR ⩽ *FRMXR*	Rankine cycle mass flow rate (FLOWR) less than upper limit (FRMXR)	C3
TEMP4 ⩾ *TEMP3*	Temperature at ④ (TEMP4) should be greater than or equal to temperature at ③ (TEMP3)	C4
QUAL4 ⩾ *1.0*	Quality at ④ (QUAL4) is superheated vapor	C5
TMAXE – TMINE ⩾ *TDELE*	TMAXE is greater than TMINE by at least TDELE	C6
TMINE – TMEP2 ⩾ *TDELC*	TMINE is greater than temperature at ② by at least TDELC	C7
CARNOT ⩾ *SYSEF1*	Ideal Carnot cycle efficiency is greater than system efficiency 1 (Sanity check 1)	C8
CARNOT ⩾ *SYSEF2*	Ideal Carnot cycle efficiency is greater than system efficiency 2 (Sanity check 2)	C9
DBTMXE ⩾ *TMAXE*	Temperatures within valid ranges for REFPROP fluid	C10

The system variable bounds ($x_j^{min} \leqslant x_j \leqslant x_j^{max}$):

500 ⩽ PMAX ⩽ 5000 (kPa)	Bound 1,2 (B1, B2)
350 ⩽ TMAX ⩽ 850 (K)	B3, B4
350 ⩽ TMAXE ⩽ 850 (K)	B5, B6

The system goals:

Goal 1: Achieve zero moisture in steam leaving the turbine (Minimize the moisture or maximize the steam quality of ①)

$RCMIT + d_1^- - d_1^+ = 0$ — Goal 1 (G1)

Goal 2: Maximize Rankine cycle efficiency

$RCEFF + d_2^- - d_2^+ = 1$ — G2

Goal 3: Maximize temperature exchanger efficiency

$TEFFEX + d_3^- - d_3^+ = 1$ — G3

Goal 4: Maximize system efficiency indicator 1

$SYSEF1 + d_4^- - d_4^+ = 1$ — G4

Goal 5: Maximize system efficiency indicator 2

$SYSEF2 + d_5^- - d_5^+ = 1$ — G5

Goal 6: Maximize heat transfer effectiveness in exchanger

$HTEFF + d_6^- - d_6^+ = 1$ — G6

Minimize

The design scenario DS to be explored

$z_{DS}(d)$ — DS

There are limitations of the method used in (Smith et al., 2015). Using the Preemptive approach, the six goals are assigned to six levels of priority. By prioritizing the goals differently, comparison may show competing goals driving the solution in different directions. Using the Archimedean approach, the six goals are grouped at the same level and linearly combined using weights. A mix of the Preemptive and Archimedean approaches is possible, but the authors of (Smith et al., 2015) did not explore this. The method proposed in this chapter, the ALWC algorithm, allows the exploration, capture, and reuse of knowledge on the mixed form of the goals for other problems. We assume no prior knowledge of the division and interrelationships of the goals or subsystems for this problem. We apply the ALWC algorithm to learn such knowledge and use the domain expertise documented in (Smith et al., 2015) to verify our findings.

6.4 The Adaptive Leveling-Weighting-Clustering (ALWC) Algorithm

6.4.1 Clustering the Goals Based on Their Interrelationship

To manage a many-goal, concurrent design problem, we explore the effective combinations of the goals $z(d)$, by identifying their interrelationships, such as correlation or orthogonality. These interrelationships inform the scalarization function or prioritization of the goals. A way to combine the goals of a many-goal problem is a design scenario (DS), such as an Archimedean approach using a weight vector, or a Preemptive approach using an order to prioritize the goals, or any mixture of the two strategies. Under a design scenario, DS_a, the merit function is denoted as $z_{DS_a}(d)$, the corresponding *satisficing* solution is x_a^s, and the deviation of Goal k is d_{ak}. For a K-goal problem, if we solve it using A design scenarios, $[DS_1, DS_2, \ldots DS_A]^T$, where A is a positive integer, we get an $A \times K$ matrix of deviations, $\mathcal{D}$; see Fig. 6.8. The interrelationship that we learn by analyzing Matrix $\mathcal{D}$ can be interpreted as an indication of the conditional correlation or conditional orthogonality,[3] that is within the feasible space $\mathcal{F}$ and under A DSs, is the correlation or the orthogonality of the K goals; see Fig. 6.9. Learning algorithms and cluster analysis methods can be expanded according to the requirements of the problems. In this chapter, the knowledge discovery and decision support relevant to the innovation, selection, and customization of learning algorithms and cluster analysis methods are not our focus, however, they can be addressed and improved using our ALWC algorithm.

The adaptiveness. To make the learning algorithm generic, we assume there is no domain knowledge as pre-knowledge on the interrelationship among the goals. In this chapter, as in each iteration, only a limited number of design scenarios are

[3] The correlation analysis and orthogonality analysis are two examples of the analyses illustrated in this chapter to learn the interrelationship among goals. There are other types of analysis that can be applied to capture the interrelationship among goals. Designers can select their own method and customize the problem based on the characteristics of their problems and their preferences.

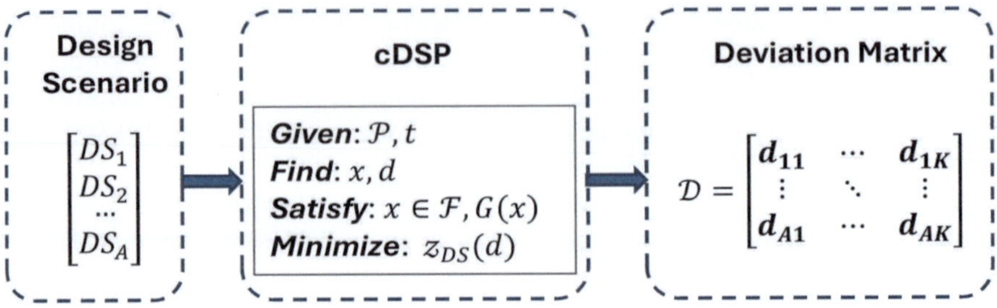

Fig. 6.8 Using multiple design scenarios to obtain a deviation matrix

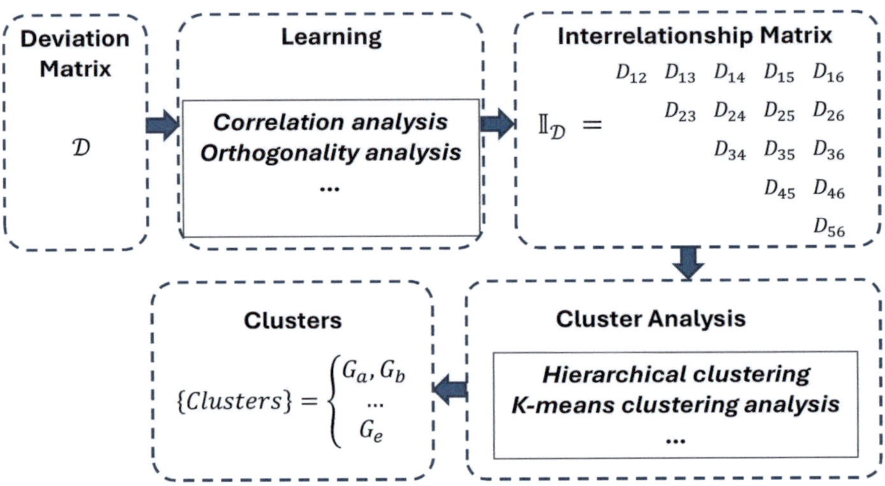

Fig. 6.9 Cluster analysis using a deviation matrix

generated to obtain solutions, hence, we learn and update the interrelationship by iteration. In this sense, our method is an adaptive method.

Learning the correlation. We learn the correlation among the deviation vector of the goals using angle-based distance. Suppose for a three-goal problem, we use two design scenarios DS1 and DS2 to learn the correlation of the goals, the scalarization of each design scenario is represented as $z_{DS_i}(d)$, $d = [d_k^-, d_k^+]^T$, where $i = 1, 2$, $k = 1, 2, 3$. The *satisficing* solution and the deviation are represented as Eqs. 6.4 and 6.5.

$$\boldsymbol{x}^{si} = \mathbf{arg}\big(\boldsymbol{min}\big(z_{DS_i}(\boldsymbol{d})\big)\big), \forall \boldsymbol{i} = 1, 2 \tag{6.4}$$

$$\boldsymbol{d}_{ik} = \left|\boldsymbol{t}_k - \boldsymbol{G}_k\big(\boldsymbol{x}^{si}\big)\right|, \forall \boldsymbol{i} = 1, 2, \forall \boldsymbol{k} = 1, 2, 3 \tag{6.5}$$

As shown in Fig. 6.10, we use a two-dimensional solution space to illustrate the three goals $G_k(x)$, $\forall k = 1, 2, 3$, and two *satisficing* solutions x^{si}, $\forall i = 1, 2$, under two design scenarios $DSi, \forall i = 1, 2$. In Fig. 6.11, we show the deviation points in two two-dimensional goal spaces. The coordinate origin O represents the utopia point where the deviation of both goals is zero, while the Point I (1, 1) is the worst point, where both goals are 0% complete. When design scenario changes from DS1 to DS2, the deviation of Goal 1 and Goal 3 change in different directions, Fig. 6.11a, whereas the deviation of Goal 1 and Goal 2 change in the same direction, Fig. 6.11b. Therefore, based on the results of these two design scenarios, Goal 1 and Goal 3 have a lower correlation and thus belong to two different clusters, while Goal 1 and Goal 2 have a higher correlation, so they belong to one cluster. The acute angle α_{ij}, between $\overline{D_1^{ij}D_2^{ij}}$ and the diagonal $\overline{OI}$ of the goal space indicates the correlation between Goal i and Goal j. We use angle-based correlation analysis to obtain the interrelationship between the two goals (Eq. 6.6). As more design scenarios are considered, the correlation among goals can be learned better. For a number of design scenarios, A, if we use $\alpha_{ij}(DS_\kappa, DS_\iota)$ to denote the acute angle between $\overline{D_\kappa^{ij}D_\iota^{ij}}$, the deviation coordinate of Goal i and Goal j under two design scenarios, DS_κ and DS_ι, and we use the average of these angle-based correlations under any two design scenarios as the correlation between Goal i and Goal j, between the two goals, it is shown as Eq. 6.7. D_{ij_cor} are the elements of the correlation matrix $\mathbb{I}_{\mathcal{D}}$, therefore, we obtain the matrix using angle-based correlation with A design scenarios as in Eq. 6.8. As we update the design scenarios in each iteration, $\mathbb{I}_{\mathcal{D}}$ is iteratively updated. Using $\mathbb{I}_{\mathcal{D}}$ to cluster the goals, we can stop iterating when the clustering results converge.

$$\boldsymbol{D_{ij\_cor}(DS_1, DS_2) = arc\big(sin\big(\alpha_{ij}\big)\big), \forall i, j = 0, \ldots K} \tag{6.6}$$

$$\boldsymbol{D_{ij\_cor}(DS_1, \ldots, DS_A) = \frac{1}{C_A^2}\sum\nolimits_{\kappa=1}^{A-1}\sum\nolimits_{\iota=\kappa+1}^{A} arc\big(sin\big(\alpha_{ij}(DS_\kappa, DS_\iota)\big)\big), \forall i, j = 0, \ldots K} \tag{6.7}$$

$$\mathbb{I}_{\mathcal{D}}\big(\boldsymbol{correlation}, [\boldsymbol{DS}_1, \ldots \boldsymbol{DS}_A]^T\big) = \big(\boldsymbol{D_{ij\_cor}(DS_1, \ldots, DS_A)}\big) \tag{6.8}$$

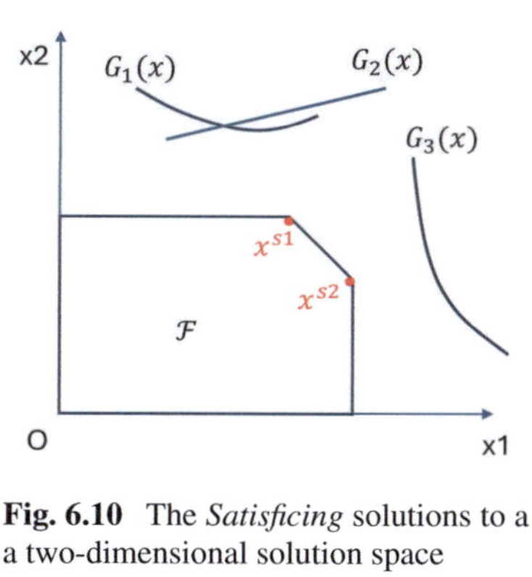

Design scenario	Solution point	Deviation		
		Goal 1	Goal 2	Goal 3
*DS*1	x^{s1}	$d_{11} = 0.7$	$d_{12} = 0.65$	$d_{13} = 0.6$
*DS*2	x^{s2}	$d_{21} = 0.4$	$d_{22} = 0.35$	$d_{23} = 0.8$

Fig. 6.10 The *Satisficing* solutions to a three-goal cDSP under two design scenarios illustrated in a two-dimensional solution space

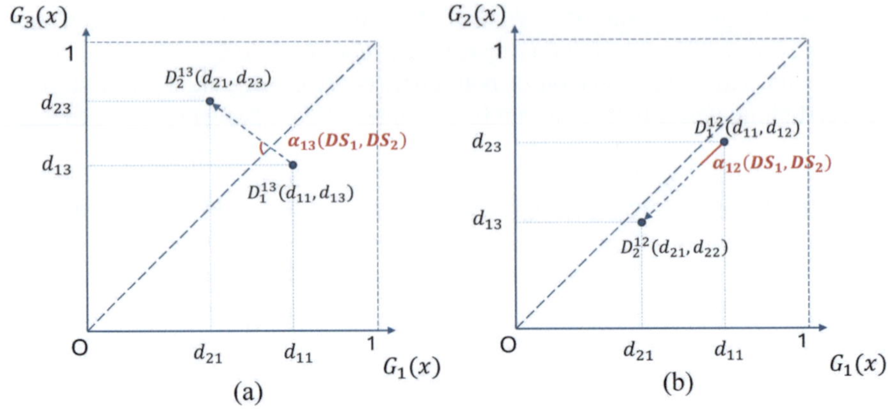

Fig. 6.11 The *Satisficing* solutions to a three-goal cDSP under two design scenarios illustrated in two two-dimensional goal spaces

Learning the orthogonality. When learning the correlation of the goals, we treat the deviation vector of the goals as statistical variables and induce the correlation of goals by analysing the correlation among the statistical variables; whereas when learning the orthogonality among the goals, we obtain the relative position of goals in geometric space, under the constraints of the feasible region $x \in \mathcal{F}$, by using the dot-product of the deviation vectors of any two goals; see Eq. 6.9. In Fig. 6.11a and b, we visualize the dot-product of the deviation vector of two goals, Goal i and Goal j, and Goal $\rangle$ and Goal $|$ using two design scenarios, DS1 and DS2. The dot-product is calculated as shown in Eq. 6.10. For Euclidean distance, $p = 2$—it is also known as the Norm-2 distance. If the dot-product is zero, the two goals are orthogonal, or perpendicular, see Fig. 6.12a; if the dot-product is a large value, the two goals are more parallel, see Fig. 6.12b. The elements, D_{ij_ort}, of the orthogonality matrix $\mathbb{I}_{\mathcal{D}}$, are obtained using orthogonality with A number of design scenarios, as shown in Eq. 6.11.

$$\{< \boldsymbol{Goali}, \boldsymbol{Goalj} > |\boldsymbol{x} \in \mathcal{F}\}$$
$$= \int_{x \in \mathcal{F}} \overline{\boldsymbol{G}_i(\boldsymbol{x})} \boldsymbol{G}_j(\boldsymbol{x}) \boldsymbol{dx} \approx \boldsymbol{D}_{\boldsymbol{ij}_{Ort}}(\boldsymbol{DS}_1, \ldots, \boldsymbol{DS}_A) = [\boldsymbol{d}_{1i}, \ldots \boldsymbol{d}_{Ai}] \cdot [\boldsymbol{d}_{1j}, \ldots \boldsymbol{d}_{Aj}]^{\boldsymbol{T}} \tag{6.9}$$

$$D_{ij_{Ort}}(DS_1, \ldots, DS_A) = \left(\sum\nolimits_{\kappa=1}^{A} d_{\kappa i}{}^{p}\right)^{\frac{1}{p}} \left(\sum\nolimits_{\kappa=1}^{A} d_{\kappa j}{}^{p}\right)^{\frac{1}{p}} cos(\theta_{ij}(DS_1, \ldots, DS_A)) \tag{6.10}$$

$$\mathbb{I}_{\mathcal{D}}\left(\boldsymbol{orthogonalitty}, [\boldsymbol{DS}_1, \ldots \boldsymbol{DS}_A]^{\boldsymbol{T}}\right) = \left(\boldsymbol{D}_{\boldsymbol{ij\_ort}}(\boldsymbol{DS}_1, \ldots, \boldsymbol{DS}_A)\right) \tag{6.11}$$

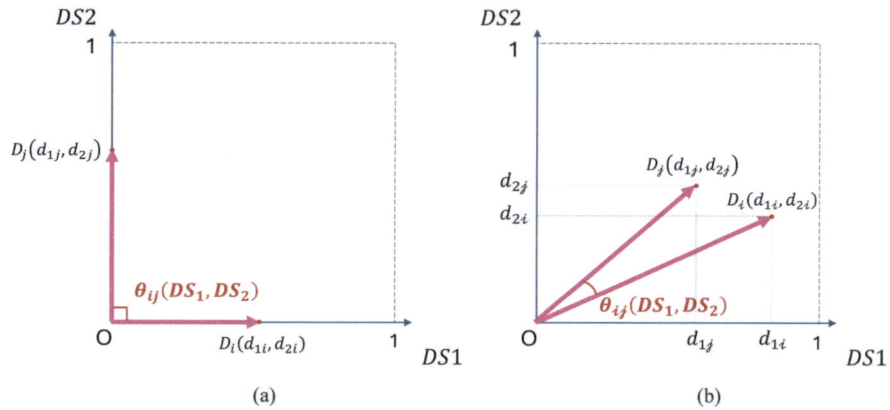

Fig. 6.12 The orthogonality between the deviation vectors of two goals using two design scenarios

Cluster analysis based on interrelationship matrix of the goals. In this chapter, we use multiple clustering algorithms and use cross validation to ensure the rationality of the clustering results. Designers have several criteria to consider when selecting appropriate clustering algorithms, such as:

- the sensitivity of the clustering results to sample size (the number of design scenarios),
- the improvement of goal achievement, and
- the diversity of the solutions by leveling the goals using a clustering result.

The selection of the clustering algorithms is worthy of study for enhancing the ALWC algorithm. If the improvement is limited in a small range, we will stop.

6.4.2 A Schematic of the ALWC Algorithm

The ALWC includes two loops; see Fig. 6.13. Leveling-Weighting-Clustering is the outer loop, and Weighting-Clustering is the inner loop. In the outer loop, leveling begins with the clustering result $\{C, Cluster\}$ from the previous iteration.[4] In the very first iteration, leveling starts with the initialized single cluster so all goals are set to a single level. In the later iterations, when more than one clusters returns from the previous clustering, each cluster is set to a level. After running all processes in this leveling setting, including the weighting and clustering in the inner loop, the levels are alternated, that is the goals in each cluster are alternately set to a

[4] C Is the number of clusters. *Cluster* is a two-dimensional array containing the clusters of the goals. E.g., $Cluster = [[G1, G2, G4], [G3, G5, G6]]$ means that there are two clusters: *Cluster 1* includes Goal 1, 2, and 4, whereas *Cluster 2* include Goals 3 5, and 6. $Cluster[i][j]$ represent the jth goal in the ith cluster, so, $Cluster[2][3] = $ G6. Array "*Level*" works in the same way.

different level. When it enters the inner loop, the goals in each level are combined using weight vectors. After being solved using the ALP, the deviation matrix $\mathcal{D}$ is obtained. Subsequently, the interrelationship matrix $\text{II}_{\mathcal{D}}$ is obtained and used to do cluster analysis, and a new $\{C, \mathit{Cluster}\}$ is returned to update the leveling for the next iteration. The inner loop stops generating more weight vectors when the $\text{II}_{\mathcal{D}}$ does not change, which means when more weight vectors do not lead to different $\text{II}_{\mathcal{D}}$. Specifically, this occurs when $\sigma(\text{II}_{\mathcal{D}}) \leq \varepsilon$, where ε is a threshold determined using heuristics. The outer loop stops leveling the goals based on the latest clustering result $\{C, \mathit{Cluster}\}$ if it does not change in the previous η iterations, where η is a positive integer determined heuristically. When the outer loop stops, convergence is reached.

6.4.3 The Algorithms in the ALWC

There are six procedures in the ALWC algorithm. The model is formulated as a compromise Decision Support Problem (cDSP) (Hajihashemi, 2023; Mistree et al., 1981, 1993) in Procedure 1.1. The number of clusters, C, is initiated as "1" in Procedure 1.2. In Procedure 2 Levelling, starting from the second iteration, the goals are levelled based on clustering results—the output of Procedure 5 in the previous iteration. In Procedure 1, the number of levels, *n*, is updated. In Procedure 2.2, each cluster take turns to be set to Level 1 to Level *n*. The algorithm follows[5]:

Algorithm 1 Procedure 2 Leveling

2.1 Updating

Given Cluster results $\{C, \mathit{Cluster}\}$, cDSP, Leveling information $\{L, \mathit{Level}\}$ // C, L are integers representing the number of the clusters and the number of levels respectively. "$\mathit{Cluster}, \mathit{Level}$" are two-dimension arrays containing the clustering result[5] and leveling information.

integer $L = C$

array $\mathit{Level} = \mathit{Cluster}$

return $\{L, \mathit{Level}\}$ and **go to** Procedure 3 Weighting

2.2 Alternating levels

if $\text{L} \geq 2$

array $\mathit{temp} = \mathit{Level}[L]$

for integer n in range $[2, L]$

  $\mathit{Level}[n] = \mathit{Level}[n-2]$

$\mathit{Level}[1] = \mathit{temp}$

return $\{\text{L, Level}\}$ and **go to** Procedure 3 *Weighting*

[5] E.g., *Cluster* = [[*G1*,*G2*,*G4*],[*G3*,*G5*,*G6*]] means that there are two clusters: *Cluster 1* include *Goal 1, Goal 2*, and *Goal 4*, whereas *Cluster 2* include *Goal 3, Goal 5*, and *Goal 6*. *Cluster*[2][3] represent the third goal in *Cluster 2*, so, *Cluster*[2][3] = *G6*. Array Level works in the same way.

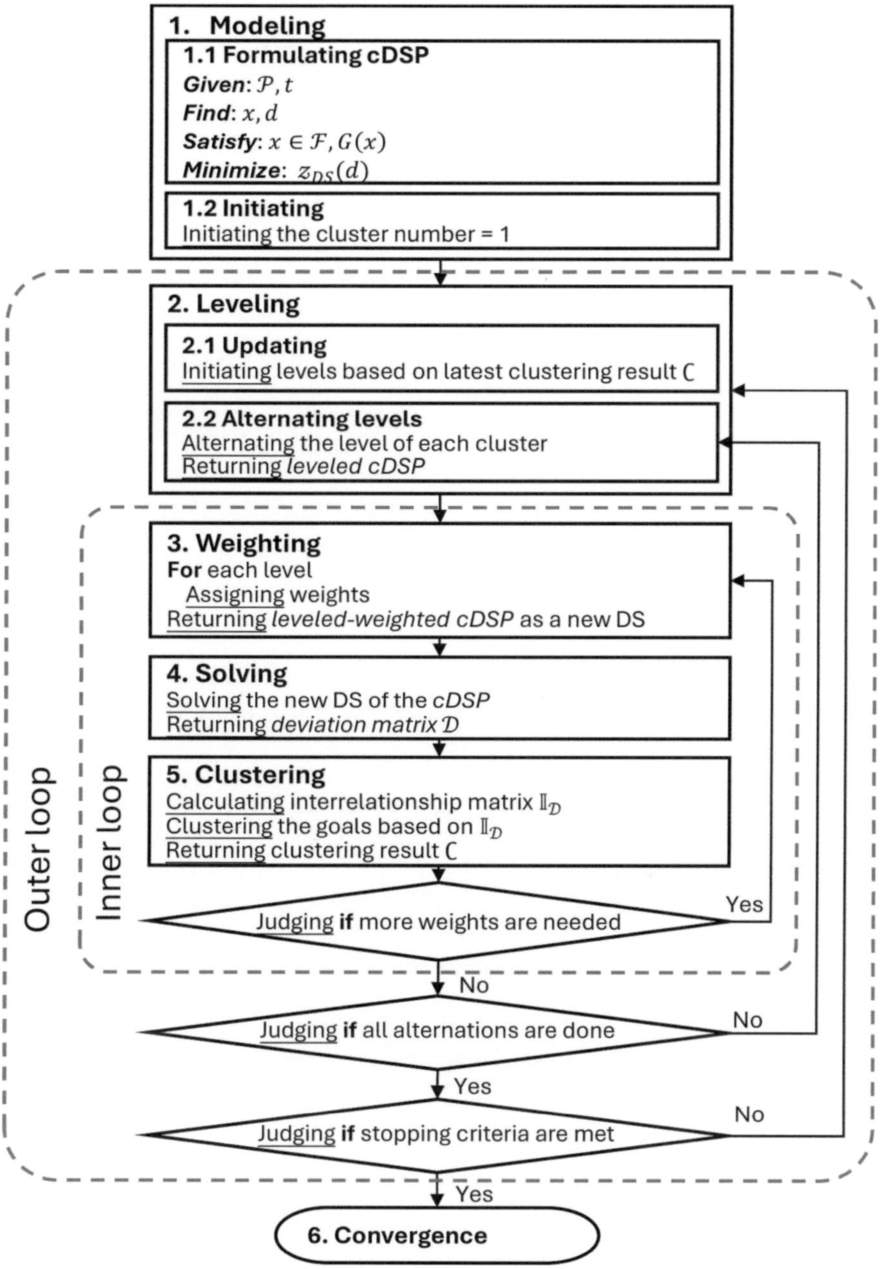

Fig. 6.13 The flowchart of the adaptive leveling-weighting-clustering (ALWC) loop

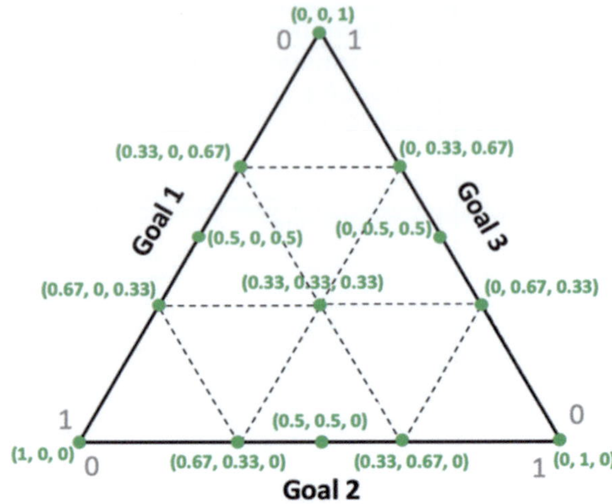

Fig. 6.14 Weight vectors of a three-goal problem with p = 3

After the goals are leveled, the goals in each level are combined using weight vectors.. It is shown that an even distribution of solutions in the solution space are not guaranteed by using an even distribution of weight vectors (Messac et al. 2002). Therefore, in Procedure 3: Weighting, the weight vectors of the goals in each level are generated in a greedy manner until the deviation matrix $\mathcal{D}$ does not increase its diversity (the variety of the solutions—see Glossary in this chapter) with the addition of more weight vectors. In this chapter, we use the standard deviation, $\sigma(\mathbb{I}_{\mathcal{D}})$, to evaluate diversity. The solution points generation method in (Seada and Deb, 2014) is improved and applied to generate weight vectors. The algorithm of weighting is a fractal algorithm. For example, if there are three goals in a level, we can use a ternary plot to illustrate how the weights are generated, see Fig. 6.14. Each edge of the triangle is an axis representing the weight of a goal. The coordinates of each spot in the triangle represent a weight vector. The sum of the coordinates of any spot is one. The algorithm for weighting follows [6,7]:

---

[6] E.g., *Weight* $= \left[[[1,0], [0,1], \left[\frac{1}{2}, \frac{1}{2}\right]], [[1,0, 0,0], [0,1, 0,0], [0,0, 1,0], [0,0, 0,1]]\right]$ means that there are two levels; *Level 1* has two goals and there are three vectors for the two goals; Level 2 has four goals and there are four weight vectors for the four goals. *Weight*[*n*] represents the weight vectors of the goals in the *n*th level; *Weight*[*n*][*m*] represents the *m*th weight vector of the goals in the *n*th level. *Weight*[*n*][*m*][*k*] represents the weight of the *k*th goal of the *m*th weight. vector in the *n*th level. In this example, *Weight*[1][3][1] $= \frac{1}{2}$.

[7] E.g., $F = [0, \frac{1}{2}, 1]$, M = 2, "*I*" can be one of the nine arrays: [0, 0], $[0, \frac{1}{2}]$, [0, 1], $\left[\frac{1}{2}, 0\right]$, $[\frac{1}{2}, \frac{1}{2}]$, $[\frac{1}{2}, 1]$, [1, 0], $[1, \frac{1}{2}]$. "*wv*" can be [0, 1], $[\frac{1}{2}, \frac{1}{2}]$, and [1, 0].

Algorithm 2 Procedure 3 Weighting

Given $\{L, Level\}$, $\{C, Cluster\}_{i-1}$, $Weight$ // "$Weight$" is a three-dimension array containing weight vectors[6]

Initiate parameters:

  integer $p = 2, i = 1$

  array $F = [\,], wv = [\,], Weight = [wv],\ \sigma(\mathcal{D}_0) = 0,\ \varepsilon = \mathcal{E}$

for integer n in range [1, L]

  integer $M = \text{length}(Level[n])$ // Assign the number of goals in Level n as M

  while $|\sigma(\mathcal{D}_i) - \sigma(\mathcal{D}_{i-1})| > \varepsilon$

    $i = i + 1$

    for integer q in range[1,p]

      for integer r in range[0,q]

        $F = F \cup \frac{r}{q}$

    for I in $product(*[F]*M)$ // "I" is a one-dimension array with three elements that are cross-combined by M number of F[7]

      if sum(I) == 1 // If the sum of the elements in "I" equals one.

        $wv = I$

        $\text{Weight}[n] = \text{Weight}[n] \cup \text{wv}$

    Call Procedure 4 and **return** $\mathcal{D}_i$

    Calculate $\sigma(\mathcal{D}_i)$

Weighting iterates while p is increasing and stops if the diversity of the deviation matrix $\mathcal{D}_i$ is not improved. We use standard deviation to represent the diversity of $\mathcal{D}_i$. To remove the variety of the ambitiousness of the target of the goals, that is the deviation range and distribution may vary from one goal to another, we normalize $\mathcal{D}_i$ into the range [0, 1] using Eq. 6.12. A part of the normalized deviations of a six-goal cDSP with 81 Iterations 81 Iterations with $L = 3, p = 2$ is shown in Table 6.4.

$$\boldsymbol{d_{ak_{norm}}} = \frac{\boldsymbol{d_{ak} - min\{d_{1k}, \ldots d_{Ak}\}}}{\boldsymbol{max\{d_{1k}, \ldots d_{Ak}\} - min\{d_{1k}, \ldots d_{Ak}\}}} \tag{6.12}$$

Therefore, the number of leveling-weighting scenarios (or design scenarios, *DS*), *A*, corresponds to each clustering result is given in Eq. 6.13.

$$\boldsymbol{A = \textbf{length}(DS) = n! \cdot n \cdot length(Weight[n])} \tag{6.13}$$

Essentially, the leveling-weighting process converts a multi-goal cDSP into multiple single-goal cDSPs. The goals are grouped into multiple clusters based on

Table 6.4 A part of the normalized deviations of a six-goal cDSP with 81 Iterations with $\mathbf{L} = 3$, $\mathbf{p} = 2$

Iteration a	Leveling-weighting scenario			$d_{a1_{norm}}$	…	$d_{a6_{norm}}$
	Level 1 (Weight)	Level 2 (Weight)	Level 3 (Weight)			
$a = 1$	Goal 2, 4 (1,0)	Goal 3, 6 (1,0)	Goal 1, 5 (1,0)	1	…	1
$a = 2$	Goal 2, 4 (0,1)	Goal 3, 6 (1,0)	Goal 1, 5 (1,0)	1	…	1
$a = 9$	Goal 2, 4 $\left(\frac{1}{2}, \frac{1}{2}\right)$	Goal 3, 6 $\left(\frac{1}{2}, \frac{1}{2}\right)$	Goal 1, 5 (1,0)	1	…	0.62
$a = 19$	Goal 2, 4 (1,0)	Goal 3, 6 (1,0)	Goal 1, 5 $\left(\frac{1}{2}, \frac{1}{2}\right)$	1	…	0.83
$a = 27$	Goal 2, 4 $\left(\frac{1}{2}, \frac{1}{2}\right)$	Goal 3, 6 $\left(\frac{1}{2}, \frac{1}{2}\right)$	Goal 1, 5 $\left(\frac{1}{2}, \frac{1}{2}\right)$	0.45	…	0.64
$a = 54$	Goal 3, 6 $\left(\frac{1}{2}, \frac{1}{2}\right)$	Goal 1, 5 $\left(\frac{1}{2}, \frac{1}{2}\right)$	Goal 2, 4 $\left(\frac{1}{2}, \frac{1}{2}\right)$	0.03		0
$a = 81$	Goal 1, 5 $\left(\frac{1}{2}, \frac{1}{2}\right)$	Goal 2, 4 $\left(\frac{1}{2}, \frac{1}{2}\right)$	Goal 3, 6 $\left(\frac{1}{2}, \frac{1}{2}\right)$	0.01	…	0.68

their correlation or orthogonality, and the clusters are converted to constraints one-by-one within a tolerance. It is assumed that goals with stronger orthogonality or less correlation perform differently under multiple design scenarios. Consequently, using a weighted sum of the highly orthogonal goals to acquire compromise solutions may waste the potential of all subsystems. To boost the performance of each subsystem in turn to *satisfice* various situations, we assign the highly orthogonal goals to different clusters and prioritize them alternatively, whereas we assign the more correlated goals in the same cluster and combined them using weight vectors.

To cluster the goals in Procedure 5, we use the interrelationship matrix $\mathbb{I}_{\mathcal{D}}$ as the input. In this chapter, the correlation analysis (Eqs. 6.6 and 6.7) and orthogonality analysis (Eqs. 6.9 and 6.10) are two options for obtaining $\mathbb{I}_{\mathcal{D}}$ for cluster analysis. We use the hierarchical clustering methods in "*hclust*" in *R*, to obtain the clustering results $\{C, Cluster\}$. By applying all methods in hierarchical clustering, different results may be acquired. The most popular result is returned as $\{C, Cluster\}$ for use in leveling of the next cycle. The algorithm for clustering follows:

Algorithm 3 Procedure 5 Clustering

Given $\mathcal{D}_i$, {C,Cluster}
Initiate $queue = [\,]$, $temp\_result = [\,]$
$\mathcal{D}_{i\_norm} \leftarrow$ Normalize $\mathcal{D}_i$ using Equation 12
Calculate $\mathbb{I}_{\mathcal{D}}$ based on their correlation (Equation 6, 7) or orthogonality (Equation 9, 10)
for each *method* in *hclust*
    $\{C,Cluster\} \leftarrow$ hclust($\mathbb{I}_{\mathcal{D}}$, $method$)
    queue.append = $\{C,Cluster\}$ // Add the latest result into an array named "*queue*"
$\{C,Cluster\} = mode(queue)$ // Select the clustering result that appears most often
return $\{C,Cluster\}$

We use the test problem, Rankine cycle thermal system design, to testify the effectiveness of the ALWC algorithm.

6.5 Unsupervised Learning: Results and Discussions

6.5.1 Clustering Result

By running the ALWC algorithm applying different interrelationship methods and clustering methods (K-means, hierarchical clustering, etc.), we iteratively obtain clustering results that are shown in Table 6.5. For each interrelationship method, although only one clustering result converges, various clustering results are found during the ALWC loop. Three clustering results identified as a result of updating the leveling and these are summarized in Table 6.6. Our aims of processing the ALWC loop include identifying the best clustering result, obtaining more solutions that better achieve the goals or improve the diversity of the deviations, and gaining knowledge on method selection and reuse them. Therefore, we use all solutions and deviations from all iterations to enlarge the solution pool for decision support.

6.5.2 Improvement in Goal Achievement Along the Design Scenario Expansion

For a many-goal problem, trade-offs between two goals often affect other goals, leading to numerous nondominated solutions and weakly dominated solutions. This makes using the concept of 'solution domination' to rank the solutions ineffective. Therefore, we use statistics to reflect whether the results are improved and

Table 6.5 The clustering results along iterations

Interrelationship	Iteration 1: 1-level			Iteration 2	Iteration 3
Angle-based correlation	6 weight vectors	21 weight vectors	71 weight vectors	22 design scenarios by alternating levels {3, [[1, 6], [2, 4], [3, 5]]}	22 design scenarios by alternating levels {3, [[1], [2, 4], [3, 5, 6]]}
	{3, [[1], [2, 4], [3, 5, 6]]}	{3, [[1, 6], [2, 4], [3, 5]]}	{3, [[1, 6], [2, 4], [3, 5]]} (returned)	{3, [[1], [2, 4], [3, 5, 6]]}	{3, [[1], [2, 4], [3, 5, 6]]} (converged)
Orthogonality	6 weight vectors	21 weight vectors		22 design scenarios by alternating levels {3, [[1, 2, 4], [3, 5], [6]]}	22 design scenarios by alternating levels {3, [[1, 2, 4], [3, 5], [6]]}
	{3, [[1, 2, 4], [3, 5], [6]]}	{3, [[1, 2, 4], [3, 5], [6]]}		{3, [[1, 2, 4], [3, 5], [6]]}	{3, [[1, 2, 4], [3, 5], [6]]} (converged)

Table 6.6 The summary of the clustering results ever returned to update the leveling

Clustering result 1	{3, [[1], [2, 4], [3, 5, 6]]}
Clustering result 2	{3, [[1, 6], [2, 4], [3, 5]]}
Clustering result 3	{3, [[1, 2, 4], [3, 5], [6]]}

enriched through iteration. In Table 6.7, we show the mean, standard deviation, best (minimum), and worst (maximum) deviations of each of the six goals, under each clustering scenario, highlighting the best case among all clustering scenarios. To make the results from different clustering scenarios comparable, we use the real deviation value instead of the normalized value. For the 1-level scenario, we use the results of the 71 weight vectors to obtain the statistics because these 71 weight vectors include the 6 and 21 weight vectors from previous iterations of the inner loop. For all the other three clustering scenarios, each with 22 design scenarios, we calculate the statistics respectively. For example, for Goal 1, among the means of each of the four clustering scenarios, the best (smallest) value 0.04 occurs in the third clustering scenario. The numbers in Table 6.7 are rounded but we use the longer floating numbers to do the comparison and highlight (emphasis) the best ones. The key take aways from Table 6.7 follow:

By leveling the goals using three clustering scenarios, the mean, standard deviation, and the worst-case result of the deviations of five goals, except for Goal 1, are improved. The worst case result for the sum of all goals is also improved. These

Table 6.7 Statistics of the results

Statistics	Clustering scenario	$d_k = max\{d_k^-, d_k^+\}$						Sum $\sum_{k=1}^{6} d_k$
		d_1	d_2	d_3	d_4	d_5	d_6	
Mean	1 level	0.05	0.81	0.36	0.81	0.90	0.02	*2.94*
	{3, [[1], [2, 4], [3, 5, 6]]}	0.05	*0.78*	0.51	*0.78*	0.92	0.01	3.04
	{3, [[1, 6], [2, 4], [3, 5]]}	*0.04*	0.84	*0.34*	0.84	*0.90*	*0.01*	2.96
	{3, [[1, 2, 4], [3, 5], [6]]}	0.05	0.80	0.45	0.80	0.91	0.01	3.01
Standard deviation	1 level	0.05	0.08	0.41	0.08	0.05	0.04	0.31
	{3, [[1], [2, 4], [3, 5, 6]]}	0.05	0.08	0.38	0.08	0.05	0.01	0.26
	{3, [[1, 6], [2, 4], [3, 5]]}	0.05	*0.05*	*0.24*	*0.05*	*0.04*	*0.01*	*0.19*
	{3, [[1, 2, 4], [3, 5], [6]]}	*0.04*	0.08	0.37	0.08	0.05	0.01	0.25
Minimum (best case)	1 level	*0.00*	*0.69*	*0.04*	*0.69*	*0.84*	*0.00*	*2.67*
	{3, [[1], [2, 4], [3, 5, 6]]}	*0.00*	0.69	0.06	0.69	0.85	0.00	2.72
	{3, [[1, 6], [2, 4], [3, 5]]}	*0.00*	0.69	0.06	0.69	0.85	0.00	2.72
	{3, [[1, 2, 4], [3, 5], [6]]}	*0.00*	0.69	0.07	0.69	0.85	0.00	2.73
Maximum (worst case)	1 level	*0.12*	0.88	0.90	0.88	0.97	0.16	3.52
	{3, [[1], [2, 4], [3, 5, 6]]}	0.12	*0.87*	*0.90*	*0.87*	*0.97*	0.02	3.31
	{3, [[1, 6], [2, 4], [3, 5]]}	0.12	*0.87*	*0.90*	*0.87*	*0.97*	*0.02*	*3.29*
	{3, [[1, 2, 4], [3, 5], [6]]}	0.12	0.87	*0.90*	0.87	*0.97*	*0.02*	*3.29*

observations indicate that by clustering and leveling the goals using the ALWC algorithm, better design scenarios are identified, reducing the deviation (or improving the achievement) of most goals.

Other statistics can be used to evaluate the iteration results from different perspectives. Designers may select or develop customized statistics based on the characteristics of their problems.

6.5.3 Reducing the Euclidean Distance to the Utopia Point

For a many-goal, concurrent design problem, goals may represent the performance of various subsystems of the design. It is possible that the improvement of one goal results in a larger sacrifice of another goal or several other goals. For example, in Fig. 6.15, as Goal 3 is improved 20% at D_2^{123} versus D_1^{123}, Goal 1 and 2 worsen by 80%. This may be desirable in some situations, but more often than not, designers prefer to avoid this situation because the overall performance of the design may not be practically enhanced by improving a single subsystem. Therefore, we use the Euclidean distance to the Utopia point of the deviations, to evaluate the overall performance of the achievement of all goals. Equation 6.14 represents the Euclidean distance to the Utopia point of the result from Design Scenario a for a K-goal problem. We use statistics to evaluate the evolution of the Euclidean distance during the iterations and summarize the statistics in Table 6.8. Using the clustering scenarios obtained during iteration, the mean, standard deviation, and the worst case of the Euclidean distance to the Utopia point are improved, whereas the best case is not improved. All the design scenarios and corresponding results from the iterations are added to the solution pool for designers to select. Designers may customize their post-solution analyses to acquire further decision support on design scenario generation and selection.

$$|\boldsymbol{OD}_a| = \left(\sum_{k=1}^{K} d_{ak}{}^2\right)^{\frac{1}{2}} \tag{6.14}$$

6.5.4 Reducing Computational Complexity

Smith and coauthors (Smith et al., 2015) explore the leveling of all six goals, and the theoretical number of leveling scenario, is $6! = 720$. In this chapter, as the goals are leveled based on clustering results, for each clustering scenario, the number of leveling scenarios is reduced to $3! = 6$. Within each level, the goals are combined using weight vectors. The stopping criteria of weight vector generation (Line 5 of Algorithm 2) help prevent designers from using too many unnecessary weight vectors. Using the angle-based correlation method (Sect. 6.4) to calculate the interrelationship matrix, we explore 142 design scenarios in three iterations and converge; and using the orthogonality method we explore 71 design scenarios in three iterations, and they converge. If we use both methods, that is $142 + 71 = 213$ design scenarios. Therefore, by using the leveling plus the weighting methods, the number of design scenarios is reduced from 720 to 213.

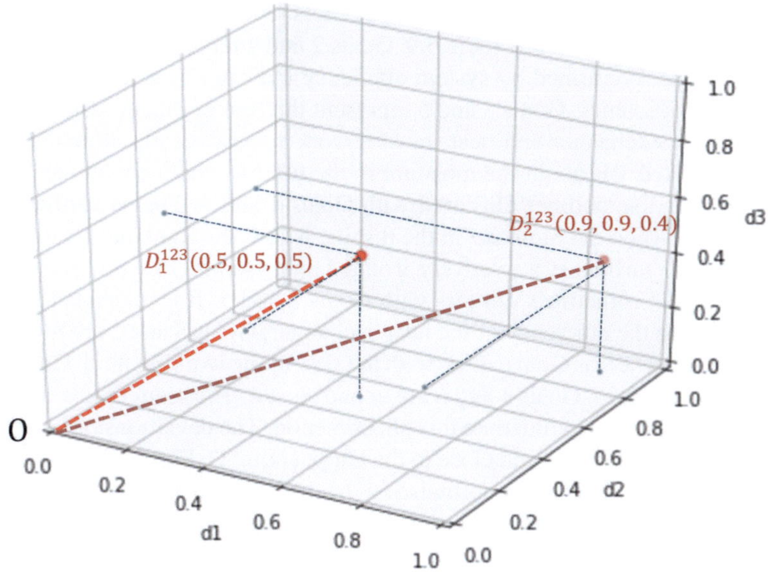

$$|OD_1^{123}| = \left((0.5)^2 + (0.5)^2 + (0.5)^2\right)^{\frac{1}{2}} = 0.87$$

$$|OD_2^{123}| = \left((0.9)^2 + (0.9)^2 + (0.4)^2\right)^{\frac{1}{2}} = 1.33$$

Fig. 6.15 An example of improving goal 3 by 20% while worsening goal 1 and goal 2 by 80% respectively

Table 6.8 Statistics of the Euclidean distance to the utopia point of the results under each clustering scenario

	Mean	Standard deviation	Minimum (best case)	Maximum (worst case)
1 level	1.56	0.08	*1.45*	1.70
{3, [[1], [2, 4], [3, 5, 6]]}	1.57	0.08	1.46	1.66
{3, [[1, 6], [2, 4], [3, 5]]}	*1.54*	*0.06*	1.46	*1.64*
{3, [[1, 2, 4], [3, 5], [6]]}	1.56	0.07	1.46	*1.64*

6.5.5 *Verification of the Results*

The results are verified using domain knowledge. Goals 2 and 4 consistently form one cluster, and Goals 3 and 5 form another. However, the clustering of Goals 1 and 6 varies with each iteration. This indicates that Goals 2 and 4 are strongly correlated, or weakly orthogonal, as are Goals 3 and 5, while the relationship between Goals 1 and 6 and other goals is not significant. The clusters represent the subsystems, which verify the clustering result; see Table 6.9. Goals 2 and 4 represent the Rankine cycle efficiency, further confirmed by system efficiency indicator 1, which aims to boost Rankine cycle efficiency. Goals 3 and 5 represent the heat exchange efficiency, with the temperature exchanger and heat transfer work synergistically affecting system performance. Goal 1 is about the moisture in the turbine, and it sometimes forms a single cluster and sometimes clustered with Goals 2 and 4. This is verified by the knowledge that when the moisture in the turbine is low, the Rankine cycle is highly efficient, but the turbine is a relatively isolated subsystem. Goal 6 is sometimes a single cluster and sometimes clusters with Goals 3 and 5. It is understandable that they all deal with the efficiency of the heat exchanger; Goals 3 and 5 are more about the temperature, whereas Goal 6 deals with the flow of the working fluid. However, in one clustering scenario, Goals 1 and 6 are in one cluster. The scatter plots of any two goals facilitate (Fig. 6.16) illustrate this phenomenon. The deviations of Goals 1 and 6 are relatively small values and close to the Utopia point O, Fig. 6.16e, compared to other two-goal plots. Even after normalizing the deviations in the range [0, 1], Goals 1 and 6 are weakly correlated. However, if such a clustering scenario is used in the next iteration, the clustering results are "back to normal." This indicates that iterative clustering and updating are necessary to expand the sample size and remove bias.

Using domain knowledge, the clustering results for the thermal system design problem are verified. For many-goal problems where the domain knowledge or expertise is missing, the ALWC algorithm can help designers identifying the interrelationships among the goals. Consequently, more design scenarios, based on the interrelationships of goal combinations, are explored. The design scenarios and goal deviations are added to the solution pool for designers to select the satisfaction of different requirements to improve the problem formulation. This allows for further analysis to better understand the interrelationship among subsystems. The ALWC algorithm can be used as a tool to partition a concurrent design problem, especially when information is incomplete.

Table 6.9 Meaning of the three clusters

Meaning	Representative goals
Rankine cycle efficiency	Goal 2: Maximize Rankine cycle efficiency Goal 4: Maximize system efficiency indicator 1
Heat exchange efficiency	Goal 3: Maximize temperature exchanger efficiency Goal 5: Maximize system efficiency indicator 2
Moisture in turbine	Goal 1: Moisture in steam leaving the turbine
Heat transfer effectiveness in exchanger	Goal 6: Heat transfer effectiveness in exchanger

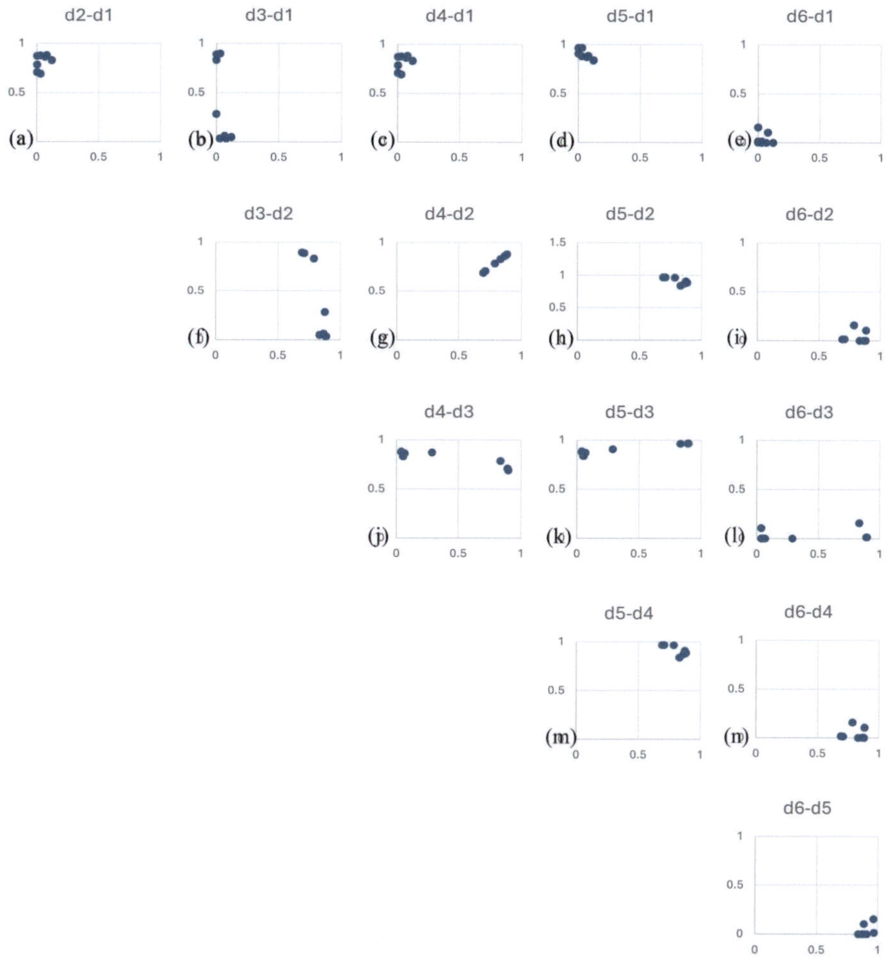

Fig. 6.16 Scatter plots of any two goals using deviations of 1-level, 21 weight vectors

6.5.6 *Closing Remarks on Using ALWC to Accelerate Learning*

In this chapter, we introduce the Adaptive Leveling-Weighting-Clustering (ALWC) algorithm for exploring multiple design scenarios. We illustrate the algorithm's effectiveness using a thermal system design test problem. The key question addressed is.

- *What is the domain-independent algorithm to capture and reuse the knowledge of a many-goal, concurrent-engineering-design problem to facilitate the exploration and selection of design scenarios?*

The answer follows: Using the ALWC algorithm, with increasing weight vectors, the interrelationships among goals, based on their deviations or achievement rates, evolve and converge. Goals are then grouped into clusters representing different subsystems based on these interrelationships.. The combinations of the goals are explored iteratively, using either a Pre-emptive or an Archimedean strategy. This facilitates assigning each cluster a different level (leveling) and combining the goals in each level using weight vectors (weighting). Through iteration more design scenarios are identified, and the corresponding solutions are obtained for designers to choose the appropriate design scenario and then improve the design. As a tool to acquire insight when domain knowledge is lacking, the combination of goals is explored to identify better solutions regarding average deviations, standard deviations, worst case scenarios, and Euclidean distance to the Utopia point, while reducing computational complexity.

The algorithms within the ALWC algorithm can be extended, modified, or customized to meet the specific requirements of a design problem. When there is insufficient expertise in the problem domain to support decision making on goals, subsystem division and tradeoffs, and design improvement, the knowledge discovered using the ALWC algorithm can be used to explore ways that may contribute to design improvement.

6.6 Role of this Chapter in This Monograph

6.6.1 Summarizing How We Complete Connecting Formulation, Exploration, and Evaluation

For a complex system composed of subsystems, when designers lack knowledge of subsystem division, the structure of the subsystems, combination status, mutual relationships, etc., the interrelationships among the subsystems cannot be effectively leveraged to improve the efficiency of the entire system. For the decision model of a concurrent engineering-design problem, the awareness of subsystems and the interrelationship among the subsystems are hidden information that can be learned by analyzing the decision model, the solution space, and the association between design scenarios and the solutions.

The essence of Hypothesis 3 (originally proposed in Sect. 3.3.3), "*learn the system nature such as interrelationships among subsystems and reorganize them based on it*" is to explore multiple design scenarios and obtain a dynamic dataset for analysis, through which the correlation, orthogonality, or other indicators that characterize the degree of contradiction or synergy among the different driving forces in the system, gradually emerge and eventually form a trend. Learning such a trend is helpful in reorganizing the subsystems or selecting the most reasonable and representative design scenarios that maximize the usage of the interrelationships among the subsystems. Such a learning process is to improve the formulation through exploration and evaluation of the model and solution space.

In this chapter, we divide the methods that deal with multi-goal (multi-objective) problems into two categories, namely, focusing on the performance of the solution algorithms and focusing on design improvement. Based on the limitations of both categories—the limitations of the Pareto front, especially in engineering-design problems, and the heavy reliance on domain knowledge, we propose the adaptive leveling-weighting-clustering (ALWC) algorithm (Figs. 6.9 and 6.13) to use the dataset that consists of (but not limited to) design scenarios, goal achievement, and solutions to learn the system nature and obtain useful information to improve the formulation. Information such as subsystems and their interrelationships is helpful in identifying the most appropriate and representative design scenarios to reorganize the subsystems and speed up the learning process.

In other words, we realize the connections among formulation, deduction, decision, and action through the activities in the circles, as shown in Fig. 6.17. The overview and flow chart of the method are illustrated in Figs. 6.9 and 6.13. By realizing the Algorithms 1–3 in Sect. 6.4.3, we implement the ALWC algorithm in the context of the Rankine cycle thermal system problem.

6.6.2 Summarizing How We Realize Type I, III, and IV Robust Design

For the test problem, designing the Rankine cycle thermal system, Type I uncertainty is identified as the various results due to the changes in the starting point. Therefore, we call the "XPLORE" module in DSIDES—I details are introduced in Sect. 2.3.3 and Glossary II.2.1 of (Hajihashemi, 2023)—to identify a local area that is feasible and has relatively good goal achievement to start solving the cDSP.

Type III uncertainty, defined as the uncertainty in model approximation (ways of combining multiple goals), is managed using the three algorithms in the ALWC algorithm: Algorithm 1 Leveling, Algorithm 2 Weighting, and Algorithm 3 Clustering; see Fig. 6.13. By exploring different leveling and weighting formats to combine the multiple goals that represent the interests of subsystems, the goals can be clustered, and the subsystems can be identified, and their interrelationships can be learned. Therefore, the combination of the goals can be updated and the most appropriate and representative design scenarios (which in this test problem is the combination of goals) can be selected. Using the selected design scenarios instead of enumerating all of them can accelerate learning and reduce computational complexity.

Type IV uncertainty, the uncertainty in using domain knowledge to simplify the model, for Test Problem 3, has two manifestations, namely, fixing decision variables based on domain expertise and selecting design scenarios using experience. The first uncertainty is managed using "fixing variables" and "XPLORE" modules. In

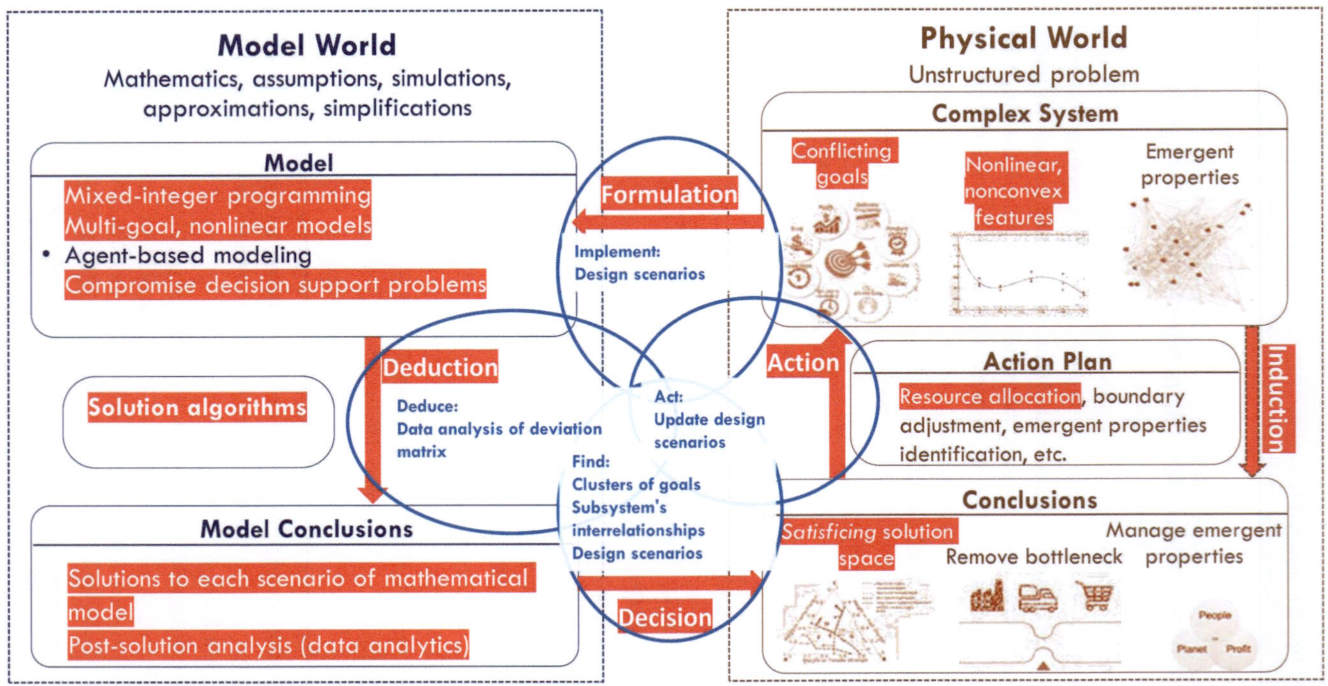

Fig. 6.17 The procedures involved in formulation-exploration-evaluation—Establish the information exchange, knowledge awareness, and instructions sharing among formulation deduction, decision, and action

DSIDES, when applying both modules to a variable, we first use an interior-point search to identify the best value for a variable and then fix it, by treating it as a deterministic parameter. If we fix more than one variable, then the best combination of the variables is returned as a vector and those variables' values are fixed by implementing the vector. This significantly reduces the computational complexity especially when there are discrete variables. The second uncertainty is managed through the leveling and clustering algorithm in the ALWC algorithm. By clustering the goals and leveling them based on their clusters, we select the design scenarios that allow boosting the system performance.

By implementing the ALWC algorithm as an extension of DSIDES, we can identify the solution space that is relatively insensitive to the Type I, III and IV uncertainty that we need to manage in a specific problem. In this way, we realize Type I, III and IV robust design. A summary of issues associated with Test Problem 3 is provided in Table 6.10.

6.6.3 Role of this Chapter in the Monograph

In this chapter, we provide a frame of reference for modeling constructs and solution algorithms for multi-goal or multi-objective problems, with a particular focus on exploring the interrelationships among subsystems and the combination of goals, which is an extension of the frame of references in Chap. 2. The Adaptive Leveling-Weighting-Clustering (ALWC) algorithm is proposed to explore the subsystems of the system and the design scenarios. A test problem, designing a Rankine cycle thermal system, is used to verify the proposed methods. It is shown that using the ALWC algorithm, the interrelationships among the subsystems can be learned, and that the most representative design scenarios that can reasonably organize the subsystems to boost the system's efficiency can be identified and applied. Research Question 3 is addressed. The "Multi-Goal and Many-Goal Management" piece of the "Design Evolution Cycle" is completed (see Fig. 6.18).

Table 6.10 Summary of test problems 3 regarding type I, II, and IV uncertainty management

RD Type	RDI-II				
			RDIII	RDIII	
				RDIV	RDIV
Method	M1: Formulation-Exploration Framework		M2: Adaptive Linear Programming with Parameter Learning (ALPPL)	**M3: Adaptive Leveling-Weighting-Clustering Algorithm (ALWC)**	M4: Scenario Planning in Agent-Based Modeling
Chapter	Ch 4		Ch 5	**Ch 6**	Ch 7
Uncertainty \ Test Problem	T1: Dam network	T2: Supply chain	T3: Hot rolling process chain	**T4: Thermal system**	T5: Promoting second season farming
Type I	*Uncertainty in timing and amount of inflow – Table 4.8*	*Uncertainty in demand side – Figure 4.25*	*Uncertainty in hyper parameter setting – Table 5.8, Figure 5.16*	***Uncertainty in parameter setting in solution algorithm (Starting point of searching) – call XPLORE in DSIDES***	Uncertainty in price (Price of agriculture products)
Type II	*Uncertainty in outflow (water release target) – Table 4.6*	*Uncertainty in supply side - Table 4.16*	*Uncertainty in user preferences – Table 5.7*		*Promotion effort and timing*
Type III			*Uncertainty in model approximation due to heuristics in approximation – Table 5.6*	***Uncertainty in model approximation (ways of combining multiple goals) – leveling-weighting-clustering algorithms (Figure 6.13)***	
Type IV				***Uncertainty in using domain knowledge to simplify the model (fixing decision variables and selecting design scenarios) – "fixing variable" and "XPLORE" modules in DSIDES, and leveling and clustering algorithm in the ALWC***	*Interventions that change the mathematical relation among promotion and result (developing local market)*

EVe – empirical verification of the method
M – method
RD – robust design
T – test problem

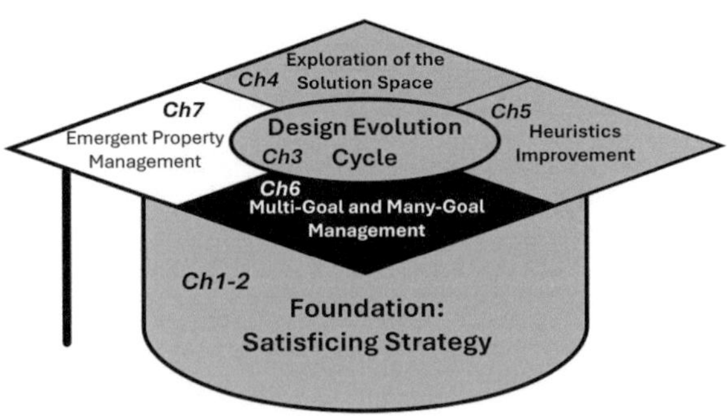

Fig. 6.18 The role of this chapter in the knowledge foundation

Glossary

Archimedean	A strategy of managing multiple goals by compromising the achievement of the goals. It is also known as the weighted sum function. Goals are weighted and combined to form a compromise goal (Guéret et al., 1999; Ignizio, 1976).
Concurrent design	An integrated design method where a designer should use various methods, technologies, and strategies (Gao et al., 2012).
Deviation	Deviations are measured from these goals both above and below the target. Unwanted deviations from this set of target values are then minimized in an achievement function.[8]
Diversity of solutions	Diversity of solutions refers to the breadth and variety of available design options. High solution diversity indicates an expanded range of innovative and adaptable choices, enabling responsiveness to evolving preferences, needs, or constraints in the design process.
Goal	In this chapter, the term "goal" is used to refer to the objective of a design problem when its target value is determined. The problem is solved by minimizing the deviation between the achieved value and the target value of the goals. In other words, the problem is solved by maximizing the goal achievement rate.
Goal achievement rate	The rate at which the target of a goal is achieved.
Many goal problems	Problems with more than three goals.
Pre-emptive	A strategy of managing multiple goals by decentralizing the problem. Also known as Lexicographic (Kortanek and Maxwell, 1969). The goals are placed at multiple levels of priority. The first level goal function will be satisfied as far as possible, and then while holding it within a tolerance, the second level goal function will be addressed, and so on, in an attempt to address all the goals across all levels (Ignizio, 1976).
Robustness	The capability of a system to be insensitive to variations or uncertainties.
Satisficing	A decision-making strategy or a cognitive heuristic entails searching through the available alternatives until an acceptability threshold is met (Byron, 1998).

[8] From Wikipedia, Goal Programming: https://en.wikipedia.org/wiki/Goal_programming#:~:text=Goal%20programming%20is%20a%20branch,criteria%20decision%20analysis%20(MCDA).&text=Deviations%20are%20measured%20from%20these,minimised%20in%20an%20achievement%20function.

Satisficing solutions	Solutions that are not necessarily optimal but good enough by minimizing the distance between what the system can achieve and what the ideal case should be (Simon, 1996).
Scalarization	The reduction of multiple goals to a single function that can be solved using a single goal. The most common scalarization methods are lexicographic ordering (Pre-emptive) and weighted sum function (Archimedean).

References

Bader, J., & Zitzler, E. (2011). HypE: An algorithm for fast hypervolume-based many-objective optimization. *Evolutionary Computation, 19*(1), 45–76.

Bandaru, S., Ng, A. H., & Deb, K. (2017). Data mining methods for knowledge discovery in multi-objective optimization: Part A-survey. *Expert Systems with Applications, 70*, 139–159.

Byron, M. (1998). Satisficing and optimality. *Ethics, 109*(1), 67–93.

Chen, W., Allen, J. K., & Mistree, F. (1997). A robust concept exploration method for enhancing productivity in concurrent systems design. *Concurrent Engineering, 5*(3), 203–217.

Choi, H.-J., Austin, R., Allen, J. K., McDowell, D. L., Mistree, F., & Benson, D. J. (2005). An approach for robust design of reactive powder metal mixtures based on non-deterministic micro-scale shock simulation. *Journal of Computer-Aided Materials Design, 12*(1), 57–85.

Deb, K., Pratap, A., Agarwal, S., & Meyarivan, T. (2002). A fast and elitist multiobjective genetic algorithm: NSGA-II. *IEEE Transactions on Evolutionary Computation, 6*(2), 182–197.

Gao, Y., Wang, F.-Y., Zhao, Z.-Q., & Xiao, Z.-Q. (2012). *Flexible manipulators: Modeling*. Academic Press.

Guéret, C., Prins, C., & Sevaux, M. (1999). Applications of optimization with Xpress-MP. Translated from the french edition of Programmation Linéaire, by S. Heipcke, Published by Dash Optimization, Ltd., Northants, UK.

Guo, L., Allen, J. K., & Mistree, F. (2024). Optimize or satisfice in engineering design? *Research in Engineering Design, 35*(3), 239–267.

Hajihashemi, S. (2023). A user-friendly wrapper for DSIDES (Decision Support in the Design of Engineering Systems). Industrial and Systems Engineering, University of Oklahoma. MS Thesis.

Hao, F., & Merlet, J.-P. (2005). Multi-criteria optimal design of parallel manipulators based on interval analysis. *Mechanism and Machine Theory, 40*(2), 157–171.

Ignizio, J. P. (1976). An approach to the capital budgeting problem with multiple objectives. *The Engineering Economist, 21*(4), 259–272.

Jaimes, A. L., Coello, C. A. C., & Barrientos, J. E. U. (2009). Online objective reduction to deal with many-objective problems. In M. Ehrgott, C. M. Fonseca, X. Gandibleux, J. K. Hao, M. Sevaux (Eds.) *Evolutionary multi-criterion optimization. EMO 2009*. (Vol. 5467). Lecture notes in computer science. Berlin, Heidelberg: Springer.

Kortanek, K., & Maxwell, W. (1969). On a new class of combinatoric optimizers for multi-product single-machine scheduling. *Management Science, 15*(5), 239–248.

Lemmon, E. W., & Huber, M. L. (2013). *Implementation of pure fluid and natural gas standards: Reference fluid thermodynamic and transport properties database (REFPROP)*. National Institute of Standards and Technology, NIST.

Macquorn Rankine, W. J. (1853). XVIII. On the general law of the transformation of energy. *The London, Edinburgh, and Dublin Philosophical Magazine and Journal of Science, 5*(30), 106–117.

Messac, A., & Mattson, C. A. (2002). Generating well-distributed sets of Pareto points for engineering design using physical programming. *Optimization and Engineering, 3*(4), 431–450.

Mistree, F., Hughes, O., & Phuoc, H. (1981). An optimization method for the design of large, highly constrained complex systems. *Engineering Optimization, 5*(3), 179–197.

Mistree, F., Hughes, O. F., & Bras, B. (1993). The compromise decision support problem and the adaptive linear programming algorithm. In M.P. Kamat (Ed.) *Structural optimization: Status and promise* (pp 247–286). Washington, DC: AIAA.

Mistree, F., Patel, B., & Vadde, S. (1994). On modeling multiple objectives and multi-level decisions in concurrent design. *Advances in Design Automation, 69*(2), 151–161

Parra, M. A., Terol, A. B., & Urıa, M. R. (2001). A fuzzy goal programming approach to portfolio selection. *European Journal of Operational Research, 133*(2), 287–297.

Rae, A. N. (1972). *A note on the solution of goal programming problems with pre-emptive priority.* Massey University.

Reynoso-Meza, G., Blasco, X., Sanchis, J., & Herrero, J. M. (2013). Comparison of design concepts in multi-criteria decision-making using level diagrams. *Information Sciences, 221*, 124–141.

Sabeghi, M., Shukla, R., Allen, J. K., & Mistree, F. (2016) Solution space exploration of the process design for continuous casting of steel. In *ASME Design Automation Conference*, Paper Number IDETC2016-59399

Schaffer, J. D. (1985). Multiple objective optimization with vector evaluated genetic algorithms. In *Proceedings of the First International Conference on Genetic Algorithms and Their Applications*, Carnegie-Mellon University, Pittsburgh, PA (pp. 93–100). Lawrence Erlbaum Associates. Inc., Publishers.

Seada, H., & Deb, K. (2014). U-NSGA-III: a unified evolutionary algorithm for single, multiple, and many-objective optimization. COIN report 2014022.

Simon, H. A. (1996). *The sciences of the artificial.* MIT Press.

Smith, W. F., Milisavljevic, J., Sabeghi, M., Allen, J. K., & Mistree, F. (2015). The realization of engineered systems with considerations of complexity. In *ASME Design Automation Conference*, Boston, MA, Paper Number DETC2015-46211.

Soltani, A. R., Tawfik, H., Goulermas, J. Y., & Fernando, T. (2002). Path planning in construction sites: performance evaluation of the Dijkstra, A∗, and GA search algorithms. *Advanced Engineering Informatics, 16*(4), 291–303.

Tang, D., Zhu, R., Tang, J., Xu, R., & He, R. (2010). Product design knowledge management based on design structure matrix. *Advanced Engineering Informatics, 24*(2), 159–166.

Vevea, J. L., & Woods, C. M. (2005). Publication bias in research synthesis: Sensitivity analysis using a priori weight functions. *Psychological Methods, 10*(4), 428.

Wang, F.-Y. (1994). On the extremal fundamental frequencies of one-link flexible manipulators. *The International Journal of Robotics Research, 13*(2), 162–170.

Wang, R., Nellippallil, A. B., Wang, G., Yan, Y., Allen, J. K., & Mistree, F. (2018). Systematic design space exploration using a template-based ontological method. *Advanced Engineering Informatics, 36*, 163–177.

Wikipedia. (2019). Rankine Cycle. Retrieved June 1, 2025, from https://en.wikipedia.org/wiki/Rankine_cycle.

Zhang, Q., & Li, H. (2007). MOEA/D: A multiobjective evolutionary algorithm based on decomposition. *IEEE Transactions on Evolutionary Computation, 11*(6), 712–731.

Zitzler, E., Laumanns, M., & Thiele, L. (2001). SPEA2: Improving the strength pareto evolutionary algorithm. TIK Report 103, Published by ETH Zurich, Computer Engineering and Networks Laboratory.

Chapter 7
Type I, II, and IV Robust Design Through Emergent Properties Identification and Interpretations

Exploring Critical Factors Through Scenario Planning in Agent-Based Modeling

Imagine you are a social entrepreneur tasked with bringing prosperity through sustainable development to a rural Indian village where life revolves around the monsoon season. In this village, villagers can only grow one annual crop because of the seasonal rains. The river's water supply dwindles when the dry season comes, leaving the fields barren. This forces many villagers to migrate to distant places for work in the dry season, disrupting family stability and reducing their quality of life.

Faced with these challenges, you envision a transformative solution: growing a second crop during the dry season. By using underground water, you aim to enable the villagers to grow an additional crop during the dry season. However, the lack of electricity to power water pumps has always been a barrier. You propose a solar-powered micro-grid solution to generate the necessary power for the community to pump underground water and support two-season farming.

To ensure the success of this initiative, you use agent-based modeling (ABM) to simulate different scenarios and understand how the villagers might respond to this change. By exploring different outcomes, you aim to identify critical factors supporting the adoption of this new practice, while respecting the community's unique socioeconomic context.

In this chapter, we explore how connecting all stages in the design evolution cycle—formulation, approximation, exploration, and evaluation—can resolve social problems, and how leveraging simulation tools such as ABM can provide robust decision support. We also explore how leveraging simulation tools such as ABM can provide robust decision support, offering social entrepreneurs and researchers insights into utilizing technology and customizable frameworks to address challenges and foster sustainable development.

L. Guo et al., *Architecting a Satisficing Strategy for Designing Complex Systems*,
https://doi.org/10.1007/978-3-032-04577-5_7

7.1 Organization of This Chapter

The organization of this chapter is illustrated in Fig. 7.1.

In Sect. 7.2 we introduce the problem of designing promotions using agent-based modeling is framed and limitations are identified. In Sect. 7.3, we describe the test problem, the motivation and challenges of learning emergent properties in promoting second-season cultivation in an island village in India. In Sect. 7.4, based on the research gaps described in Sect. 1.5 and Method 4 proposed in Sect. 3.3.4, we introduce a framework for scenario planning in agent-based modeling for exploring critical factors in detail. In Sect. 7.5, the proposed method is applied to the second-season cultivation promotion simulation problem for critical factor identification and learning about emergent properties. In Sect. 7.6, we summarize the role of this chapter. In Table 7.1 we summarize what is contained in this chapter in terms of "what, how, and why".

The plan for specifying and answering Research Question 4 in the context of the test problems is shown in Table 7.2. In this chapter, the Proposed Method 4 (M4), the scenario planning framework in agent-based modeling, is empirically verified (EVe4) using a test problem, designing the promotion of second-season cultivation in a rural community (T4). Research Question 4 (RQ4) is defined in the context of the test problem (SQT4) and answered (AQ4) by using M4. The empirical validation and theoretical validation are in Chaps. 8 and 9.

In this chapter, Research Question 4 (RQ4), "*What is a method that allows passing the information through multiple scales of a system?*" is defined for the test problem, designing the promotion of second-season cultivation in a rural community (T4), as follows (SQT4). It is then further broken into two sub-questions indicating their tasks and their answers.

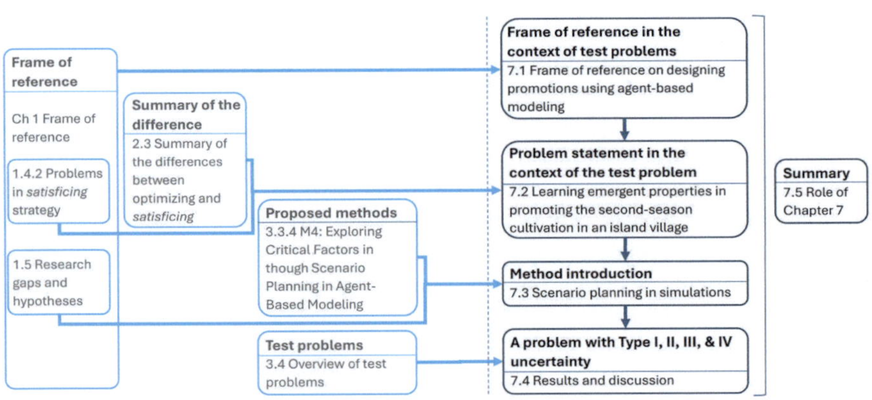

Fig. 7.1 Organization of this chapter

Table 7.1 Summary interpretation of the problems investigated in this chapter

Elements	What?
Components	The formulation, approximation, exploration, and evaluation stages of the design evolution cycle
Connectors	The proposed scenario planning framework for agent-based modeling allows designers to recognize the possibility of connecting all four stages in the design loop and learning knowledge in one stage to replace heuristics applied in another
Form	How?
Component roles	Provides one possible conceptual framework for connecting the four stages
Properties	The coherence, consistency, and mutual constraints of the four stages—heuristics, hypotheses, and scenarios—are used, tested, and verified or denied so that better settings are identified and applied
Relationship	The relationships among the activities—scenarios installation, evaluating and quantifying the model performance associated with the scenarios, and use of association to evolve the design
Rationale	Why?
Motivation	Provide the feasibility of realizing all four types of robust design by using simulation and scenario planning
Assumptions	The complex systems that consist of (or can be decomposed as) agents can be designed, learned, and evolved using systematic simulations and scenario planning
Interpretation	Tentatively understanding how agents' behaviors can be generated, monitored, predicted, and directed to benefit a system

SQT4: What method facilitates recognizing emergent properties in a system and identifying critical factors to reach system goals under these emergent properties?

SQT4.1: What critical factors influence collective behavior under interventions?

SQT4.2: How can we identify the critical factors and select an appropriate scenario to achieve an expected result?

We hypothesize that interactions among the formulation, exploration, approximation, and evaluation stages of designing should be studied and the mechanisms for information sharing and intervention among the four stages should be established; see Fig. 7.2.

Table 7.2 Plan of specifying Research Question 4 (RQ4) and empirically verifying the scenario planning framework in agent-based modeling (M4)

Chapter	Ch 1	Ch 2	Ch 3	Ch 4–7		Ch 8	Ch 9
				Ch 4–6	Ch 7		
Actions	RG H	RD RQ SH	TVe M	EVe 1–3 SQT 1–3 AQ 1–3	EVe4: Use a test problem to design the promotion in a social network to verify SH4 and demonstrate M4	CQ EVa	TE
					SQT4: What method facilitates recognizing emergent properties in a social network and identifying critical factors to reach promotion goals under the emergent properties? SQT4.1: What critical factors affect the collective behavior under interventions? SQT4.2: How can we identify the critical factors and select the appropriate scenario to reach an expected result?		
					AQ4: Using the proposed scenario planning framework, designers can perform simulations, identify critical factors in their systems, and select specific scenarios that accommodate different site-specific input values or domain-dependent knowledge to reach their goals		
Nomenclature	AQ—answer research questions CQ—closure of the answers to research questions Eva—empirically validate hypotheses Eve—empirically verify hypotheses H—give hypotheses M—introduce methods RD—tie to roust design RG—give research gaps RQ—pose research questions SH—specify hypotheses SQT—specify research questions in the context of test problems TVe—theoretically verify hypotheses TE—theoretically extend the research						

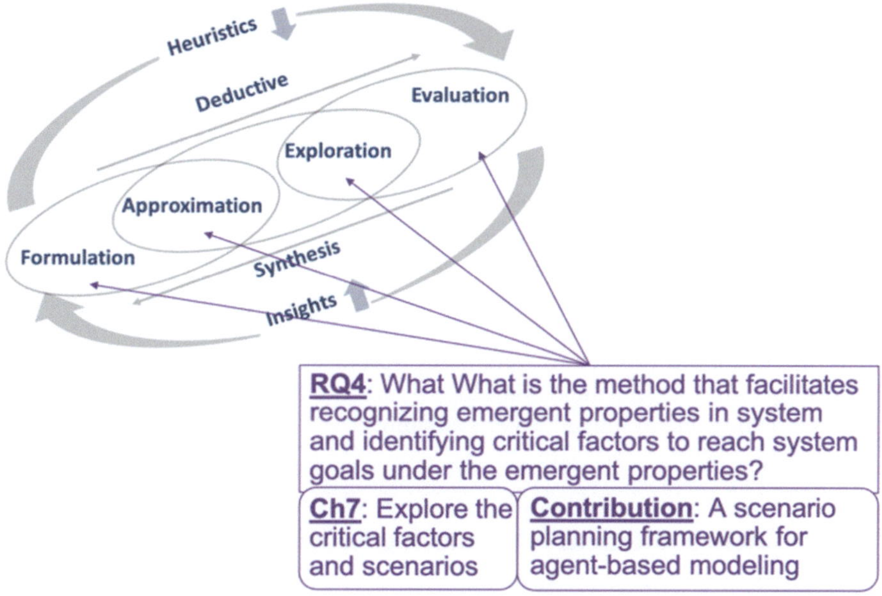

Fig. 7.2 Specified Research Question 3 and the relevant stages to be connected in the design evolution cycle

7.2 Frame of Reference on Designing Promotions Using Agent-Based Modeling

Changing people's lifestyles when new technology is available is challenging, especially in relatively closed and underdeveloped areas (Stewart, 2016; Wilhelm, 2000). Modeling methods and simulations can be used to predict people's acceptance and adoption of a new lifestyle or technology. However, the simulation verification and the utility of the method often need to be considered (Chitungo and Munongo, 2013; Parker et al., 2003). Due to unavoidable errors and flaws in modeling (Box, 1976) and due to the complexity of the model environment, methods such as sensitivity analysis (Abreu and Ralha, 2018; Guo et al., 2019) and Monte Carlo analysis (Lumbroso and Davison, 2018) are often used to manage uncertainties and provide decision support in changing environments. For the design of sociotechnical systems, where the initial data is lacking or difficult to verify and quantify (Afshari and Peng, 2015) and when the future is uncertain, we need a method to explore the variability of the model results and identify sensitive factors that contribute to variabilities (Abreu and Ralha, 2018).

From the literature on simulating social behaviors and providing decision support for policy-making, agent-based modeling (ABM) is a practical tool (Qiu et al., 2018). There are many examples of the use of ABM to observe patterns in collective behavior, identify critical factors, and intervene in the system by changing critical factors. These include exploring the extent of influence, or radius of influence, among neighbors concerning promotion actions (Opiyo, 2019), identifying critical factors that contribute to population dynamics (Qiu et al., 2018), estimating the effects of policy interventions on investment in new equipment (Al Irsyad et al., 2019), and providing decision support for managing potential labor reproduction (Rossoshanskaya, 2019). In Table 7.3, we list some representative publications related to capturing social behavior and leveraging critical factors to serve a promotion goal. We conclude, there needs to be more literature on decision support based on scenario planning. Therefore, we provide scenario-planning-based decision support to social entrepreneurs to help them achieve socio-economic goals in various situations. We use a test problem of promoting second-season cultivation: growing a second crop each year in an underdeveloped, rural Indian village previously limited to monocropping due to limited water availability.

7.3 Problem Statement—Promoting Second-Season Cultivation in an Island Village in India

– ***Test Problem 4: On applying scenario planning in agent-based modeling to a social system design problem***

Our objective is to give decision support to social entrepreneurs promoting second-season cultivation in Kudagaon, a relatively isolated and underdeveloped village in Odisha, India. Kudagaon is a village on an island in a river (Fig. 7.3). There are 85 households in the village, and each household has some farmland. The region experiences a rainy season (or monsoon season) and a dry season. Most households practice one-season cultivation, growing rice or vegetables in the rainy season using the water from the river. In the dry season, the river's water level drops significantly, preventing most families from farming because of water scarcity. Therefore, two-season cultivation is impossible, and the villagers cannot increase their family savings through it. Kudagaon has sufficient underground water for farming in the dry season, provided electric power is available. However, villagers need to purchase or rent equipment to pump the underground water and transport it to their farmland. Since most families lack adequate savings, many cannot afford the equipment. This situation presents a dilemma.

When a household grows crops only once a year, the main laborers migrate during the dry season to obtain daily wages. These daily wages are the only source of income for their families during the dry season. This is referred to as "migration income." Migration income is usually less than the income earned from second-season cultivation, assuming sufficient water. Additionally, the well-being of families with migration workers (referred to as "migration families") in terms of family stability and social status is worse than that of families who practice two-season cultivation (referred to as "cultivation families". During the rainy season, families who

Table 7.3 Some representative applications of Agent-Based Modeling (ABM) for new technology acceptance and policy impact (Guo et al., 2020)

Author and year	Problem description	Method	Results	Contribution
Opiyo (2019)	Study neighborhood influence and social pressures on temporal diffusion of solar home systems	Agent-based modeling with survey data	Visibility of newly installed SHS and increasing influence radius leads to growth in SHS installations	The survey method is helpful to acquire relatively quantifiable data for a social problem
Qiu et al. (2018)	Simulate urban land development and population dynamics	An agent-based and spatial genetic algorithm framework (PDULD)	Government policies dominate the process of land development	Community cohesion theory is introduced into the model; historical data are used to verify the results
Al Irsyad et al. (2019)	Estimate the effects of four solar energy policy interventions on photovoltaic (PV) investments, government expenditure, economic output, etc.	Uses hybrid energy-agent-based modeling	The results call for a PV donor gift policy, improved production efficiency, after-sales services, and rural financing institutions	Integrate input–output analysis, environmental factors, and socioeconomic characteristics of households in Indonesia
Rossoshanskaya (2019)	Simulate labor potential reproduction; among the scenario forecasts, provide decision support on management actions	Agent-based modeling with multi-agents and multi-scenarios	The integrated agent-based model of labor potential reproduction at the municipal level	The model is filled with sociological and statistical data and has a user-friendly interface

practice one-season cultivation on farmlands away from the river may still consider growing crops in the dry season based on the specific climate and market situation each year. They can lease equipment and water pumps to access underground water for second-season cultivation, but the cost and risk are high. If they anticipate gaining more profit from second-season cultivation than from migration work, they may stay and grow second-season crops instead of migrating. Currently, the probability of a household that practices one-season cultivation switching to two-season cultivation

Fig. 7.3 The satellite map of Kudagaon (Guo et al., 2020)

is as low as 5%. This number is provided by our colleagues, the social entrepreneurs at SunMoksha, based on historical data and interviews with the villagers.

The social entrepreneur plans to provide farmers with initial investments as loans to construct public infrastructure, enabling the generation of electrical power and access to underground water, as well as transporting water and electricity to make farming possible in the dry season. The social entrepreneur plans to provide farmers with initial investments as loans to construct public infrastructure so that electrical power can be generated and underground water can be obtained, as well as transporting the water and electricity to make farming possible in the dry season. Underground water and electricity are provided to farmers as utilities using affordable, tiered pricing. After profiting from second-season cultivation and accumulating savings, farmers repay the loans within several years. In Fig. 7.4, we illustrate how the social entrepreneur's initial investment helps make second-season cultivation feasible.

Acceptance of second-season cultivation is crucial for the success of this project. Therefore, in addition to loans and technical support, promoting second-season cultivation among farmers is important. The benefits of second-season cultivation include improving a family's economic and social status with higher and more stable income and keeping the family together. However, not all villagers realize these benefits, and

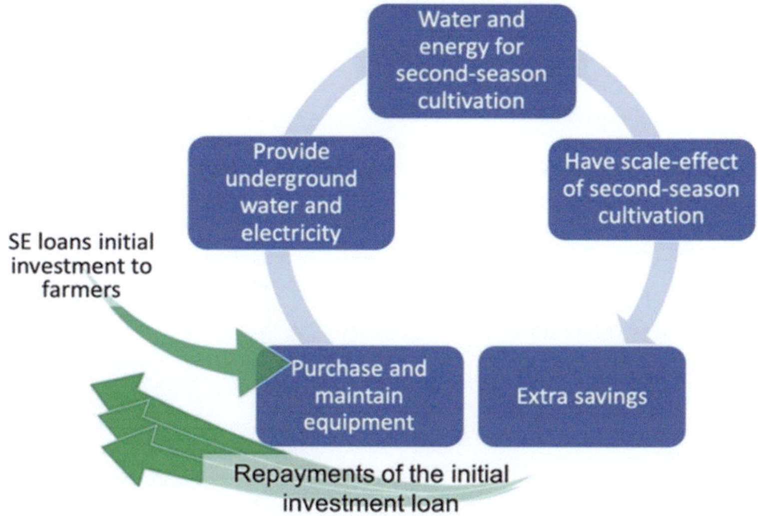

Fig. 7.4 The SE's plan for facilitating second-season cultivation (Guo et al., 2020)

there is a considerable reluctance to change their mode of production, lifestyle, and source of income.

The conceptual compromise Decision Support Problem of the project is:

Given

Factors [Scenarios]

Find

Critical Factors [Appropriate scenarios];
$Output_{Factor_i} = f\left(Scenario_j\right)$

Satisfy

Simulation goals
Simulation constraints

Minimize

Scenario combination

Based on the dilemma and the social entrepreneur's difficulties, we summarize the social entrepreneur's possible actions and targets in Table 7.4. Each target is defined to improve the villagers' social or economic status or both. We hypothesize that increasing the number of two-season cultivation households can create a scale effect due to stable market demand and lower unit costs of storage and transportation, thereby increasing villagers' income and enhancing their economic status. The social entrepreneur's job is to boost villagers' economic and social status in both the short

Table 7.4 The social entrepreneur's target based on the current situation (Guo et al., 2020)

Current problem	Reason or dilemma	SE's action	Target	Category
Only β_1 of the households anticipate a better profit from second-season cultivation, so they grow twice a year $\beta_1 \approx 5\%$	Farmers cannot afford the equipment to acquire underground water for farming in the dry season and previously electricity has been unavailable	Having promotion activities[1]; providing underground water;	Raising the second-season cultivation rate from β_1 to β_2 in the promotion year	Short-term economic status—increase profit level
			Maintaining the second-season cultivation rate of β_3 after 2 years following the promotion year	Long-term economic status—stability and reliability of income sources
$(1 - \beta_1)$ of the households do not anticipate a better profit the second-season cultivation, so they grow only once a year and migrate in the dry season			Reducing the migration rate from $(1 - \beta_1)$ to $(1 - \beta_2)$ in the promotion year	Short-term social status—the confidence to make an improvement
			We are maintaining the migration rate at $(1 - \beta_3)$ 2 years after the promotion year	Long-term social status—the family stability and sense of security and self-sufficiency

and long term. In the short term, the social entrepreneur aims to improve the two-season cultivation rate from β_1 to β_2, and in the long term, the social entrepreneur wants to improve β_1 to β_3. β_1 is a static rate based on historical data. β_2 and β_3 are targets that the social entrepreneur aims to reach. Different scenarios exist for obtaining β_2 and β_3. In this chapter, we offer suggestions on promotion efforts and their durations to reach each scenario of β_2 and β_3.

In Sect. 7.4, we introduce modeling and scenario development, discuss the results of the scenario planning in Sect. 7.5, and summarize the role of this chapter in this monograph in Sect. 7.6.

[1] The promotion activities include holding community discussions, visiting each household in person, giving training on second-season cultivation and strategies to conserve water, etc. As Kudagaon is a small village with only 85 households, SEs can visit and interview each household.

7.4 Modeling and Scenario Development

7.4.1 *Build the Architecture and Set the Baseline Scenario of the Agent-Based Model*

We simulate the households' behavior using agent-based modeling (ABM) to capture the factors that impact the promotion effects. AnyLogic 8 PLE software is used for the simulation. Kudagaon is a single community with a relatively flat social hierarchy, so each household is defined as an agent, and a single type of agent is used. Social influence and the influence of each family's neighbors are randomized within the same numerical range, although they vary from household to household.

The project's duration is 3 years. We simulate the households' behavior for 4 years. To establish a baseline, no intervention occurs in the first year to simulate the villagers' behavior before the project is launched. This pre-project simulation allows us to verify the model using current actual data. The social entrepreneurs' promotion begins in the second year, during the rainy season. The third and fourth years are post-promotion, allowing for tracking the long-term effects of the social entrepreneur's promotion.

States and transitions between states. An agent (household) has 11 possible states, as follows.

$S_1 for1S$	Growing once a year before social entrepreneur's promotion
*MigrationWork*1	Migrating during the dry season, before the promotion
$S_1 ToBePromoted$	Growing once a year in the social entrepreneur's promotion year
BeingPromoted	Being promoted to grow twice a year by social entrepreneurs
*MigrationWork*2	Migrating during the dry season after being promoted
$S_1 for1SAfter$	Growing once a year after the social entrepreneur's promotion year
$S_1 for2Ss$	Growing the first season crops for a two-season cultivation year
*Season*2	Doing a second season of growing in the two-season cultivation year
Harvest	Harvesting the second-season crops
Profit	Gaining at least the expected profit or even more from second-season cultivation
NoProfit	Gaining less profit than expected from second-season cultivation

The transitions between states are described in Table 7.5. The flowchart of the agents' state transition is shown in Fig. 7.5. The trigger condition and the duration of each transition are illustrated in Fig. 7.6.

Table 7.5 Transitions between different states for an agent (Guo et al., 2020)

Transition	Meaning
$S_1\ for\ 1S \overset{\beta_1}{\rightarrow} Season2$	Before social entrepreneur's promotion, with probability β_1, a one-season cultivation household will change to two-season cultivation $\beta_1 = 5\%$
$S_1\ for\ 1S \overset{(1-\beta_1)}{\rightarrow} MigrationWork1$	Before social entrepreneur's promotion, with probability $(1 - \beta_1)$, a one-season cultivation household will migrate during the dry season
$BeingPromoted \overset{\beta_2}{\rightarrow} Season2$	During the social entrepreneur's promotion year, with probability β_2, a household being promoted will change to two-season cultivation $\beta_2 \gg \beta_1$
$BeingPromoted \overset{(1-\beta_2)}{\rightarrow} MigrationWork2$	During the social entrepreneur's promotion year, with probability $(1 - \beta_2)$, a household being promoted will migrate during the dry season
$Harvest \overset{\alpha}{\rightarrow} Profit$	With probability α, a household will reach at least the anticipated profit through second-season cultivation and decide to do two-season cultivation next year
$Harvest \overset{(1-\alpha)}{\rightarrow} NoProfit$	With probability $(1 - \alpha)$, a household will not reach their anticipated profit through second-season cultivation and decide to do one-season cultivation next year
$S_1\ for\ 1SAfter \overset{\beta_3}{\rightarrow} Season2$	After the social entrepreneur's promotion year, with probability β_3, a one-season cultivation household will change to two-season cultivation $\beta_3 = \beta_2 \cdot \alpha$ (1)
$S_1\ for\ 1SAfter \overset{(1-\beta_3)}{\rightarrow} MigrationWork2$	After the social entrepreneur's promotion year, with probability $(1 - \beta_3)$, one-season cultivation households will migrate during the dry season

7.4.2 Scenario Development

A factor refers to any source of uncertainty, including the model structure, initial conditions, and input parameters. In this project, based on the basic knowledge from the social entrepreneur, we define the network structure, social entrepreneur's promotion effort, promotion duration, villagers' anticipation of profit, and actual profit as factors to be explored. Different scenarios are developed for each factor through conversations between an employee of SunMoksha, Ayushi Sharma, and research assistants in the Systems Realization Laboratory, Lin Guo and Vishnu Kamala, based on their experience, domain knowledge, and assumptions. Through scenario planning, the sensitivity of the results to each factor is analyzed, and critical factors

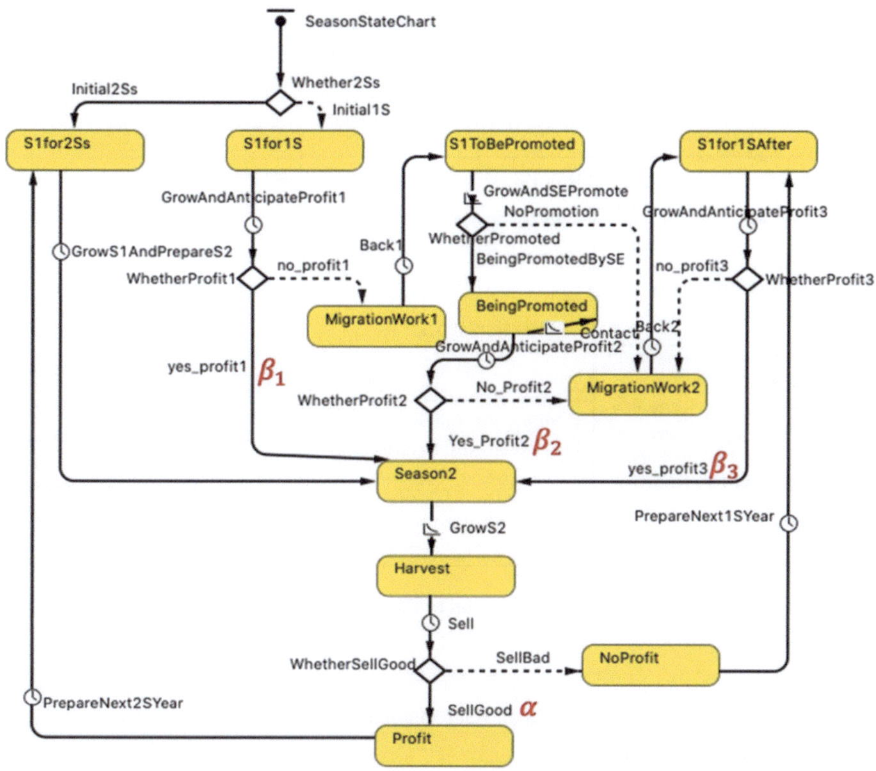

Fig. 7.5 The flowchart of the agents' state transitions (Guo et al., 2020)

are identified. In Table 7.6, we list the scenarios related to each factor. The expected outcome of scenario planning and how to obtain these outcomes are given in Table 7.7.

7.5 Results and Discussions

We explore the network type, promotion effort and duration, anticipation level of profit, and actual profit level. The results indicate that acceptance of second-season cultivation is insensitive to network type but is sensitive to promotion efforts. The factors that particularly affect the short-term results are promotion duration and villagers' anticipation; the factor that affects the long-term results is the actual profit that second-season cultivation produces. In Table 7.8, we summarize the results of the scenario planning. We detail our exploration process and observations in Sects. 7.5.1–7.5.3.

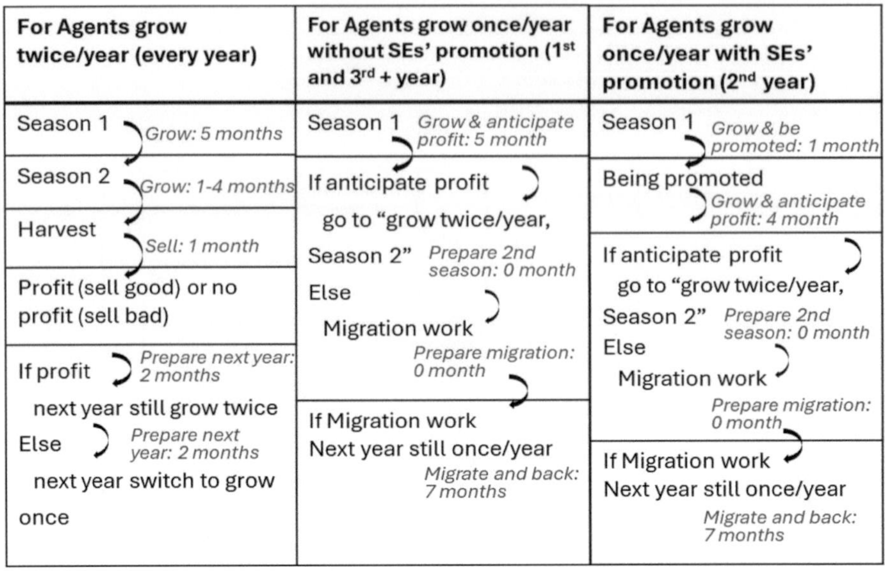

Fig. 7.6 The trigger conditions and the duration of the transitions between the states of the agents (Guo et al., 2020)

Table 7.6 Scenarios for testing each factor (Guo et al., 2020)

Factors	Scenarios
Network structure	Distance-based Scale-free
Promotion effort	Different percentages of households being promoted—10, 50, 75, 100%
Promotion duration	Different promotion durations—1, 2, 3, 4 months
Anticipation	Different positive anticipations of second-season profit—75, 95%
Profit	Different percentages of households who obtain a higher profit—75, 95%

7.5.1 *Exploring Network Type and Promotion Effort and Their Interaction Effects*

The social entrepreneur aims to reduce the migration population and increase the profit population. In Figs. 7.7 and 7.8, we present the results of two network types when the social entrepreneur's promotion reaches every household and 50% of households, respectively. In each graph, the horizontal axis represents time in months, and the vertical axis represents the number of households. The smooth, thick line represents the number of migration households during the dry season in the first year, the thin line with dots represents households that gain expected profit from second-season cultivation, and the thin line with squares represents the number of migration

Table 7.7 The expected outcome of the scenario planning

Expected outcome	The way to obtain the outcome		
Simulation results		Short-term effect	Long-term effect
	Improved economic state	The number of households that gain expected profit by growing twice a year in the *promotion year*	The number of households that gain expected profit by growing twice a year in *the fourth year*
	Improvement of social status	The number of households that migrate during the dry season during the *promotion year*	The number of households that are directly or indirectly promoted
Critical factors	Identify whether the simulation results are sensitive to scenario changes for each factor		
Variability of the simulation results	By changing the scenarios of the critical factors, obtain the range of the output of the model		
Identify how the critical factors affect the simulation results	Capturing qualitative and quantitative relationships among critical factors		

households during the dry season in and after the promotion years (the second, third and fourth year).

We explore different influence radii of the distance-based network, 50 m (Fig. 7.7a) and 100 m (Fig. 7.7b). We also explore different average degrees of the scale-free network, 5 (Fig. 7.7c) and 10 (Fig. 7.7d), which is the average number of connections in each household. We also explore two promotion efforts, reaching all households or 50% of households (the households can promote each other through the network), with a one-month duration.

Observations from exploring the type of network: Comparing Fig. 7.7a, b with c, d **we conclude that** neither the short-term nor the long-term results are sensitive to network types.

Observations from exploring the influence radius: Comparing Fig. 7.8a with b we conclude that if the village is a distance-based network, when the influence radius is larger, the short-term effect is slightly better, whereas the long-term effect is insensitive to the influence radius.

Observations from exploring the average degree of the network: Comparing Fig. 7.8c with d we conclude that when a village is a scale-free network, neither the short-term nor the long-term results are sensitive to the average degree, the average number of connections of each household.

Observations from exploring the interaction effect of network type and promotion effort: Only the short-term results are sensitive to promotion effort, whereas

Table 7.8 Summary of the scenario planning results

Factor	Controllable or not	Scenario	Meaning	Observation	Whether critical or not
Network type[a]	No	Distance-based	One community with strong neighborhood influences but weak or zero distant interactions	The scale-free network has slightly better results in the promotion year, but in the long-term, network type does not affect the promotion result; see Figs. 7.5 and 7.6	No
		Scale-free	One community has asymmetric influences between any two connected households, and the influence does not depend on distance		
Promotion effort	Yes—the social entrepreneur can control the promotion effort by reaching out to different numbers of households	Reaching different numbers of households: 10, 30, 50, 75, 100%	The promotion condition and target can be different from village to village, so searching for an appropriate rate for each village is necessary to reduce the promotion cost	The promotion result is always improved by increasing the promotion effort. The short-term effects are more sensitive to the promotion effort than the long-term effect; see Figs. 7.8 and 7.9 and Table 7.9	Yes, it is more critical for short-term effects

(continued)

Table 7.8 (continued)

Factor	Controllable or not	Scenario	Meaning	Observation	Whether critical or not
Promotion duration	Yes—social entrepreneur can perform the promotion for different lengths of time	The promotion duration may last from 1 to 4 months during the rainy season	A short promotion means a relatively short direct promotion but a long indirect promotion. This is suitable for a village with a strong mutual influence among households and vice versa	For short-term effects, long promotion has much better results; for long-term effects, short promotion and long promotion show similar results; see Fig. 7.10 and Table 7.10	Only critical for short-term effects
Villagers' anticipation and actual profit—β_2 of households anticipate a better profit through second-season cultivation, α of them gain actual profit as anticipated	Initially, $\boldsymbol{\alpha}$ and $\boldsymbol{\beta}_2$ are uncontrollable, but as the project goes on, the social entrepreneur can control them by improving productivity, developing more market demand, etc.	$\boldsymbol{\beta}_2 = 95\%$ $\boldsymbol{\alpha} = 95\%$	Optimistic profit anticipation; good economy, weather conditions, or soil fertility	The short-term result is more sensitive to profit anticipation; The long-term result is more sensitive to actual profit; see Fig. 7.11	Profit anticipation is critical for short-term effects; real profit is critical for long-term effect
		$\boldsymbol{\beta}_2 = 95\%$ $\boldsymbol{\alpha} = 75\%$	Optimistic profit anticipation; acceptable economy, weather condition, or soil fertility		
		$\boldsymbol{\beta}_2 = 75\%$ $\boldsymbol{\alpha} = 95\%$	Conservative anticipation: good economy, weather conditions, or soil fertility		

[a] A social entrepreneur needs to determine a village's network type through observation and statistics. The indicators to determine the network type may vary from village to village

Table 7.9 Promotion effort exploration—migration household in the promotion year and the end-of-project year with different network scenarios

Type and setting of the network	Promotion effort									
	10%		30%		50%		75%		100%	
	2nd year	4th year	2nd year	4th year	2nd year	4th year	2nd year	4th year	2nd year	4th year
Distance-based network With influence radius: 50 m	67	37	51	30	40	27	24	20	12	11
Distance-based network With influence radius: 100 m	66	37	51	28	40	27	26	20	9	13
Scale-free network The average number of connections for a household is 10	68	40	62	37	44	26	27	22	12	12

Table 7.10 Migration population (households) during the 4 years with different promotion durations

Promotion duration (month)	1st year	2nd year (promotion year)	3rd year	4th year
1	80	1–10	1–11	1–9
2	80	10	1–9	1–10
3	80	0–1	1–9	2–9
4	80	0–2	1–8	2–9

the long-term results are insensitive to it because, in the long term, the households promote each other sufficiently.

In Fig. 7.9, we illustrate the number of migration households during the promotion year (the short-term effect) and 2 years after the promotion year (the long-term effect). As the promotion rate increases, the marginal improvement in the short term is greater than in the long term. In Table 7.9, we list the interaction effects of influence radius and promotion effort. Even if the social entrepreneur promotes to every household, 11–13 households will still migrate during the dry season.

Exploring promotion effort and its interaction with influence radius: Considering various promotion efforts, the simulation results remain insensitive to the types of networks and the network setting. The migration population decreases as promotion efforts increase. To save cost, a social entrepreneur can select the appropriate promotion effort (given a certain target for population migration) instead of promoting to every household, and economic and social status can be increased.

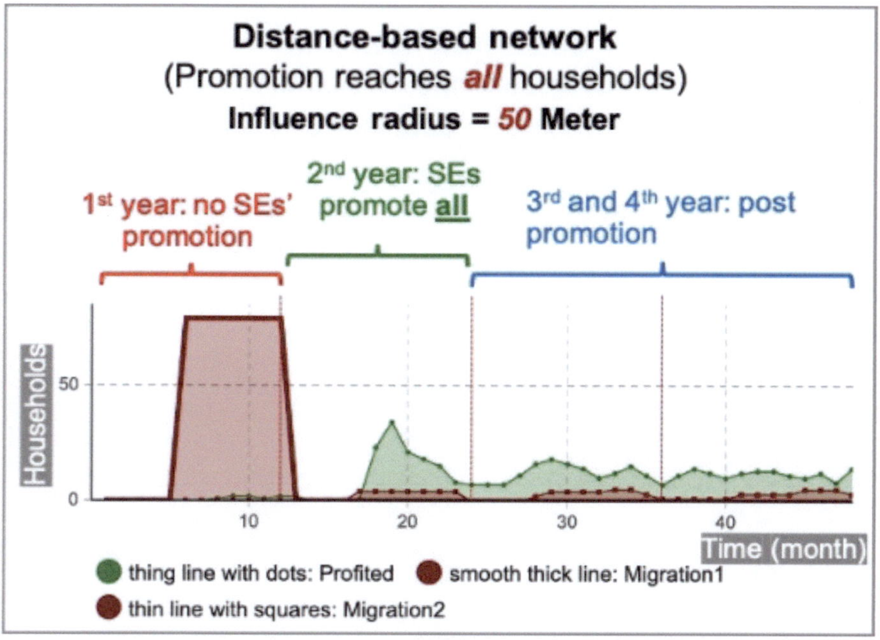

(a)

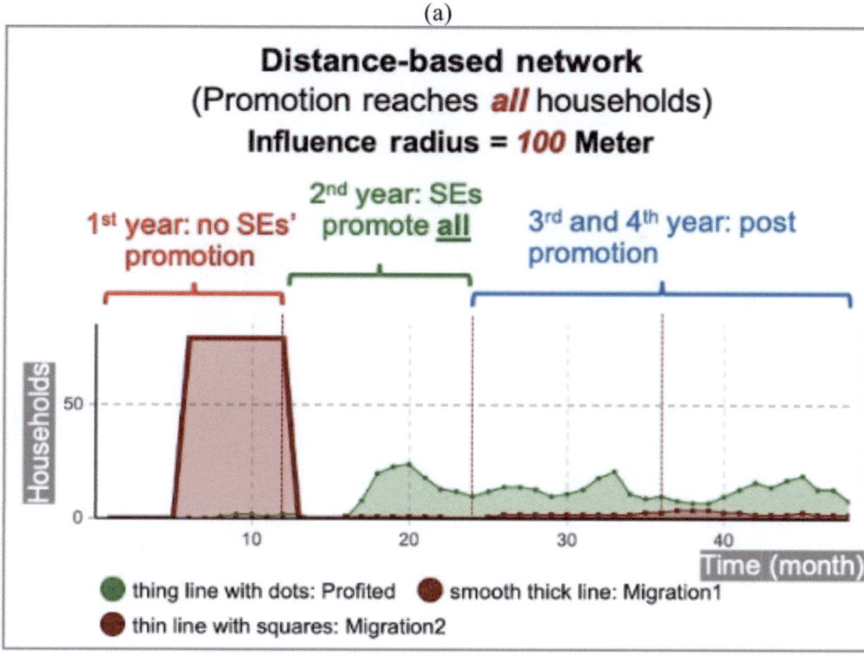

(b)

Fig. 7.7 Results for two network types when promotion reaches Reaches all households

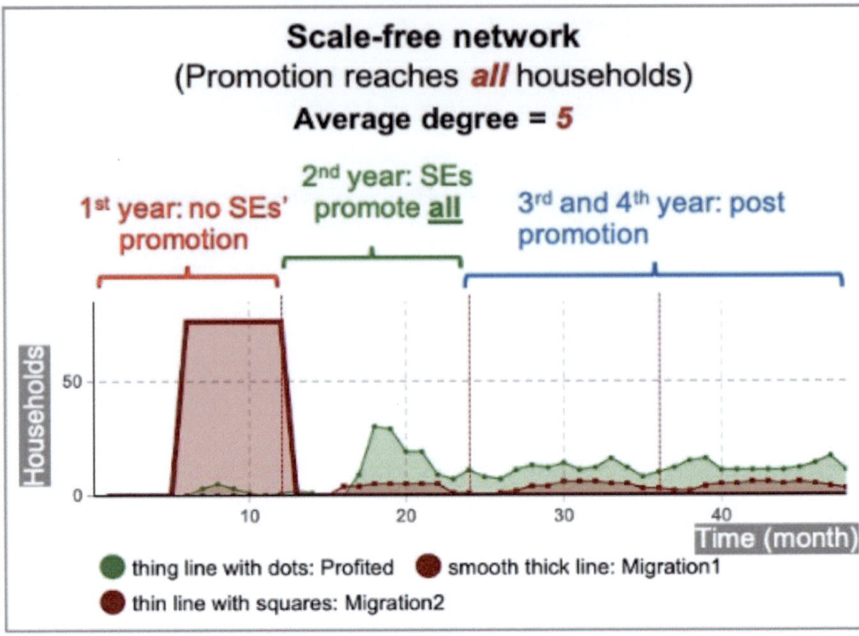

(c)

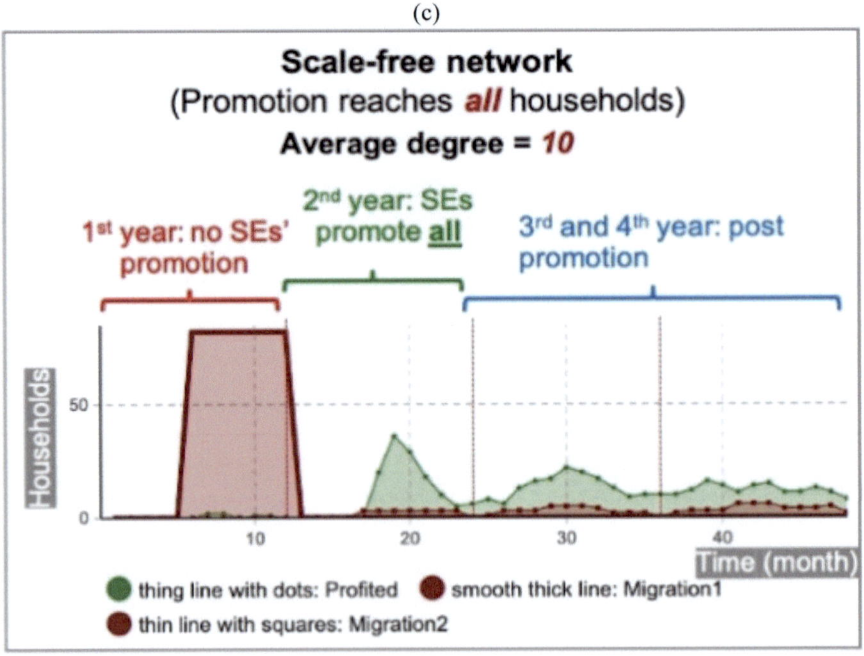

(d)

Fig. 7.7 (continued)

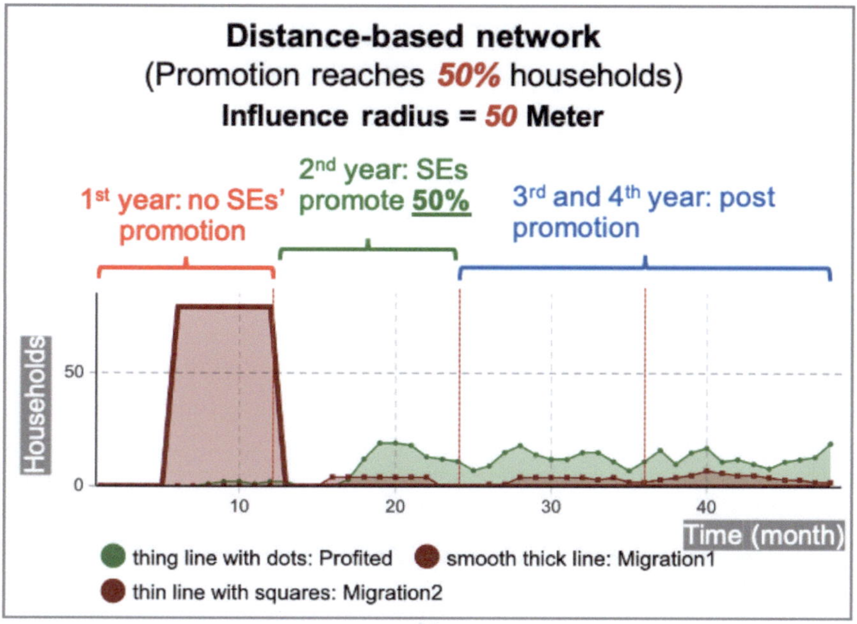

(a)

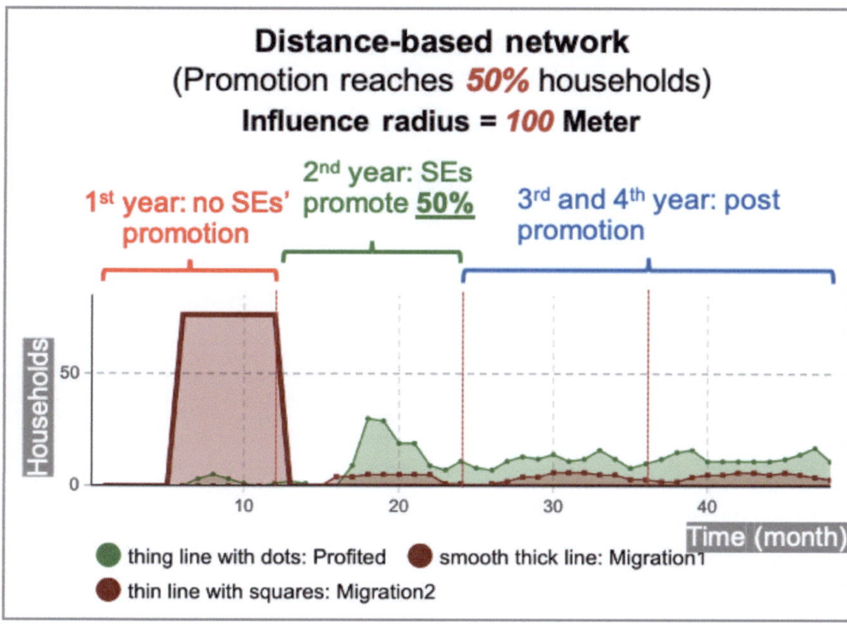

(b)

Fig. 7.8 Results for two network types when promotion reaches 50% of households

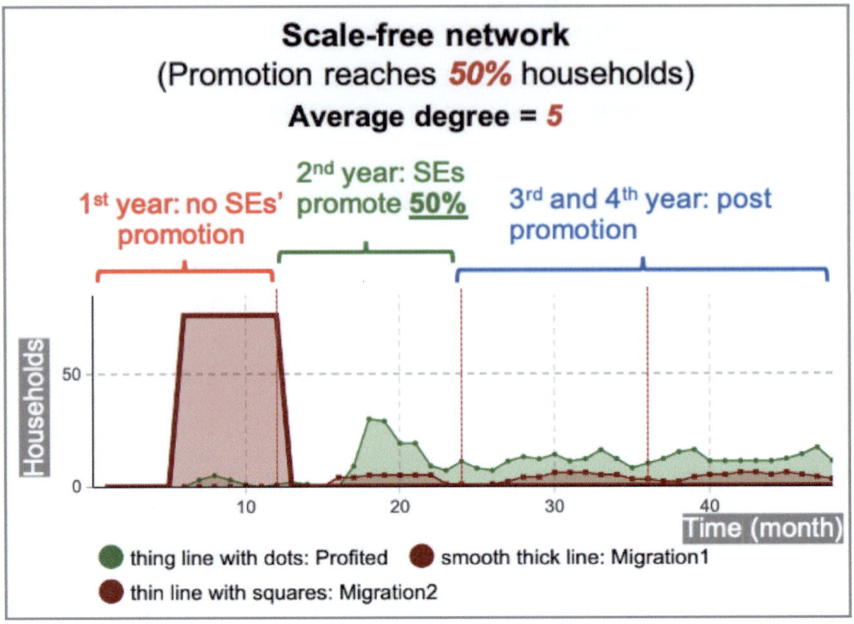

(c)

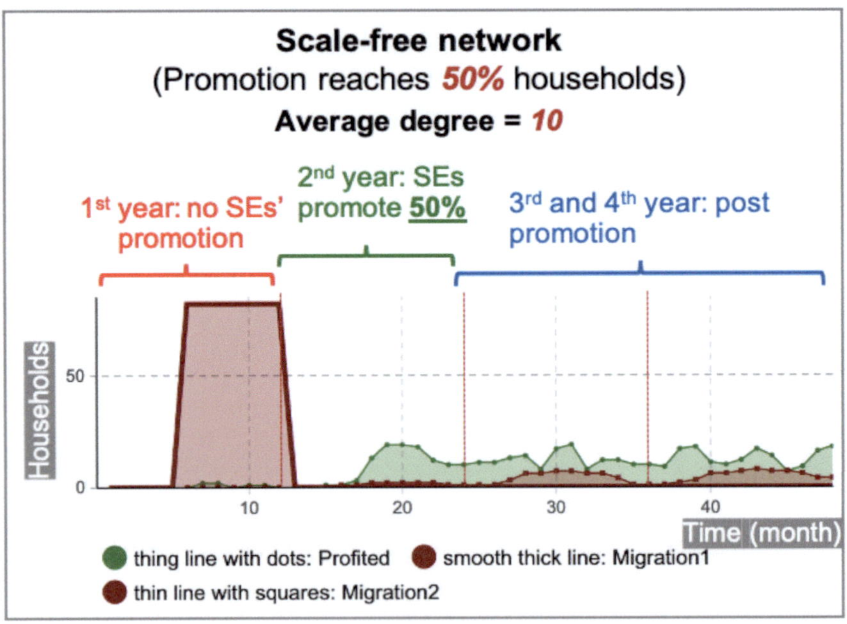

(d)

Fig. 7.8 (continued)

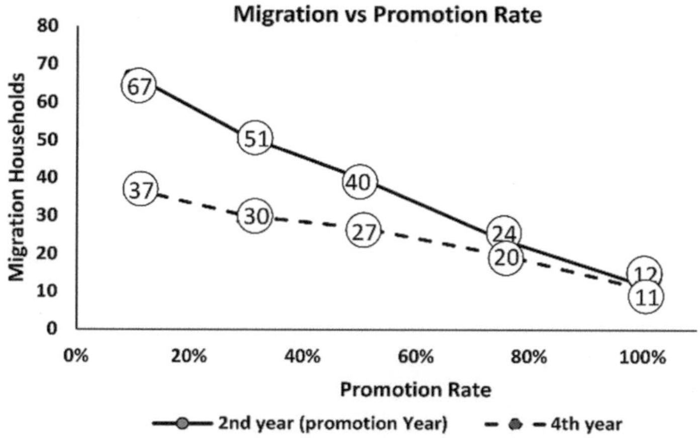

Fig. 7.9 The migration households with different promotion efforts

7.5.2 Exploring the Promotion Duration

In Fig. 7.10, we show the promotion results with different durations: 1 month—Fig. 7.10a, 2 months—Fig. 7.10b, 3 months—Fig. 7.10c, and 4 months—Fig. 7.10d. Since the results are insensitive to the type and setting of the network, a distance-based network with a 75-m influence radius is selected. The migration population for different scenarios is summarized in Table 7.10. We observe that extending the promotion results in larger profits in both the promotion year and the following years. This is because a longer promotion results in a larger population being promoted directly (by the social entrepreneur) and indirectly (by villagers). Direct promotion, combined with indirect promotion, reinforces the villagers' acceptance of second-season cultivation. However, when extending the promotion from 3 to 4 months, the results do not improve. This indicates that the villagers' capacity for accepting an idea through promotion has an upper limit; therefore, overwhelming promotion does not lead to a higher social acceptance, and a three-month promotion yields the best results for this village.

Exploring promotion duration: Extending the social entrepreneur's promotion allows more interactions between the social entrepreneur and the villagers, triggering more indirect promotion among villagers. Therefore, it enhances the acceptance of second-season cultivation. However, the value of increasing the duration of promotion has an upper limit. In this project, the limit is 3 months—after 3 months, additional promotion does not help.

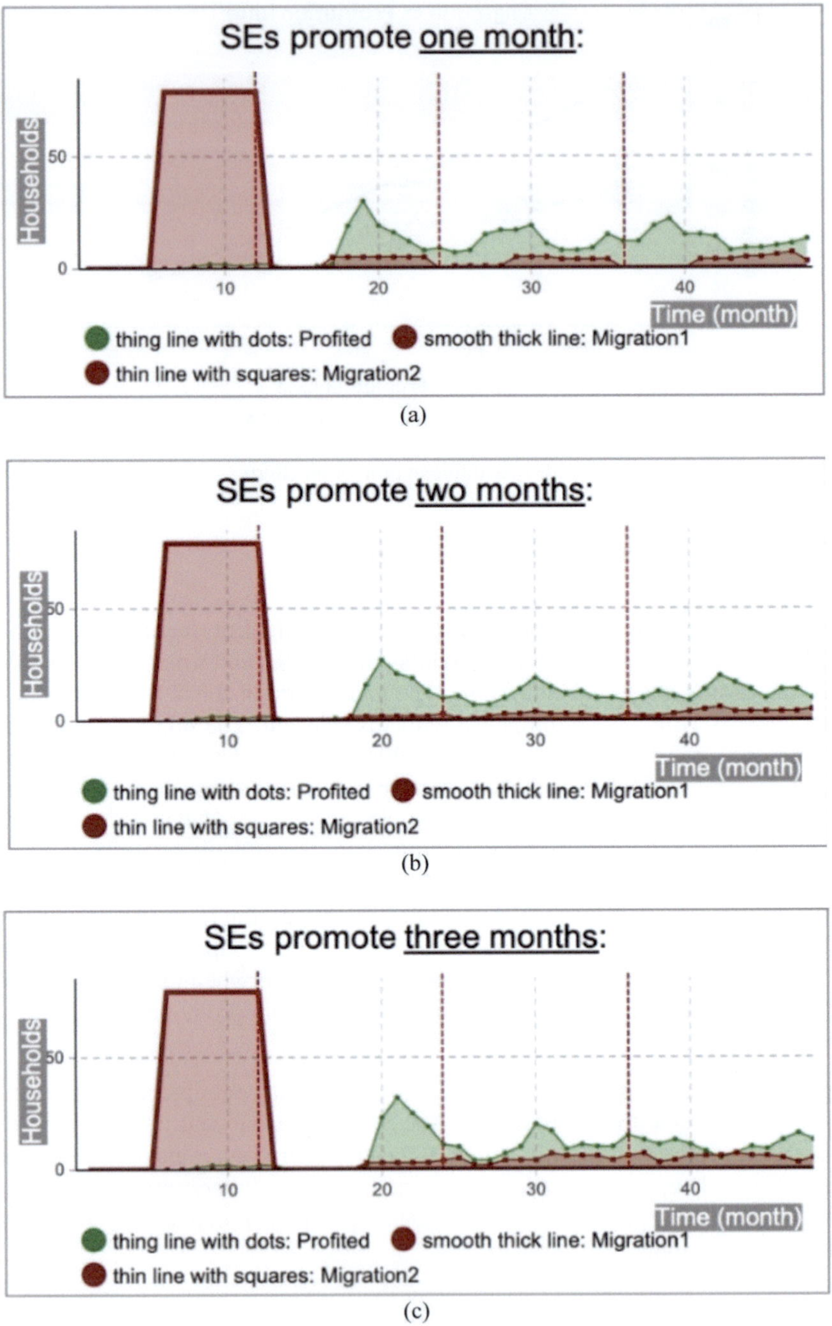

Fig. 7.10 Simulation results for different promotion durations—using a distance-based network with a 75-m influence radius

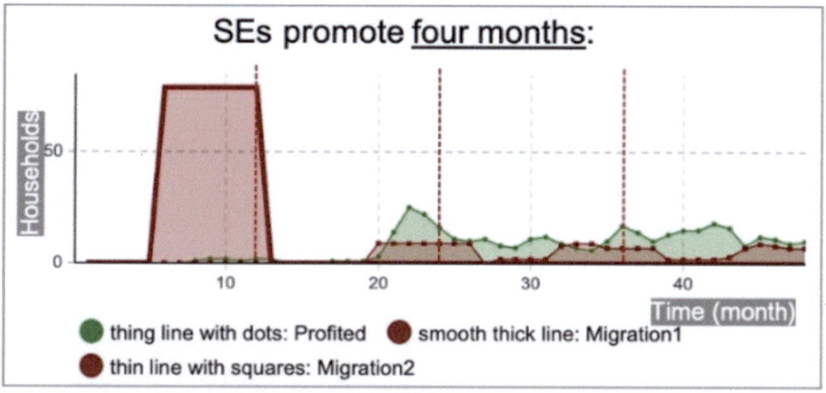

(d)

Fig. 7.10 (continued)

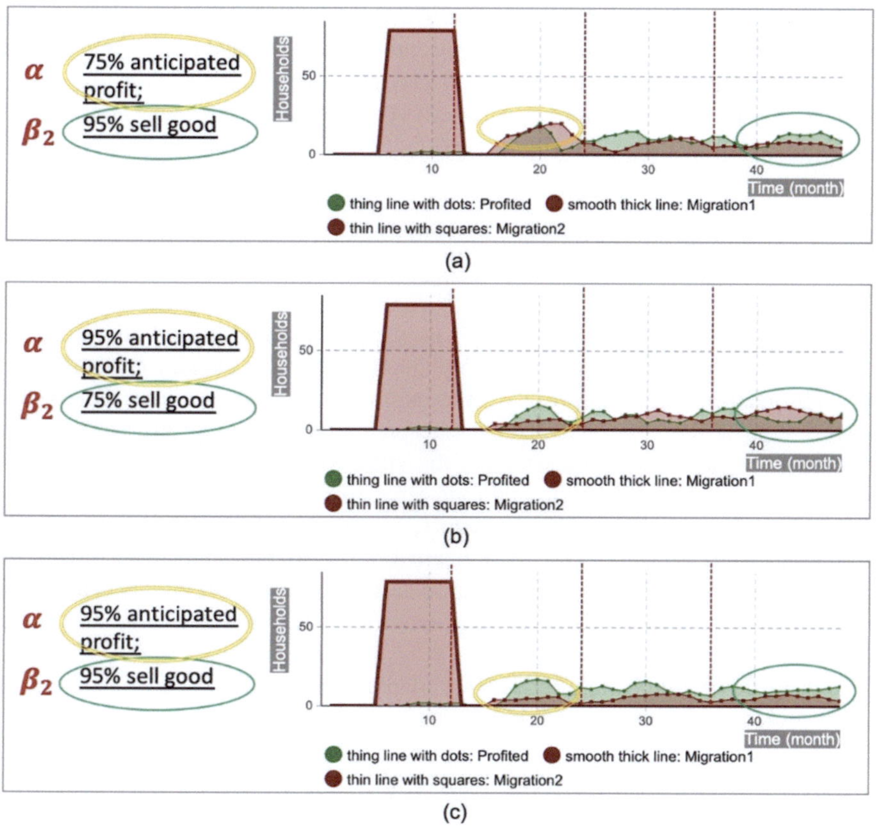

Fig. 7.11 Simulation results of three scenarios of anticipation β_2 and profit α

7.5.3 Anticipation and Profit Exploration

Two additional factors are the villagers' anticipation of profit and the actual profit they gained in the previous year. These factors are important because the social entrepreneur needs to set targets for them when planning promotion efforts. Therefore, it is necessary to identify the relationship: (1) between the villagers' anticipation and their actual improvement of actual economic and social status, and (2) between the villagers' profit and their decision on whether to do two-season cultivation the next year.

We explore three combinations of two values of α and β_2. Since $\beta_3 = \beta_2 \cdot \alpha$, we do not need to set a value for β_3. We use a distance-based network with a 75-m influence radius and the social entrepreneur promotes 50% households. The long-term effect is determined by the short-term effect and the actual gain. In Fig. 7.11, we show the results from the three scenarios for α and β_2.

Exploring the relationship between the anticipated profit and the actual profit: the short-term result is more sensitive to anticipated profit, while the long-term result is more sensitive to actual profit. A social entrepreneur can select the appropriate target for anticipated profit when promoting and taking other actions such as improving farmland productivity, expanding market share, or reducing inventory and logistics costs to improve farmers' actual profit and reach the desired migration rate target.

7.5.4 Closing Remarks

In this project, agent-based modeling is used to simulate villagers' acceptance of second-season cultivation, and scenario planning is used to identify critical factors that significantly affect the results. We explore scenarios for four factors: the network type of the village, the social entrepreneur's promotion efforts, the social entrepreneur's promotion duration, and the villagers' anticipated profit and actual profit concerning the short-term and long-term effects on villagers' economic and social status. Among all the explored factors, the duration of the social entrepreneur's promotions and villagers' anticipation are critical to the short-term effects. In contrast, the villagers' actual profit is critical to the long-term effects. The social entrepreneur can select the appropriate scenario to achieve economic and social goals.

To make this scenario planning process adaptable for other socio-technical system design projects, we summarize it in Fig. 7.12 and describe it as follows.

- Identify the factors in the system with uncertainties.
- Determine whether each factor is controllable or uncontrollable.
- For controllable factors, identify their scenarios and the meaning of each scenario based on data, domain expertise, or assumptions, and identify or suggest how to set each scenario.

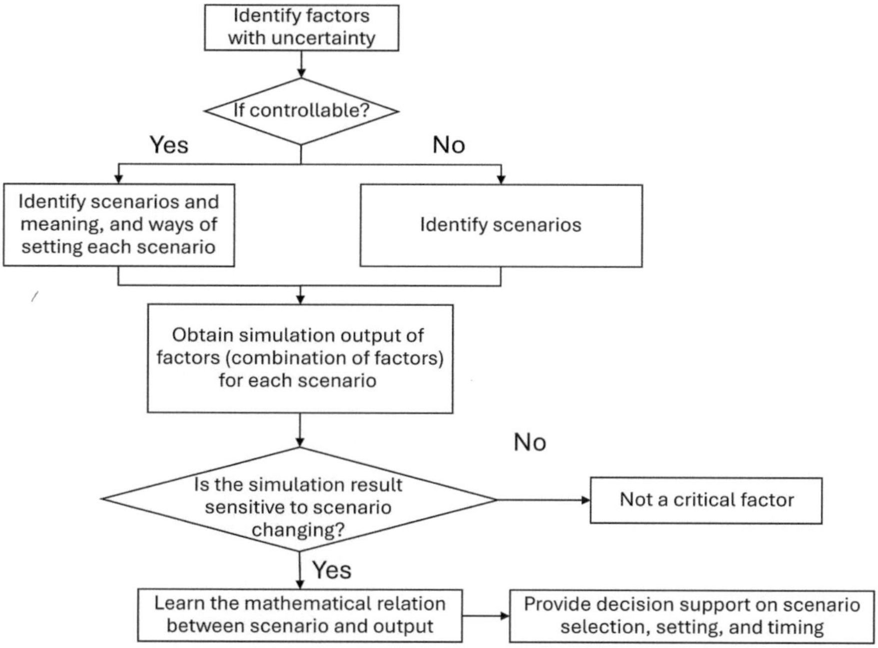

Fig. 7.12 Scenario planning for identifying critical factors in simulation

- For uncontrollable factors, identify possible scenarios, connect each scenario with system performance, and predict the impact of each scenario on system performance.
- Analyze the sensitivity of the simulation output to each factor and the necessary combinations of multiple factors.
- Identify critical factors—if the simulation output is sensitive to a factor or a combination of multiple factors, then that factor or combination of multiple factors is a critical factor.
- Identify the quantitative relations among each scenario of the critical factors and the simulation output.
- Provide decision support to the system designer by giving all the scenario-output relations.

7.6 Role of This Chapter in This Monograph

7.6.1 Summarizing How We Connect Formulation, Approximation, Exploration, and Evaluation

For a wicked problem, such as designing interactions with a social system to achieve a certain result, designers need more knowledge and data to formulate the problem. To address the problem, the following questions need to be answered: What is the boundary of the problem? What can be controlled and leveraged, and what cannot? What factors should be designed as decision variables, and what parameters have uncertainty? What are the functional relationships or cause-and-effect relations among variables, and how do they evolve over time or in response to other changing factors? For a social system, even with knowledge of each individual's preferences and behavior rules, designers may not fully understand the collective behavior of the community because the collective behavior is nonlinear with the individual behavior, and the emergent properties may change the environment, culture, and nature of the system. Hence, in this chapter, we use simulation scenario planning to identify the critical and controllable factors and the mathematical relationship among the factors (Fig. 7.12) and use it to capture the emergent properties. With such knowledge, the appropriate scenarios and interventions are selected to achieve the simulation goals.

The essence of Hypothesis 4, "*capture and quantify emergent properties through scenario planning in simulations*," is to explore the critical and sensitive factors that significantly affect the simulation results, quantify the relationship between the settings of those factors and the results regarding simulation goals, and update such learned mathematical relations in the model formulation and approximation as the emergent properties. By evaluating whether some interactions imposed on the system can control the controllable factors to achieve specific simulation results, either in the short term or over a longer time frame, designers gain the knowledge on how to improve further the formulation (such as the architecture of the simulation model) and approximation (ways to control the controllable factors and incorporate uncertainties of sensitive factors).

In other words, we realize the connections among formulation, decision, and action through the activities in the circles, as shown in Fig. 7.13. The method's flowchart is illustrated in Fig. 7.12. By realizing the proposed method for the test problem, promoting second-season cultivation in a village in India, we capture and model the emergent properties of the social system by managing the critical and sensitive factors.

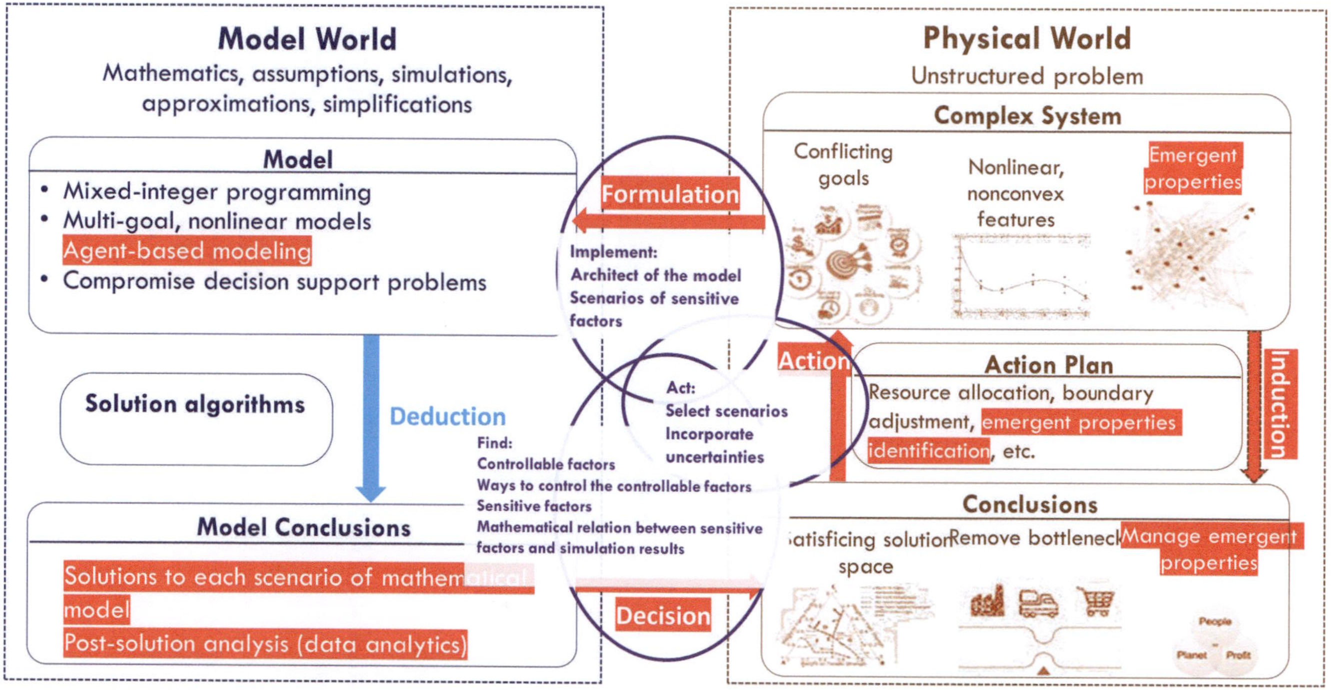

Fig. 7.13 The procedures involved in formulation-approximation-exploration-evaluation—establish the information exchange, knowledge awareness, and instructions sharing among formulation, decision, and action

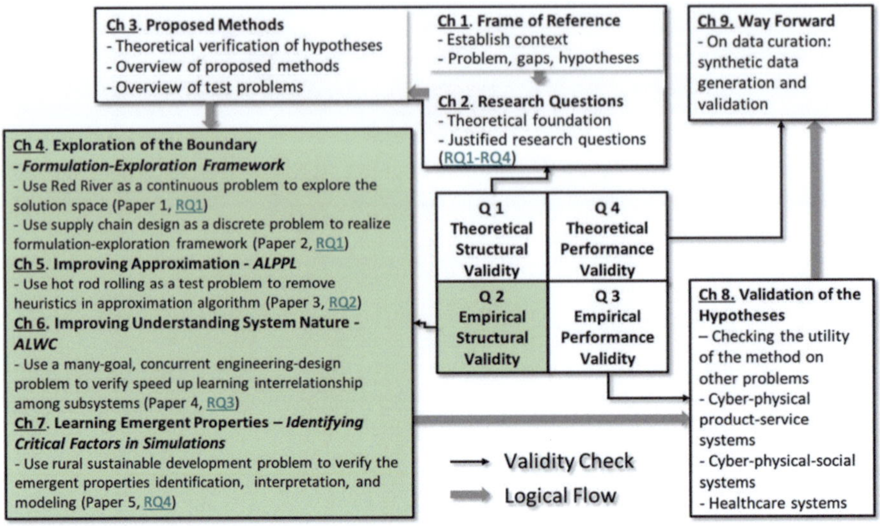

Fig. 7.14 Finishing empirical structural validity in Chaps. 4, 5, 6, and 7

Fig. 7.15 The role of this chapter in the Knowledge Foundation

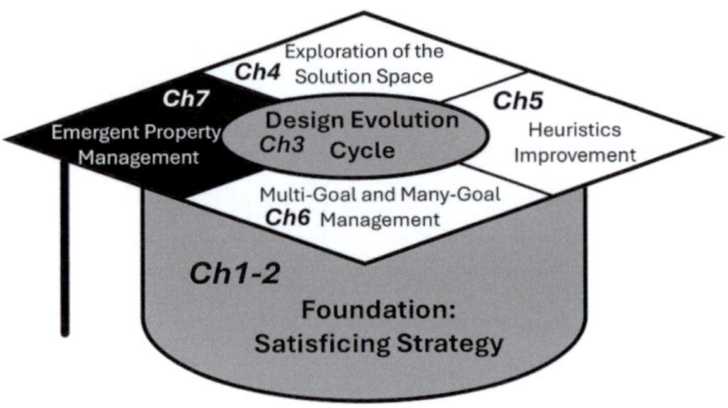

7.6.2 *Summarizing How We Realize Type I, II, and IV Robust Design*

For the test problem, promoting the second-season cultivation in a village, Type I uncertainty is identified as the price of the agricultural products and the way it affects the profit α. We incorporate the uncertainty as different scenarios with changes in the value of α (Table 7.8 and Fig. 7.5). As a result, the simulation results vary with α; see Fig. 7.11a, c. In the short term, prices are uncontrollable by farmers or social entrepreneurs, but in the long term, they can intervene through market development, technology adoption, or budget control. As the time frame for this project is 4 years, we take product price as a factor that cannot be directly controlled, so it is a parameter, and the price fluctuation is treated as Type I uncertainty.[2]

Type II uncertainty is represented in our test problem as the unknown and uncertain impact of different promotion efforts and promotion durations on the short-term and long-term promotional effects. The scenarios we explore and their physical meaning are listed in Table 7.8. The results of various scenarios of promotion efforts are shown in Figs. 7.8 and 7.9 and summarized in Table 7.9. The results of various scenarios of promotion duration are shown in Fig. 7.10 and summarized in Table 7.10.

Type IV uncertainty is recognized as the unknown feature and uncertainty in the villagers' estimation management. Villagers' original estimation of gaining more profit by growing the second-season crops than migrating (profit estimation) without underground water, β_1, is known. However, their profit estimation with underground water, β_2, depends on the actual profit α and their original profit estimation β_1. In other words, villagers' actual profit from the previous year and the promotion results comprehensively impact their estimations for the next year. Therefore, the uncertainty of β_2 results from managing Type I and II uncertainty. That is why we define it as Type IV uncertainty. This uncertainty is described in Table 7.8. The results are shown in Fig. 7.11b, c.

By implementing the proposed scenario planning method (Fig. 7.12), we can identify the solution space relatively insensitive to the Type I, II, and IV uncertainty we need to manage in a specific problem. This way, we realize Type I, II, and IV robust design. See the summary in Table 7.11 as the closing remarks of Table 3.3 regarding the robust design realization and uncertainty management for Test Problem 4.

7.6.3 Role of This Chapter

In this chapter, the frame of reference for designing promotions using agent-based modeling is given, as an extension to the frame of reference which is provided in Chap. 2. A method and framework for identifying critical factors through scenario planning in agent-based modeling is proposed to identify critical factors and quantify the scenarios associated with simulation results for modeling emergent properties. The proposed methods are verified using a test problem: promoting second-season cultivation in an island village in India. Different scenarios are identified and explored to provide decision support to social entrepreneurs. Agent-based modeling (ABM) is used to simulate villagers' acceptance of second-season cultivation, growing two crops a year instead of one. We explore the possibility of second-season cultivation to improve the villagers' socioeconomic status in the short and long term. The proposed method of capturing and using critical factors influencing individuals' behavior in a community can be used in other projects. Research Question 4 is

[2] The four types of uncertainty are discussed in Chap. 1, Sect. 1.3.

Table 7.11 Summary of test problem 4 regarding Type I, II, & IV uncertainty management

RD Type	RDI-II				
			RDIII	RDIII / RDIV	**RDIV**
Method	M1: Formulation-Exploration Framework		M2: Adaptive Linear Programming with Parameter Learning (ALPPL)	M3: Adaptive Leveling-Weighting-Clustering Algorithm (ALWC)	**M4: Scenario Planning in Agent-Based Modeling**
Chapter	Ch 4		Ch 5	Ch 6	**Ch 7**
Uncertainty \ Test Problem	T1: Dam network	T2: Supply chain	T3: Hot rolling process chain	T4: Thermal system	**T5: Promoting second-season farming**
Type I	*Uncertainty in timing and amount of inflow – Table 4.8*	*Uncertainty in demand side – Figure 4.25*	*Uncertainty in hyperparameter setting – Table 5.8, Figure 5.16*	*Uncertainty in parameter setting in solution algorithm (Starting point of searching) – call XPLORE in DSIDES*	***Uncertainty in price (Price of agriculture products) – α, Table 7.7, Figure 7.11 (a) and (c)***
Type II	*Uncertainty in outflow (water release target) – Table 4.6*	*Uncertainty on the supply side - Table 4.16*	*Uncertainty in user preferences – Table 5.7*		***Promotion effort and timing – Table 7.8-7.10, Figure 7.8-7.10***
Type III			*Uncertainty in model approximation due to heuristics in approximation – Table 5.6*	*Uncertainty in model approximation (ways of combining multiple goals) – leveling-weighting-clustering algorithms (Figure 6.13)*	
Type IV				*Uncertainty in using domain knowledge to simplify the model (fixing decision variables and selecting design scenarios) – "fixing variable" and "XPLORE" modules in DSIDES, and leveling and clustering algorithm in the ALWC*	***Interventions that change the mathematical relation among promotion and result (developing local market) – β_2, Table 7.8, Figure 7.5, Figure 7.11 (b) and (c)***

EVe – empirical verification of the method
M – method
RD – robust design
T-test problem

addressed. The "Emergent Property Management" piece of the "Design Evolution Cycle" is completed (see Figs. 7.14 and 7.15).

Role of *Chaps. 4* ***to this chapter***: From Chap. 4 to this chapter, Quadrant 2 of the Research Questions is addressed (see Fig. 7.13). Empirical structural validity is established by testing the hypotheses and proposed methods using test problems.

Glossary

Critical Factor	Designers identify a critical factor as a variable they control whose fluctuations significantly influence the system's output.
Scenario Planning	Designers use this method to identify and implement multiple scenarios of variable factors so that they can analyze different system outputs and associate desired results with appropriate scenarios.
Sensitive Factor	Designers define a sensitive factor as a variable whose fluctuations significantly impact the system's output.
Social Entrepreneur	An individual who develops and implements innovative solutions that address social, cultural, or environmental challenges. They prioritize social impact over profit and ensure financial sustainability.
Wicked Problem	Designers understand a wicked problem as a complex issue that resists straightforward definition and resolution. It involves multiple stakeholders, evolving requirements, and no clear endpoint.

References

Abreu, C. G., & Ralha, C. G. (2018). An empirical workflow to integrate uncertainty and sensitivity analysis to evaluate agent-based simulation outputs. *Environmental Modelling and Software, 107*, 281–297.

Afshari, H., & Peng, Q. (2015). Modeling and quantifying uncertainty in the product design phase for effects of user preference changes. *Industrial Management and Data Systems, 115*(9), 1637–1665.

Al Irsyad, M. I., Halog, A., & Nepal, R. (2019). Estimating the impacts of financing support policies towards photovoltaic market in Indonesia: A social-energy-economy-environment model simulation. *Journal of Environmental Management, 230*, 464–473.

Box, G. E. (1976). Science and statistics. *Journal of the American Statistical Association, 71*(356), 791–799.

Chitungo, S. K., & Munongo, S. (2013). Extending the technology acceptance model to mobile banking adoption in rural Zimbabwe. *Journal of Business Administration and Education, 3*(1), 51–79.

Guo, L. S., Chen, J. K. A., & Mistree, F. (2019). A framework for designing the customer order decoupling point to facilitate mass customization. *ASME Journal of Mechancial Design, 143*(2), Article 022002.

Guo, L., Mohebbi, S., Das, A., Allen, J. K., & Mistree, F. (2020). A framework for the exploration of critical factors on promoting two-season cultivation in India. *ASME Journal of Mechanical Design, 142*(12), 124503.

Lumbroso, D., & Davison, M. (2018). Use of an agent-based model and Monte Carlo analysis to estimate the effectiveness of emergency management interventions to reduce loss of life during extreme floods. *Journal of Flood Risk Management, 11*, S419–S433.

Opiyo, N. (2019). Impacts of neighbourhood influence on social acceptance of small solar home systems in Rural Western Kenya. *Energy Research and Social Science, 52*(C), 91–98.
Parker, D. C., Manson, S. M., Jensen, M. A., Hoffmann, M. J., & Dealman, P. (2003). Multi-agent systems for the simulation of land-use and land-cover change: A review. *Annals of the Association of American Geographers, 93*(2), 314–337.
Qiu, R., Zhu, W., Zhang, J., & Staenz, K.: (2018). Modelling and simulating urban residential land development in Jiading New City, Shanghai. *Applied Spatial Analysis and Policy, 11*(4), 753–777.
Rossoshanskaya, E. A. (2019). Integrated agent-based model of labor potential reproduction of a municipal formation. *Econ Soc Changes Facts Trends Forecast, 12*(1), 124–137.
Stewart, F. (2016). *Technology and underdevelopment*. Springer.

Chapter 8
Validation of the Hypotheses in Realizing Design Evolution

Summary and Validation of Model Evolution

Engineers often customize design methods for specific problems and tasks, making it relatively straightforward to demonstrate internal correctness (verification). However, establishing the general applicability of these methods to a broader class of problems (validation) is more complicated. Key challenges include defining the scope and characteristics of the problem class, identifying modulable and adjustable functions, rules, mechanisms, and hyperparameters of the design method, and determining how to adapt the method for new and similar problems. In this chapter, we systematically address these questions, providing a framework for validating the proposed design evolution cycle and incorporated methods, and demonstrating their adaptability across various engineering design contexts and scenarios.

In this chapter, we summarize the empirical performance validity (Chap. 1, Sect. 1.1, Fig. 1.2, Quadrant 3, the bottom right quadrant) of the design evolution methods. We provide closing remarks on the answers to the research questions and validate the hypotheses. In this chapter, we recap the contributions and the application scope of the proposed methods and introduce additional examples.

8.1 Organization of This Chapter

In this chapter, as illustrated in Fig. 8.1, we summarize our contributions in Sect. 8.2. We clarify the application scope of each method in Sect. 8.3. We introduce other examples in Sect. 8.4 and highlight the role of this chapter in Sect. 8.5.

In Fig. 8.1 we illustrate how information from various chapters feed into different section of this chapter. In Tables 8.1 and 8.2, we illustrate the interconnections between earlier chapters and this chapter. The single arrows are for theoretical connections, whereas the double arrows are for empirical connections. In Sect. 8.2, we review the strength of the *satisficing* strategy, summarize the answers to each research question—RQ1 addressed in Chap. 4, RQ2 in Chap. 5, RQ3 in Chap. 6, and RQ4

L. Guo et al., *Architecting a Satisficing Strategy for Designing Complex Systems*,
https://doi.org/10.1007/978-3-032-04577-5_8

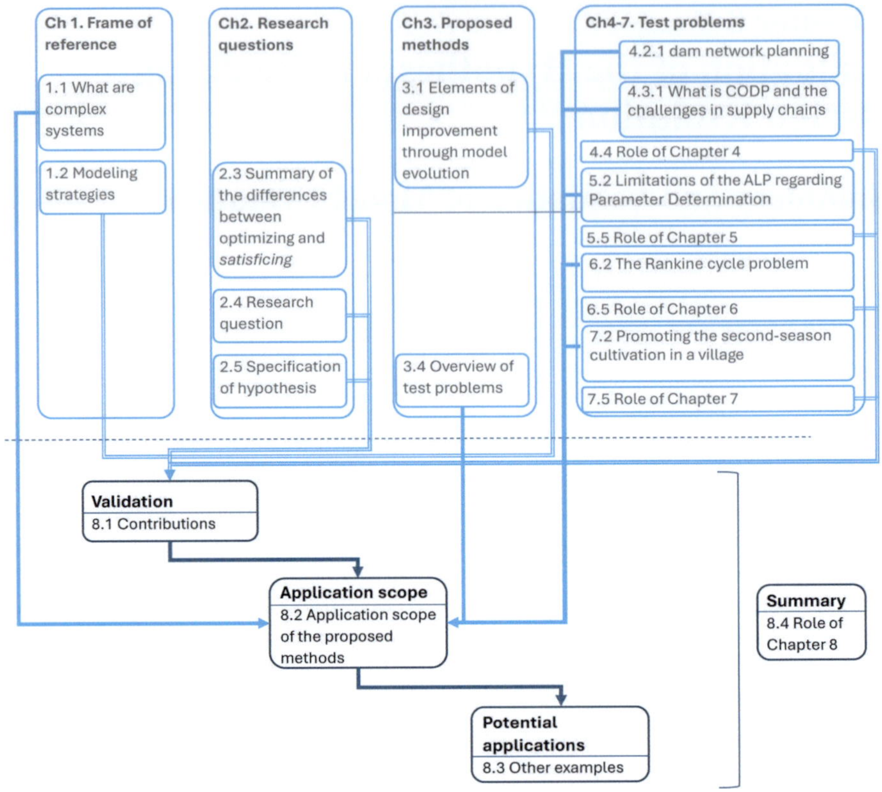

Fig. 8.1 Organization of this chapter

in Chap. 7, validate the hypotheses, and reflect on the tasks we completed to create knowledge. In Sect. 8.3, we review the characteristics and challenges associated with designing complex systems and clarify the application scope of the proposed methods, leading to additional applications in Sect. 8.4.

8.2 Contributions Documented in This Monograph

In this monograph, we address the challenges of designing complex systems, categorized into three major areas: (i) incomplete and inaccurate information, particularly in the early stages of design, (ii) overly rigid convergence conditions for nonlinear, nonconvex, multi-goal, and multi-stage optimization problems, which significantly reduce the probability of reaching optimal solutions, and iii) environmental and inherent uncertainties that often render mathematical solutions infeasible.

Table 8.1 Summary interpretation of the problems investigated in this chapter

Elements	What?
Components	Validation square
Connectors	Validity check and logic flow through quadrants of the validation square
Form	*How?*
Component roles	Provides one possible conceptual framework for connecting the four stages for the design evolution cycle—formulation, approximation, exploration, and evaluation
Properties	The coherence, consistency, and mutual constraints of the four stages—heuristics, hypothesis, and scenarios are used, tested, and verified or denied so that better settings can be identified and applied
Relationship	The relations among the activities—scenarios installation, evaluating and quantifying the model performance associated with the scenarios, and use of association to evolve the design
Rationale	*Why?*
Motivation	Provide the feasibility of realizing all four types of robust design by using the simulation and scenario planning
Assumptions	The complex systems that consist of (or can be decomposed as) agents can be designed, learned, and evolved using regular simulations and scenario planning
Interpretation	Tentatively understanding how agents' behaviors can be generated, monitored, predicted, and directed to benefit the system

Table 8.2 The support for this chapter in previous sections

Ch1 Sections	Ch2 Sections	Ch3 Sections	Ch4-7 Sections	Supporting Relation	Ch8 Sections
1.2				Select the appropriate strategy – *satisficing*	8.1
	2.3			The uniqueness and benefits of our strategy	
	2.4			Answer research questions	
	2.5			Validate hypotheses	
		3.1		Finish tasks	
			4.4 5.5 6.5 7.5	Contributions of the proposed methods	
1.1				Characteristics of complex systems and challenges in design	8.2 8.3
		3.4		Features of examples of complex system	
			4.2.1 4.2.3 5.2 6.2 7.2	Problem statement in the context of test problems	

To effectively manage these challenges, we propose using design evolution cycle as a conceptual and comprehensive tool.

We use the design evolution cycle as both a concept and a tool to help designers navigate complex system design. We interlink multiple stages of the design process to achieve this evolution. For example, in Chap. 4 we connect formulation and evaluation by using sensitivity analysis results (exploration) to identify model bottlenecks and then reallocate resources to eliminate them (formulation). In Chap. 5 we interlink approximation, exploration, and evaluation by exploring the solution space (exploration) to develop rules for assessing hyperparameter performance (evaluation) and by tuning hyperparameters for robustness under different approximation scenarios (approximation). We establish these interconnections by linking tasks and activities across stages, exchanging information, and dynamically updating parameter settings, activity rules, or key performance indicators as new information emerges. Specifically, we transfer information between stages to inform or revise decisions, and we establish mathematical relationships among parameters, actions, and heuristics at different stages.

We demonstrate that by integrating formulation, approximation, exploration, and evaluation in the design of complex systems, designers can manage four distinct types of uncertainty—specifically, through:

(i) the Three-Step Exploration Method to perform sensitivity analysis and update the model boundary to improve robustness (Sect. 4.2),
(ii) the Formulation-Exploration framework to identify and remove bottlenecks to better goal achievement (Sect. 4.3),
(iii) the Adaptive Linear Programming algorithm with Parameter Learning (ALPPL) to ensure the solution robustness to hyperparameter settings (Chap. 5),
(iv) the Adaptive-Leveling-Weighting-Clustering Algorithm (ALWC) to intelligently and dynamically combine multiple goals based on their interrelationship (Chap. 6), and
(v) scenario planning in agent-based modeling to capture and leverage emergent properties (Chap. 7).

The four distinct types of uncertainty in the test problems include:

Type I—Noise factors or uncontrollable parameters (Taguchi, 1993), such as precipitation in a river basin (Sect. 4.2) or demand fluctuation in a supply chain (Sect. 4.3).

Type II—Control factors or variations in design variables (Chen et al., 1996, 1999), such as productivity variations in a supply chain (Sect. 4.3) or hyperparameter settings in a solution algorithm (Chap. 5).

Type III—Variations in mathematical model structure arising from inherent complexity (Choi et al., 2005a, 2005b), such as the dynamic orthogonality between goals in the solution space (Chap. 6).

Type IV—Uncertainties in the process chain, introduced when managing the first three types (Choi et al., 2005a, 2005b), including unexpected emergent properties resulting from interventions, such as unexpected collective behavior patterns as a result of a social entrepreneur's promotion of a new technology in a village (Chap. 7).

We summarize what has been done regarding addressing the research questions and verifying the hypotheses. In Tables 4.2, 5.2, 6.2, and 7.2, we provide full versions of Table 8.2 showing how we address Research Questions 1, 2, 3, and 4, respectively.

Table 8.3 Summary of addressing the research questions and verifying the hypotheses—– with section number

Chapter	Actions
Ch1	RG – 1.5.1; H – 1.5.2
Ch2	RDI-II – 2.4.1; RDIII – 2.4.1; RDIV – 2.4.1; RQ1 – 2.4.2; RQ2 – 2.4.2; RQ3 – 2.4.2; RQ4 – 2.4.2; SH1 – 2.5; SH2 – 2.5; SH3 – 2.5; SH4 – 2.5
Ch3	TVe1 – 3.2.1; TVe2 – 3.2.2; TVe3 – 3.2.3; TVe4 – 3.2.4; M1 – 3.3.1; M2 – 3.3.2; M3 – 3.3.3; M4 – 3.3.4
Ch4	EVe1 – 4.2.3/4.3.3/4.4.1; SQT1 – 4.1/4.3.2; AQ1 – 4.2.6/4.3.2; VRDI-II – 4.4.2/5.5.2
Ch5	EVe2 – 5.2/5.5.1; SQT2 – Table 5.2; AQ2 – 5.5.1; VRDIII – 5.5.2/6.5.2
Ch6	EVe3 – 6.3/6.5.1; SQT3 – Table 6.2; AQ3 – 6.5.1; VRDIII – 6.5.2/7.5.2
Ch7	EVe4 – 7.3/7.5.1; SQT4 – Table 7.2; AQT4 – 7.5.1
Ch8	CO – 8.1/8.4; EVa – 8.2/8.3
Ch9	TE
Nomenclature	AQ"X" – answer research question "X" CQ – closure of the answers to research questions EVa – empirically validate hypotheses EVe"X" – empirically verify hypothesis "X" H – give hypotheses M"X" – introduce method "X" RD"X" – tie to Type "X" robust design RG – give research gaps RQ"X" – pose Research Question "X" SH"X" – specify hypothesis "X" SQT"X" – specify research question "X" in the context of test problems "X" TE – theoretically extend the research TVe"X" – theoretically verify hypothesis "X VRD"X" – validate Type "X" robust design

In Table 8.4, we summarize the research gap, hypothesis, four types of robust design, etc., in Chaps. 1 to 3 (Table 8.3).

8.2.1 Summarizing the Satisficing Strategy

From Chaps. 1 to 3, we describe the theoretical foundation of the satisficing strategy—Q1 of the validation square; see Fig. 8.2.

Table 8.4 Addressing research question 1 and verifying the hypotheses

Chapter	Ch1	Ch2			Ch3		Ch4–7	Ch8	Ch9
Actions	RG: How can designers realize design evolution using a satisficing strategy to manage chaos in the physical world, reduce the risk of losing an optimal solution, and discover domain-independent knowledge to update metaheuristics? H: By connecting the multiple design stages and passing information through them, designers can improve their decision models through iterations	RDI-II: What mechanisms and procedures enable the exploration of the solutions relatively insensitive to Type I and Type II uncertainty?	RQ1: What is a method used to evolve the model boundary?	SH1: Explore the sensitivity of the segments of the model boundary and improve accordingly	TVe1: Explore the boundary of the system by connecting formulation with exploration	M1: Formulation-Exploration framework	EVe, SQT, AQ	C1, EVa	TE
			RQ2: What is a method to evolve the model to update metaheuristics?	SH2: Explore the sensitivity of the segments of the model boundary and improve accordingly	TVe2: Improve the approximation by connecting approximation, exploration, and evaluation	M2: Adaptive Linear Programming Algorithm with Parameter Learning (ALPPL)			
		RDIII: What mathematics in the design method allows exploring solutions that are relatively insensitive to Type III uncertainty?	RQ3: What is a method to speed up learning the system nature?	SH3: Learn the system's nature, such as the interrelationships among subsystems, and reorganize them based on it	TVe3: Speed up the learning system nature by connecting formulation, exploration, and evaluation	M3: Adaptive Leveling-Weighting-Clustering algorithm (ALWC)			

(continued)

Table 8.4 (continued)

Chapter	Ch1	Ch2			Ch3		Ch4–7	Ch8	Ch9
		RDIV: What is a method that allows sensing and managing Type IV uncertainty?	RQ4: What method allows information to be passed through multiple system scales?	SH4: Capture and quantify emergent properties through scenario planning in simulation	TVe4: Learn emergent property by connecting formulation, approximation, exploration, and evaluation	M4: Scenario planning in agent-based modeling			
Nomenclature	AQ "X"—answer research question "X" CQ—closure of the answers to research questions EVa—empirically validate hypotheses EVe "X"—empirically verify hypothesis "X" H—give hypotheses M "X"—introduce method "X" RG—research gaps RD "X"—tie to Type "X" robust design RQ "X"—pose Research Question "X" SH "X"—specify hypothesis "X" SQT "X"—specify research question "X" in the context of test problems "X" TE—theoretically extend the research TVe "X"—theoretically verify hypothesis "X" VRD "X"—validate Type "X" robust design								

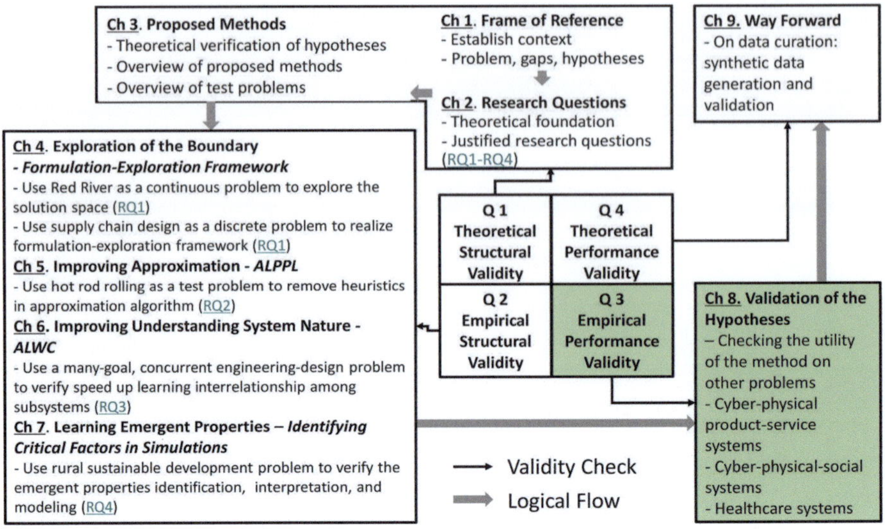

Fig. 8.2 Finishing empirical performance validity in this chapter

In Chap. 1, we frame the content of this monograph from two perspectives: (1) design methods and strategies (Tables 1.4, 1.5 and 1.6), and (2) challenges in the model-based realization of complex systems—there is intellectual disconnection between the model world and physical world (Fig. 1.8). Methods in literature can be classed into two categories regarding whether the researchers are aware of or admit such intellectual disconnection—the optimizing literature and satisficing literature. In the optimizing literature, in which, authors formulate a problem exactly using optimization model, and aim at finding the single optimal solution, assuming that models are complete and accurate, authors seek solutions that meet the necessary and sufficient Karush–Kuhn–Tucker (KKT) conditions (Guo et al., 2024). These conditions are sensitive to errors, incompleteness, and variations in any segment of the decision model. In the satisficing literature, acknowledging that models can be flawed and incorporating segments with varying fidelity, authors aim for "good enough" solutions, also known as satisficing solutions, that are relatively insensitive to errors, incompleteness, and uncertainties. Sometimes, the satisficing solutions are near-optimal solutions. In this monograph, we propose identifying satisficing solutions by satisfying the necessary KKT conditions without requiring the sufficient KKT conditions. However, designers must rely on domain knowledge and metaheuristics, which can make design knowledge unreproducible. Therefore, we identify the research gap to be addressed in this monograph:

How can designers realize design evolution using a satisficing strategy to manage chaos in the physical world, reduce the risk of missing an optimal solution, and discover domain-independent knowledge to update metaheuristics?

We hypothesize that *by running the different stages of the design in iterations and connecting multiple stages and passing information through them, designers can improve their decision models iteratively*, which in this monograph is defined as "design evolution."

In Chap. 2, we pose research questions and specify the hypothesis. By analyzing the assumptions underlying the use of the KKT conditions for seeking optimal solutions, we conclude that an optimization strategy requires designers to accept at least three key assumptions, or ensure that their problem formulation and solution algorithm satisfy these three requirements, to guarantee model convergence to an optimal solution:

(1) models are perfect representations of the physical world,
(2) all equations of a decision model are continuous and differentiable, and
(3) the degree of convexity of at least one non-zero linear combination scenario of all constraints is higher than the degree of convexity of the objective function.

However, only the second assumption is required when using a *satisficing* strategy (Sect. 2.1.4). Even if it is not, by using the adaptive linear programming (ALP) algorithm, a problem with coupled-decision feature (selection decision support problem or discrete variables) can be approximated into one or a series of accumulated continuous, differentiable problems, ensuring that the second assumption is met.

Through five toy problems in Sect. 2.2, we illustrate how the results of using these two strategies differ. The differences are in all four stages of the design loop:

Formulation: using goals with target values as the right-hand side value, instead of objectives without a right-hand side value);

Approximation: using the ALP algorithm, which incorporates the second-order sequential linearization and accumulated linearization;

Exploration: combining interior-point search (to explore robust starting point) and vertex search (Dual Simplex); and.

Evaluation: allowing violations and post-solution analysis.

Based on these discussions, we choose *satisficing* as our strategy to manage complex-system design problems and run the design evolution cycle iteratively. Therefore, we justify the research questions regarding requirements (four types of robust design, Table 2.15) and tasks (four stages of the design loop, Table 2.16).

In Chap. 3, we theoretically verify the hypotheses to address the research questions through checking the internal consistency of the proposed methods. We demonstrate the feasibility of completing the research tasks connecting multiple stages of the design loop through information sharing and performance evaluation. By running the design evolution cycle and connecting the stages, we update the metaheuristics in the early stage of a design and manage additional types of uncertainty. We propose the methods that allow designers to realize the design evolution cycle in Chap. 3 (Fig. 3. 3). To explore and update the mathematical boundary and physical boundary of the design problems to improve the robustness, we propose connecting the formulation and exploration using the Three-Step Method and Formulation-Exploration framework (M1, Fig. 3.8). To improve the robustness of the approximation method under the *satisficing* strategy, we propose using parameter learning to improve the adaptive linear programming algorithm—the adaptive linear programming algorithm with parameter learning (ALPPL) (M2, Fig. 3.9). To get aware of subsystems and explore the interrelationships among them, we employ unsupervised learning to understand the orthogonality and correlation among the goals in the solution space and update the goal combination based on it—the adaptive leveling-weighting-clustering (ALWC) algorithm (M3, Fig. 3.10). To understand the emergent properties of a design, such as identifying critical factors in scenario planning of a decision model, we standardize the scenario planning process in simulation—scenario planning through agent-based modeling (M4, Fig. 3.11).

8.2.2 Summarizing the Test Problems

In Chaps. 4 through 7, we demonstrate the proposed methods M1-M4 using six test problems: T1.1, T1.2, T2, T3, T4, and T5.

T1.1—Dam network planning. We propose a Three-Step Method to plan the monthly water release from each reservoir in a dam network to satisfy three user groups. The boundary of the reservoirs has evolved to make the system to be less sensitive to uncertainties in water inflows.

In this research project, we focus on water management within a fourteen-dam network along the Red River, which borders Oklahoma and Texas (Figs. 4.3 and 4.4). The goal is to address the water demands of multiple user-groups, including municipal, agricultural, and wildlife, under conditions of uncertainty. Municipal demand, influenced by collective human behavior, is predictable but variable. Agricultural and wildlife demands fluctuate due to precipitation variability, ecosystem dynamics, and climate change.

Method: We satisfy multiple user-groups' water demand through two ways: First, dynamic water management, which is adjusting monthly water storage and release for each reservoir. Second, pool reallocation, which is redistributing water within reservoir pools, including the pools in each reservoir (Fig. 4.6).

To identify and resolve inefficiencies, we use the Three-Step Method (M1) to run the design evolution cycle by connecting the Formulation and Exploration stages. In each iteration, we pinpoint bottlenecks (represented as active constraints or bounds with zero slack or surplus) and over-resourced areas (represented as improvable or redundant constraints or bounds with over-sized slack or surplus). Through iterative adjustments, water storage is reallocated from over-resourced pools to bottleneck pools, effectively resolving constraints.

Results: Over three iterations, we successfully reduce unmet water demand across user-groups,. reduce the number of active and improvable constraints and bounds by reallocating the pool sizes and adding buffers to the mathematical models versus the physical system. We improve the water supply and the robustness of the system.

T1.2—CODP positioning for a supply chain. We expand the Three-Step Method into the Formulation-Exploration framework and use it to position the customer order decoupling point (CODP) for a three-echelon single-product supply chain across different product life cycle phases (PLC). We remove the bottleneck in iterations by evolving the CODP and relevant parameters.

In this research project, we aim to improve the operational efficiency and the configurability to realize mass customization for the single product through dynamically positioning the CODP across the phases in the PLC. The goals include ensuring profit, service level, and equity across echelons. The decisions that we need to make include the CODP position, the service level of each echelon, and the reliability of each echelon.

Method: We reach the three goals through two ways—positioning CODP in different phases of PLC and determine the service level and reliability to compromise the three goals. We expand the Three-Step Method into the Formulation-Exploration framework to run the design evolution cycle. The expanded part is using DSIDES to formulate the problem into a compromise Decision Support Problem and using the ALP to approximate and solve the problem, which dealing with the nonlinear feature.

Results: Over the iterations, we identify and remove bottlenecks in the supply chain, position the CODP dynamically in different phases of PLC. In this way, we can reach to a near- mass production operational efficiency while providing relatively high configurability.

T2—Improving the hot rod process chain. We incorporate parameter learning into the adaptive linear programming algorithm. The improved algorithm, the adaptive linear programming algorithm with parameter learning (ALPPL), facilitates designers evaluate the performance of the approximation regarding multicriteria and learning the association between parameter settings and approximation performance in order to update the metaheuristics applied in parameter setting.

Hot rolling is a multi-stage manufacturing process where reheated billets, slabs, or blooms undergo thermo-mechanical processing, leading to microstructural evolution, including grain refinement and phase transformations (e.g., austenite to ferrite, pearlite). The final microstructure, determined by cooling rates, final austenite grain size, and chemical compositions, defines the mechanical properties (yield strength, tensile strength, hardness) of the steel rods. Traditional plant trials for developing new steel grades are costly and time-intensive, necessitating a simulation-based design approach to optimize conflicting property goals. Using a goal-oriented inverse design method (Nellippallil et al., 2017), the target microstructure (high ferrite fraction, low pearlite interlamellar spacing, and small ferrite grain size) is identified to meet desired mechanical properties. This problem focuses on exploring the solution space of cooling rates, final austenite grain size, and chemical compositions using ALPPL to achieve the target microstructure, ensuring the steel rod's mechanical requirements are met, while primarily testing the solution algorithm rather than detailing material and process design. The hot rod rolling problem has nonlinear and nonconvex features, multiple goals, coupled decision making, and various levels of fidelity of sub-systems. The challenge of the hot rod rolling process design is that the hyperparameter in the solution algorithm may severely affect the solution space.

Method: To realize the design with limited computational resources, problems with the features above need to be linearized and then solved using solution algorithms for linear programming. The adaptive linear programming (ALP) algorithm is an extension of the Sequential Linear Programming algorithm where a nonlinear compromise decision support problem (cDSP) is iteratively linearized, and the resulting linear programming problem is solved with satisficing solutions returned. The reduced move coefficient (RMC) is used to define how far away from the boundary the next linearization is to be performed, and historically, it has been determined based on heuristics. The choice of RMC significantly affects the efficacy of the linearization process and, hence, the rapidity of finding the solution. In this research project, we propose a rule-based parameter-learning procedure to vary the RMC at each iteration, thereby significantly increasing the speed of determining the ultimate solution. To demonstrate the efficacy of the ALP algorithm with parameter learning (ALPPL), we use the integrated design of a hot-rolling process chain for the production of steel rods. Using the proposed ALPPL, we can incorporate domain expertise to identify the most relevant criteria to evaluate the performance of the linearization algorithm, quantify the criteria as evaluation indices, and tune the RMC to return the solutions that fall into the most desired range of each evaluation index.

Results: Compared with the old ALP algorithm using the golden section search to update the RMC, the ALPPL improves the algorithm by identifying the RMC values with better linearization performance without adding computational complexity. The insensitive region of the RMC is better explored using the ALPPL—the ALP only explores the insensitive region twice, whereas the ALPPL explores four times throughout the iterations. With ALPPL, we have a more comprehensive definition of linearization performance—given multiple design scenarios, using evaluation indices (EIs) including the statistics of deviations, the numbers of binding (active) constraints and bounds, the numbers of accumulated linear constraints, and the number of iterations. The desired range of evaluation indices (DEI) is also learned during the iterations. The RMC value that brings the most EIs into the DEI is returned as the best RMC, which ensures a balance between the accuracy of the linearization and the robustness of the solutions. For our test problem, the hot rolling process chain, the ALP returns the best RMC in twelve iterations considering only the deviation as the linearization performance index, whereas the ALPPL returns the best RMC in fourteen iterations considering multiple EIs. The complexity of both the ALP and the ALPPL is $O(n^2)$. The parameter-learning steps can be customized to improve the parameter determination of other algorithms.

T3—Rankine cycle thermal system design. We use unsupervised learning to identify subsystems based on the interrelationships among the goals and reorganize them to boost the system's performance. Learning the tradeoffs among subsystems and providing scenarios for users to realize certain tradeoffs can make the system more robust.

In this research project, we address the issue of solving problems with multiple components, multiple objectives, and target values for each objective. There are limitations in managing these multi-component, multi-goal problems such as the need for domain expertise to combine or prioritize the goals.

Method: We propose a domain-independent method, Adaptive Leveling-Weighting-Clustering (ALWC), to manage the exploration of design scenarios of multi-goal, engineering-design problems. Through clustering analysis on the randomly generated solution points, we identify the orthogonality among the goals. The goals that are more "parallel" are grouped in the same cluster, whereas the goals that are more orthogonal are grouped in different clusters. We use weights to combine the goals in the same cluster, and Lexicography to manage the goals in different clusters.

Results: Using ALWC, designers explore combinations and priorities of the goals based on their interrelationships. Through iteration, design scenarios are obtained with higher goal achievements and an improved understanding of the interrelationship among subsystems. This is achieved without increasing computational complexity. This knowledge is helpful for multi-component design. The ALWC method is demonstrated using a Rankine cycle-based thermal-system design problem.

T4—Learning emergent properties of a social system under interventions. We use scenario planning in agent-based modeling to identify the critical and sensitive factors in the interactions with the system and the interactions among the factors in the system so corresponding actions can be taken to reach the target of the system goals.

For a remote and isolated village where the farmers only grow crop using the surface water during the monsoon season and migrate during the dry season, a social entrepreneur wants to promote farmers to use solar power to pump the underground water during the dry season to do a second-season farming. In this way, the farmers can have a better life with a more stable income. This will not only promote the new energy solution but also change the farmers' lifestyle. The challenge is how to predict the collective behavior and leverage it to better promote the new energy solution.

Method: To promote a new mode of production and a changed lifestyle in a relatively closed, underdeveloped community, critical interventions should be explored, and, as appropriate, implemented. Different scenarios are identified and explored so that decision support can be provided to social entrepreneurs (SEs). Here, agent-based modeling (ABM) is used to simulate villagers' acceptance of second-season cultivation, growing two crops a year instead of one. We explore the possibility of second-season cultivation to improve the villagers' social-economic status in both the short term and the long term.

Results: We explore the network type, promotion effort and duration, and anticipation level of profit and actual profit level. The results indicate that the acceptance of second-season cultivation is insensitive to network type but is sensitive to promotion effort. The factors that particularly affect the short-term result are promotion duration and villagers' anticipation; the factor that affects the long-term result is the real profit that second-season cultivation actually produces. The proposed method of capturing and making use of critical factors in influencing individuals' behavior in a community can be used in other projects.

8.2.3 *Summarizing the Answer to the Research Questions*

In summary, the answer to the research questions is new knowledge: connecting multiple stages in the design evolution cycle. In Table 8.5, we summarize the new knowledge, how it helps answer each research question, differences we make, and the relevant publications. Table 8.5 is another format of the illustration in Fig. 3.2. The differences between "before the new knowledge" and "after the new knowledge" are major contributions documented in this monograph. The essence of why we can make a difference is that we choose the *satisficing* strategy and fill the research gaps.

8.2.4 *Summarizing the Four Types of Robust Design*

In summary, to achieve the four types of robust design, we manage the four types of uncertainty in designing complex systems using various methods and mechanisms. In different systems, each type of uncertainty may have different representations. The four types of uncertainty in each test problem, the new knowledge to achieve the four types of robust design, and the corresponding tables and figures with more details are summarized in Table 8.6, which is an alternative representation of the illustration in Fig. 3.17.

8.3 Application Scope of the Proposed Methods

8.3.1 *Application Scope of the Design Evolution Cycle*

The design evolution cycle can be used for designing complex systems with multiple types of uncertainty. Designers can evolve the system's boundary, handle both discrete and continuous variables, use heuristics to proceed and improve the design by updating them, reorganize and dynamically combine the subsystems based on their interrelationships, and learn and manage emergent properties.

The typical features of problems that make the design evolution cycle particularly effective include nonlinear (especially nonconvex) constraints or objectives, multiple and conflicting goals, multi-stage design requirements, limited information in the early stages of design, performance sensitivity to hyperparameter settings, and emergent properties that introduce unexpected features.

Table 8.5 New Knowledge in this Monograph

Chapter	Ch4		Ch5	Ch6	Ch7
Research Question (RQ)	RQ1: What is a method used to evolve the model boundary?		RQ2: What is the method to evolve the model to update metaheuristics?	RQ3: What is the method to speed up learning the system's nature?	RQ4: What method allows information to be passed through multiple system scales?
Answer to RQ – new knowledge	The Three-Step Exploration Method	The Formulation-Exploration Framework	The Adaptive Linear Programming Algorithm with Parameter learning (ALPPL)	The Adaptive Leveling-Weighting-Clustering Algorithm (ALWC)	Scenario planning workflow for Agent-Based Modeling
Before the New Knowledge	Solutions are on the boundary of the mathematical model (formed by constraints or bounds) as well as the boundary of the physical system, so they are sensitive to the variation in the system boundary; No suggestion on model improvement regarding both goal completion and robustness of the solution; We are improving the goal completion results by sacrificing the model robustness and vice versa		Heuristics in the algorithm (Golden Section Search in ALP) have not been evaluated and improved; No criteria other than the goal completion for solution quality evaluation there is no mechanism for replacing heuristics with insight obtained from calculation and analysis	The priority and combination of multiple goals rely on domain expertise It is not easy to visualize the results for models with more than three goals; The tradeoff scenarios are limited	The emergent property of a system under interventions are not captured and incorporated into the model; There is no mechanism to identify critical and sensitive factors and quantify them with respect to system goals
After the New Knowledge	Solutions are on the boundary of the mathematical model but away from the physical boundary, so they are relatively insensitive to the variation in the boundary (formed by constraints or bounds); Designers can suggest model improvement regarding both goal completion and robustness of the solution; Designers can improve the goal completion without sacrificing the model's robustness, and vice versa		Heuristics in the algorithm are evaluated using criteria with consideration of goal completion, robustness, and computational complexity; Heuristics are replaced with insight obtained from post-solution analysis	Decision support on the priority and combination of the multiple goals can be provided to designers without any domain knowledge; Visualization of more than three goals is possible; More tradeoff scenarios are provided	Designers can capture and incorporate emergent properties of a system under interventions into the model; Designers can identify critical and sensitive factors and quantify them with respect to system goals

(continued)

Table 8.5 (continued)

Chapter	Ch4		Ch5	Ch6	Ch7
Research Question (RQ)	RQ1: What is a method used to evolve the model boundary?		RQ2: What is the method to evolve the model to update metaheuristics?	RQ3: What is the method to speed up learning the system's nature?	RQ4: What method allows information to be passed through multiple system scales?
Answer to RQ – new knowledge	The Three-Step Exploration Method	The Formulation-Exploration Framework	The Adaptive Linear Programming Algorithm with Parameter learning (ALPPL)	The Adaptive Leveling-Weighting-Clustering Algorithm (ALWC)	Scenario planning workflow for Agent-Based Modeling
Publication	Guo, L., Zamanisabzi, H., Neeson, T. M., Allen, J. K., Mistree, F., (2018). "Managing Conflicting Water Resource Goals and Uncertainties in a Dam-Network by Exploring the Solution Space," Journal of Mechanical Design, 141(3): 031,702	Guo, L., Chen, S., Allen, J. K., Mistree, F., 2021,"A Framework for Designing the Customer Order Decoupling Point to Facilitate Mass Customization," *ASME Journal of Mechanical Design* 143(2):022,002	Guo, L., Nellippallil, A. B., Smith, W. F., Allen, J. K., Mistree, F., "Adaptive Linear Programming Algorithm with Parameter Learning for Managing Engineering-Design Problems," under review, ASME Design Automation Conference, St. Louis, MO, USA, Paper number: 19643	Guo, L., Milisavljevic-Syed, J., Allen, J. K., Mistree, F., "Formulation-Exploration Model for Managing Many-Goal Engineering Design Problems	Guo, L., Mohebbi, S., Das, A., Allen, J. K., and Mistree, F. (2020). A Framework for the Exploration of Critical Factors on Promoting Two-Season Cultivation in India. Journal of Mechanical Design, 142(12):124,503

Table 8.6 Summary of thelr types of robust design

Questions regarding RD Type	RDI-II: What mechanisms and procedures enable the exploration of the solutions relatively insensitive to Type I and Type II uncertainty?		RDIII: What mathematics in the design method allows exploring solutions that are relatively insensitive to Type III uncertainty?	RDIV: What method allows sensing and managing Type IV uncertainty?	
Realization of RD – Proposed Method	M1: Formulation-Exploration Framework		M2: Adaptive Linear Programming with Parameter Learning (ALPPL)	M3: Adaptive Leveling-Weighting-Clustering Algorithm (ALWC)	M4: Scenario Planning in Agent-Based Modeling
Chapter	Ch 4		Ch 5	Ch 6	Ch 7
Uncertainty / Test Problem	T1.1: Dam network planning	T1.2: CODP positioning for a supply chain	T2: Improving the hot rod rolling process chain	T3: Rankine cycle thermal system design	T4: Learning emergent properties of a social system under interventions
Type I	Uncertainty in timing and amount of inflow – Table 4.8	Uncertainty in demand side – Figure 4.25	Uncertainty in hyperparameter settings – Table 5.8, Figure 5.16	Uncertainty in parameter setting in solution algorithm (Starting point of searching) – call XPLORE in DSIDES	Uncertainty in price (Price of agriculture products) – α, Table 7.8, Figure 7.11 (a) and (c)
Type II	Uncertainty in outflow (water release target) – Table 4.6	Uncertainty on the supply side - Table 4.16	Uncertainty in user preferences – Table 5.7		Promotion effort and timing – Tables 7.8-7.10, Figure 7.8-7.10
Type III			Uncertainty in model approximation due to heuristics in approximation – Table 5.6	Uncertainty in model approximation (ways of combining multiple goals) – leveling-weighting-clustering algorithms (Figure 6.13)	
Type IV				Uncertainty in using domain knowledge to simplify the model (fixing decision variables and selecting design scenarios) – “fixing variable” and “XPLORE” modules in DSIDES, and leveling and clustering	Interventions that change the mathematical relation among promotion and result (developing local market) – β_2, Table 7.8, Figure 7.5, Figure 7.11 (b) and (c)

M – method
RD – robust design
T-test problem

The application fields of the design evolution cycle are not limited to engineering design, although the concept of satisficing and model evolution was initially inspired by numerous engineering-design challenges. In fact, *any problem* exhibiting the aforementioned features falls within the scope of the design evolution cycle's applicability. The design evolution cycle is an open framework, and multiple stages may have different representations for connecting them. The methods proposed in this monograph are four examples of many implementations for connecting the stages and evolving the design. The contributions, potential contributions, and scope of application of each method (M1-M4) are summarized hereafter.

8.3.2 Application Scope of M1—Formulation-Exploration Framework

The three-step method can be used to enhance the robustness of a mechanical engineering system. In mechanical design, when multiple goals conflict with each other and design preferences evolve over time, designers can use step 1 to explore the weight space and dynamically establish rules on weight selection. To enhance the system's robustness, a reasonable buffer is added to ensure that the solution does not approach the physical boundary too closely, making it relatively insensitive to uncertainties. Our three-step method is a way of doing this. The advantage of our method is that it boosts the potential of the physical system while enhancing its robustness; hence, we neither sacrifice system robustness for better performance nor the opposite. Additionally, the satisficing space provides designers with a relatively insensitive design space and options with awareness of the system output, which is useful in concurrent and multistage design.

The Formulation-Exploration framework can be used to design mass-customized products at mass-production costs. It is especially useful for continuously improving a system with conflicting goals, evolving design preferences, and multiple stakeholders pursuing common interests.

8.3.3 Application Scope of M2—Adaptive Linear Programming Algorithm with Parameter Learning (ALPPL)

ALPPL can be applied to multi-goal engineering-design problems, especially when goals conflict with one another, the priority of the goals evolve with the environment, and the model's output must be insensitive to errors and variations. Rule-based parameter learning can be used to improve algorithms, especially when there are no customizable criteria for evaluating algorithm performance or when algorithm performance is highly sensitive to some critical parameters that are determined with heuristics or human intuition.

8.3.4 Application Scope of M3—Adaptive Leveling-Weighting-Clustering Algorithm (ALWC)

For concurrent engineering design problems with conflicting and evolving requirements across multiple disciplines that are unclear to designers in the modelling stage, managing the interactions and coupling effects between subsystems using the combination forms of the goals is challenging. Using the ALWC algorithm, designers can capture knowledge of the division, interaction, and coupling scenarios among the subsystems. Therefore, they can concurrently leverage tradeoffs among conflicting requirements.

8.3.5 Application Scope of M4—Scenario-Planning for Simulations

This proposed scenario-planning process allows designers in various fields to perform simulations to identify critical system factors and select specific scenarios that accommodate different site-specific input values or domain-dependent knowledge to achieve their goals.

8.4 Generalization of the Methods—Other Examples

8.4.1 Network Planning for Improving Hospital Visiting Process

– An Expansion of Dam Network Planning (Sect. 4.2)

Patients visiting hospitals may experience long wait times due to operational inefficiencies and bottlenecks. Additionally, long wait times decrease patient satisfaction and happiness. Healthcare systems can be modeled in various ways, including agent-based models, which can track individual patients and their movements through a hospital, and network flow models, which can model larger groups of patients and their interaction within a hospital. While agent-based models can detect bottlenecks by evaluating patient flow speed, network flow models can easily detect bottlenecks by applying an abstraction layer to the healthcare network. In network flow models, patients flow along the capacitated edges of a network while receiving treatment at the nodes, making it easy to identify bottlenecks by examining the flow in and out of nodes. These bottlenecks manifest in metrics used to evaluate the model run, including efficiency and wait times. Data regarding the capacities of the edges for a network flow model are derived from an agent-based model of a primary care clinic case study. Data describing patient flow through network facilities is difficult to find, so synthetic data is generated using an adaptation of the random subspace method. Ensemble runs of the network flow model are created to account for uncertainty in synthetic data, culminating in a distribution analysis for various metrics. By changing the topology of the network flow model, bottlenecks are removed, increasing efficiency and decreasing patient wait times. Furthermore, the network flow model is sensitive to the constraints of the random subspace, so care should be taken when initializing the model.

There are several potential contributions in this test problem. Synthetic data is generated via the Random Subspace Method, and data populates the various ensemble models. Bottlenecks readily appear in network flow models and are identified and assessed for potential removal. Moreover, modifying the topology of the model when wait times are long can partially remove a bottleneck at a node, leading to shorter wait times and increased efficiency. If necessary, leveraging model parameters from multiple types of models (agent-based model, system dynamics, and queueing model) can allow designers to cross validate the parameter and scenario settings through multiple models. For example, the statistics such as average patient wait times, median wait times, and standard deviation of wait times, etc., remain similar among various models, indicating that network flow models can represent the same system effectively.

An opportunity for future work is to apply the network flow model to a situation where only limited information is known and few parameters can be controlled, such as the topology and the number of patients flowing into the system. The goal is to achieve a certain level of efficiency. This approach shares several principles with bagging and training neural networks. The efficiency of the system is a function that depends on all the underlying variables while being subject to certain constraints. By taking the gradient of this function, we can determine the direction to adjust the underlying variables to achieve sufficient efficiency. In this more general solution, a larger solution space is considered, allowing for a satisficing solution set to the given problem.

8.4.2 Leveraging Social Drivers in Rural Development

– An Expansion of Learning Emergent Properties of a Social System Under Interventions

Development in rural regions often depends on policy actions by the government or social entrepreneurs. Generally, these policies are designed to motivate changes through economic incentives, but this is only a short-term solution. For policies to be sustainable and successful, long-term results must be produced, and progress should not always rely on continuous funding. This poses a challenge in measuring the long-term effects of short-term policies on individual behavior. We hypothesize that qualitative factors such as social drivers are the primary motivators for long-term change and that Agent-Based Modeling can be used to simulate and predict villager behavior to model the effects of policy on social interaction and behavioral patterns.

In this monograph we offer as a proof of concept that economic, social, and environmental drivers should be considered in modeling behavioral reactions to policy, not just economic drivers, and that agent-based modeling effectively simulates the propagation of their effects. Most critically, the difference between the socio-economic impact of a driver, such as access to electricity, which only alters productivity, versus a driver, like family life which influences the adoption of those in a villager's social network, shows that social impacts not only affect behavior but changes in them across network interactions can have a more drastic impact than the more commonly considered purely economic factors, which is a vital shift for development of rural communities.

Therefore, the methods and conclusions documented in this monograph can be applied in numerous applications where economic considerations have been the primary concern but have underlying social implications. Further work is needed to model social drivers within networks across disciplines, from how people react in healthcare to how education can more effectively engage students. Even complex systems such as welfare, policing or the judicial system, which generally focus on cost and data-driven deployment of resources, can be evaluated from this perspective, considering the social network roles of those administering these programs and systems and those using or impacted by them.

8.4.3 Knowledge Management in Designing Cyber-Physical Product-Service Systems

– An Expansion of Positioning Codp in a Supply Chain

The automation and intelligence highlighted in Industry 4.0, 5.0, and 6.0 place higher requirements for reasonable tradeoffs between humans and machines for decision-making governance. However, in the context of Industry 4.0, the vision for decision support for design engineering is still unclear. Additionally, the corresponding methods and system architectures are lacking to support the realization of value-chain-centric complex engineered systems design lifecycles. Hence, we identify decision support demands for complex engineered systems designs in the Industry 4.0 era, representing integrated design problems at various stages of the product value chain. In response, in this monograph, the architecture of a Knowledge-Based Design Guidance System (KBDGS) Chaps. 3 and 4 for cloud-based decision support is presented that highlights the integrated management of complexity, uncertainty, and knowledge in designing decision workflows, as well as systematic design guidance to find satisfying solutions with the iterative process "formulation-refinement-exploration-improvement." The KBDGS can be designed to facilitate diverse multi-stakeholder collaborative decisions in end-to-end cloud services. Our aim, is to provide design guidance to facilitate knowledge discovery, capturing, and reuse in the context of decision-centric digital design, thus improving the efficiency and effectiveness of decision-making. This, in turn, improves the efficiency and effectiveness of decision-making and supports the evolution of decision support in the field of design engineering for the Industry 4.0; see Jiao et al., (2021) and Allen et al., (2023).

8.5 Role of This Chapter in This Monograph

In this chapter, the contributions documented in this monograph are summarized in Sect. 8.2, where we review the theoretical foundation, recap the test problems, conclude the answers to the research questions and how we achieve the four types of robustness. The application scope of the design evolution cycle and each method is discussed in Sect. 8.3. Other examples which are extensions of the methods and test problems in this monograph are presented in Sect. 8.4.

In summary, in Chapter 8, we provide the empirical performance validity (see Fig. 8.2, also, the same as Fig. 1.2). The empirical performance validity is assessed by reviewing the differences made by using the proposed methods and clarifying their scope of application.

References

Allen, J. K., Nellippallil, A. B., Ming, Z., Milisavljevic-Syed, J., & Mistree, F. (2023). Designing evolving cyber-physical-social systems: Computational research opportunities. *ASME Journal of Computing and Information Science in Engineering, 23*(6), Article 060815.

Chen, W., Allen, J. K., Tsui, K.-L., & Mistree, F. (1996). A procedure for robust design: Minimizing variations caused by noise factors and control factors. *ASME Journal of Mechanical Design, 118*(4), 478–485.

Chen, W., Simpson, T. W., Allen, J. K., & Mistree, F. (1999). Satisfying ranged sets of design requirements using design capability indices as metrics. *Engineering Optimization, 31*(5), 615–619.

Choi, H.-J., Austin, R., Allen, J. K., McDowell, D. L., Mistree, F., & Benson, D. J. (2005b). An approach for robust design of reactive powder metal mixtures based on non-deterministic micro-scale shock simulation. *Journal of Computer-Aided Materials Design, 12*, 57–85.

Choi, H.-J., Allen, J. K., Rosen, D., McDowell, D. L., & Mistree, F. (2005). An inductive design exploration method for the integrated design of multi-scale materials and products. In *ASME design automation conference*. Paper Number DETC2005-85335.

Guo, L., Allen, J. K., & Mistree, F. (2024). Optimize or satisfice in engineering design? *Research in Engineering Design, 35*(3), 239–267.

Jiao, R., Commuri, S., Panchal, J., Milisavljevic-Syed, J., Allen, J. K., Mistree, F., & Schaefer, D. (2021). Design engineering in the age of industry 4.0. *ASME Journal of Mechanical Design, 143*(7), 070801.

Nellippallil, A. B., Song, K. N., Goh, C.-H., Zagade, P., Gautham, B., Allen, J. K., & Mistree, F. (2017). A goal-oriented, sequential, inverse design method for the horizontal integration of a multistage hot rod rolling system. *ASME Journal of Mechanical Design, 139*(3), Article 031403.

Taguchi, G. (1993). Robust technology development. *Mechanical Engineering-CIME, 115*(3), 60–63.

Chapter 9
Closing Remarks—Advancing Design Evolution in Other Disciplines

A Vision for Future Research in the Realization of Model-Based Complex Systems

In this chapter we frame the design of complex systems as an exercise in architecting a satisficing strategy. We use that strategy to address uncertainty, promote adaptability, and respond to evolving requirements. Designers now leverage technologies such as RFID, the Internet of Things, and AI, including large language models, to guide industry beyond the Industry 4.0 paradigm and into successive generations. Designers find that these tools expand available options while also introducing new layers of complexity in decision making and system evolution.

We introduce a framework, concepts, methods, and algorithms that address contemporary challenges while anticipating and managing future complexities. As AI technologies advance, we transform design to unlock new imagination and possibilities. We observe that open innovation and open-source development reshape how users contribute to and participate in the evolution of complex systems. By embracing these advancements, we move toward a future where design is more adaptive, collaborative, and intelligent.

We illustrate the necessity of model-based evolution in complex system design by examining a range of challenges throughout this monograph. From the intricate design of a Rankine cycle to the challenges of sustainable development in a village, from handling water releases satisfying the diverse needs of river basin stakeholders to adaptively fine-tuning the solution space of a hot-rod rolling process by judiciously adjusting the algorithm's hyperparameters, these stories underscore the importance of managing complexity through a structured, iterative process. We confirmed the effectiveness of these strategies through verification across chapters and validation in Chap. 8, both in specific applications and in broader contexts.

We began by exploring the motivation behind satisficing methods, driven by the need to pragmatically manage complexity in design evolution. Unlike optimization, in which we seek a single best solution, satisficing embraces adaptability—delivering solutions that balance multiple competing objectives while mitigating uncertainty. We delved into the theoretical underpinnings and practical applications, seeing how these approaches excel where optimization alone cannot, particularly in nonlinear,

L. Guo et al., *Architecting a Satisficing Strategy for Designing Complex Systems*,
https://doi.org/10.1007/978-3-032-04577-5_9

multi-goal problems. Throughout this monograph, we have developed and applied a satisficing strategy for designing complex systems. We propose the design evolution cycle that includes formulation, approximation, exploration, and evaluation, as the core of our approach. We rely on the cycle's structured framework to manage uncertainty, refine models, and ensure adaptability. By architecting a satisficing strategy, we enable sustained viability in system design, balancing competing objectives and constraints.

In this chapter, we reflect on our overarching goal: establishing a decision-based approach that enables us to navigate vast solution spaces with limited prior domain knowledge while ensuring robust, adaptable, and effective designs. We recognize that principles for robust design in complex systems extend beyond engineering. We can adapt these principles to various disciplines as a blueprint for advancing design evolution in new fields. In this chapter, we summarize key insights, reinforcing the importance of these methodologies and outlining their potential to evolve and inspire innovation in other domains.

9.1 Organization of This Chapter

In this chapter, we begin in Sect. 9.2 as we revisit our motivation, key contributions, verification, validation, and related publications, and we show these links in Fig. 9.1. In Sect. 9.3 we build on this foundation and outline the current research thrusts and applications while we frame the "what, how, and why" questions presented in Table 9.1. In Sect. 9.4 we extend our outlook as we group the thrusts into focused future investigations and cross disciplinary opportunities that connect our design evolution cycle to broader societal contexts. We culminate with a vision for future research in Sect. 9.5.

9.2 Summary of This Monograph

9.2.1 Motivation of Design Evolution Using Satisficing Strategy

Building on the overview in Sect. 9.1, we recognize that researchers use models to approximate the real world, and we observe sensitivity in some solutions to limitations in those decision models.

An optimal solution in a mathematical model may fail in the physical system (Fig. 9.2). Although a solution may work as a near-optimal, feasible answer within the mathematical framework, it can falter in practice due to incomplete real-time information. For instance, neglecting a key constraint in ship design might yield a submarine (Fig. 9.3); or, with poor-quality data, designers may overfit noisy inputs,

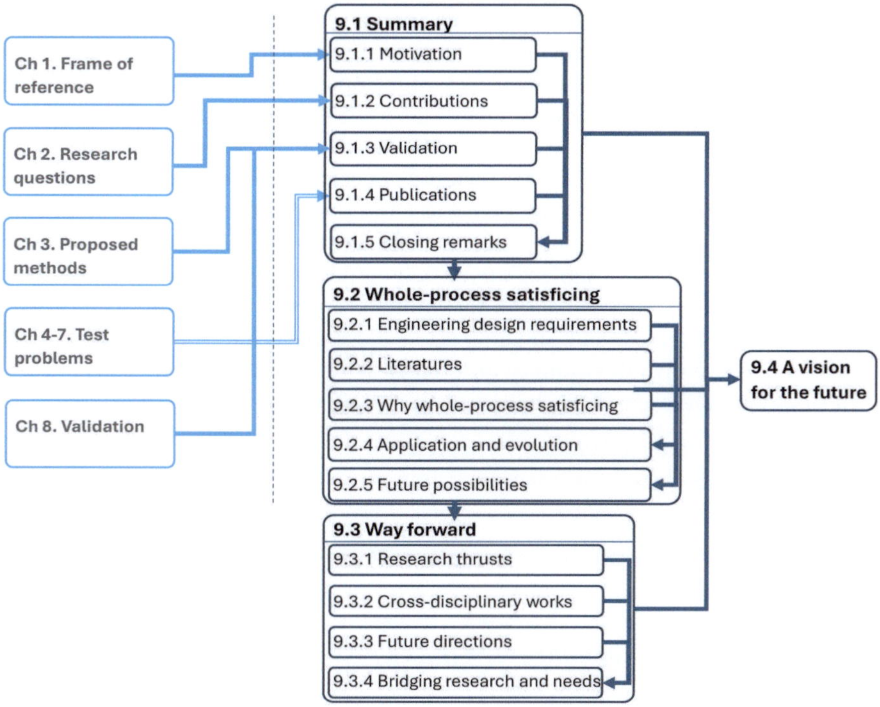

Fig. 9.1 Organization of this chapter

Table 9.1 Summary interpretation of the problems investigated in Chap. 3

Elements	What?
Components	Summaries of what we have done in this monograph and anticipation of the research and applications way forward
Connectors	The broader impact of the methods and applications in this monograph
Form	*How?*
Component roles	Summarizing the monograph and anticipating the future based on it
Properties	Conclusion and "I statement"
Relationship	Logic between what we can do and what we will do
Rationale	*Why?*
Motivation	Provide support to assist designers in taking the design evolution cycle forward and identifying more applications
Assumptions	Every complex system may encounter four types of uncertainty
Interpretation	Reemphasizing how the design evolution cycle can be enriched and useful to other problems

leading to flawed decisions (Fig. 9.4). In complex systems with emergent behaviors, outcomes remain unpredictable even if designers and participants understand the complexity, such as the growing day-night power gap from widespread solar adoption, known as the "California Duck" (Fig. 9.5). These are typical examples of George Box's famous statement "All models are wrong, but some are useful."

Based on these observations, we conclude that an intellectual disconnect exists between the model world and the physical world, as illustrated in Fig. 9.6. In the model world, we prefer: (i) models to be linear, continuous, and convex, (ii) many control factors (decision variables) but few requirements (goals and constraints), (iii) confirmed and known combinations for multiple goals, (iv) certainty—or at least predictable uncertainties, and, in summary, (v) complete information. In contrast, the physical world operates entirely differently.

In summary, there are three challenges in designing complex systems:

(i) Incomplete and inaccurate information
(ii) Excessive restrictive optimality conditions
(iii) Uncertainties.

Incomplete and inaccurate information—Given a nonlinear, multi-objective problem:

$$\boldsymbol{x}^* = \underset{x \in \mathcal{F}}{\boldsymbol{argmin}} \boldsymbol{F}(\boldsymbol{x}) = [\boldsymbol{f}_1(\boldsymbol{x}), \boldsymbol{f}_2(\boldsymbol{x}), \ldots, \boldsymbol{f}_k(\boldsymbol{x})] \tag{9.1}$$

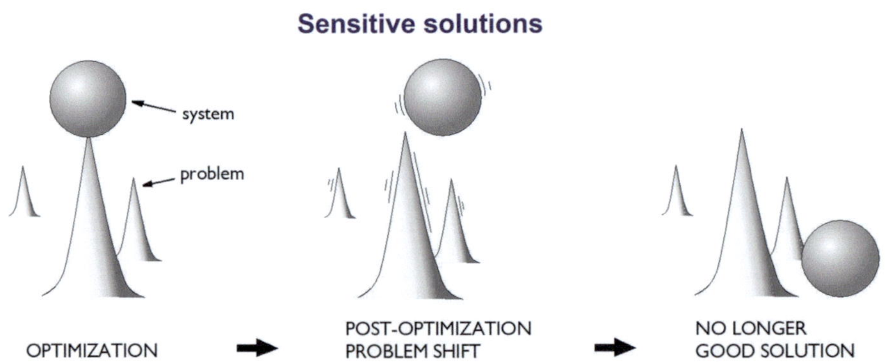

Fig. 9.2 Optimal solutions are sensitive to model error or inaccuracy

Fig. 9.3 Design failure due to missing a constraint

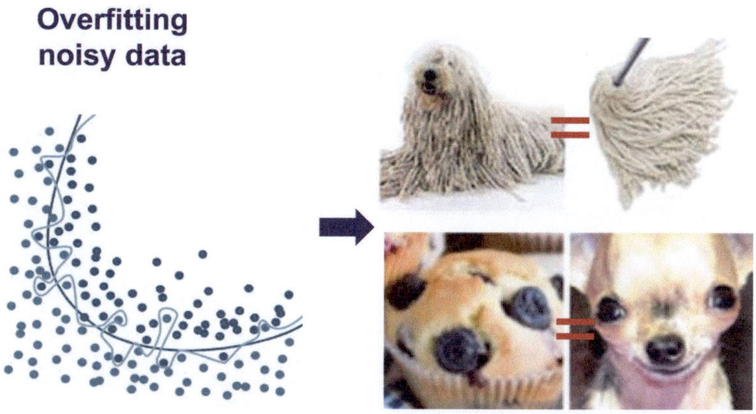

Fig. 9.4 Model failure due to lacking sufficient, high-quality data[1]

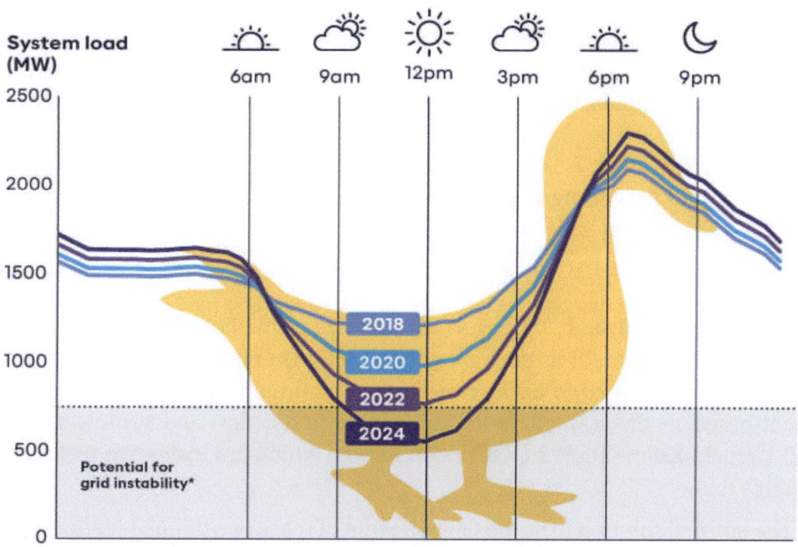

Fig. 9.5 California duck due to solar panel adoption[2]

$$\text{where } \mathcal{F} = \left\{ \boldsymbol{x} \in \mathbb{R}^n | \boldsymbol{g}_i(\boldsymbol{x}) \geq 0, \forall \boldsymbol{i} = 1, \ldots, \boldsymbol{l}; \boldsymbol{h}_j(\boldsymbol{x}) = 0, \forall \boldsymbol{j} = \boldsymbol{l} + 1, \ldots, \boldsymbol{m} \right\} \tag{9.2}$$

---

[1] https://www.metamaven.com/chihuahua-vs-muffin-comparison-computer-vision-apis/.

[2] https://essolar.com/blogs/es-solar/eliminating-the-duck-curve-in-california-with-solar-and-A-battery.

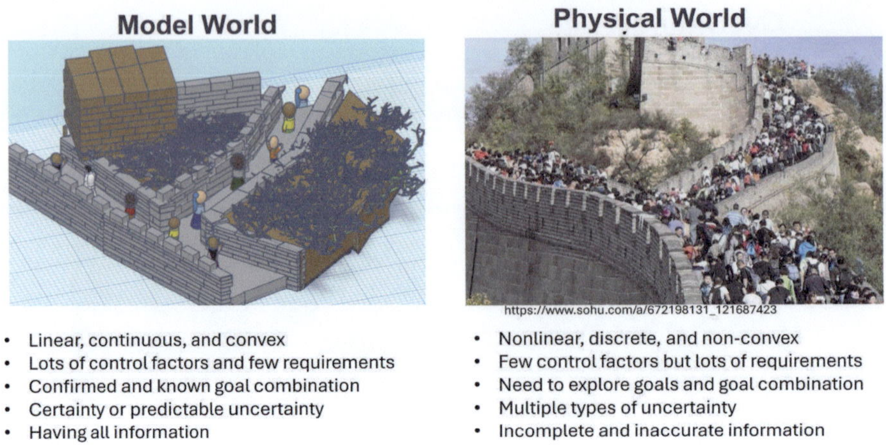

Fig. 9.6 Intellectual disconnection between model world and physical world[3]

- $m \gg n$—Sometimes, the number of equations (including objectives and constraints) far exceeds the dimensionality (the number of decision variables), meaning the simplex tableau becomes a tall, narrow matrix that is difficult to solve.
- $\forall i, j \in \{1,2,\ldots,k\}, i \neq j$, the interdependence between $f_i(x)$ and $f_j(x)$ are unknown, and the orthogonality (Eq. 9.3) changes across $\mathcal{F}$.

$$(\boldsymbol{x}) = \nabla \boldsymbol{f}_i(x)_{\mathrm{T}} \nabla \boldsymbol{f}_j(\boldsymbol{x}) \tag{9.3}$$

As Albert Einstein said, "So far as the theories of mathematics are about reality, they are not certain; so far as they are certain, they are not about reality."

Excessive restrictive optimality conditions—optimality conditions, especially for gradient-based methods, include the necessary (first-order) and sufficient (second-order) Karush–Kuhn–Tucker (KKT) conditions, which are excessive restrictive, as it requires that.

(i) The optimal solution to be the tangent point of these two things—defined as "(a)" and "(b)" here: (a) the linear combination (using the of the Lagrange multipliers) of all equality constraints and the surface of all inequality constraints (hereafter, we refer to "(a)" as "the feasible surface" for short), and (b) a contour of the objective, which is the necessary KKT condition; see Figs. 9.7 and 9.8—the first order condition can be me and cannot be met, respectively.

(ii) The convexity of "(a)" is no less than the convexity of "(b)". Here, we define the convexity as the diagonal elements of the Hessian Matrix as defined in Mistree et al., (1981); three situations may be involved into the sufficient condition—see Figs. 9.9, 9.10, and 9.11. However, some other situations may not meet the sufficient condition, for example, see Fig. 9.12.

[3] https://www.sohu.com/a/672198131_121687423.

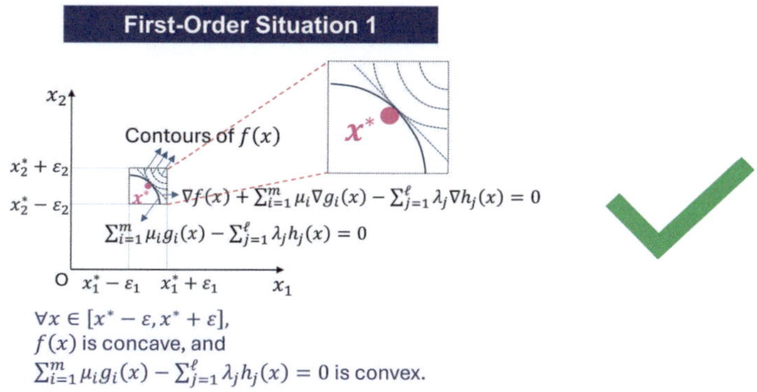

Fig. 9.7 Meeting the Necessary (First-Order) KKT Condition Means the Optimal Solution is a Tangent Point of **a** the linear combination (using the Lagrange multipliers) of all equality constraints and the surface of all inequality constraints (hereafter, we refer to "(a)" as "the feasible surface" for short), and **b** a contour of the objective

Fig. 9.8 Not Meeting the Necessary (First-Order) KKT Condition Means One Cannot Find a Tangent Point of **a** the feasible surface, and **b** a contour of the objective

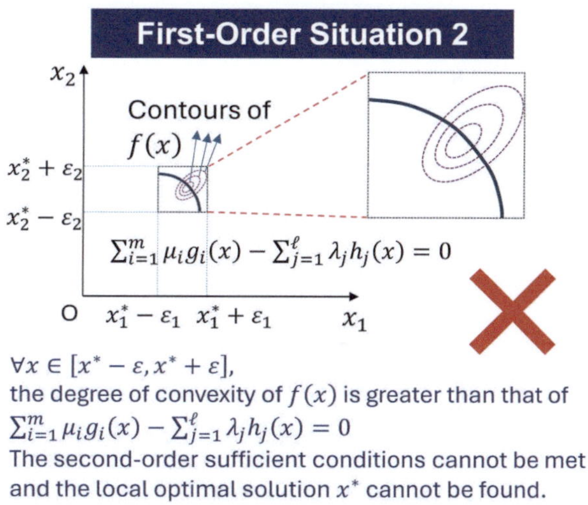

When using optimization, one has to formulate a model that only in the situations in Figs. 9.9, 9.10, and 9.11, which can be over-simplified. In contrast, when we apply a satisficing strategy, we can identify 'good enough' solutions to realistic but complex problems, as shown in Fig. 9.12. As Herbert Simon said, "The decision maker has a choice between an optimal decision from an *imaginary simplified world*, or decisions that are 'good enough', that *satisfice*, for a world *approximating* the complex real one more closely."

Uncertainties—The probability of maintaining an optimal solution x^* under uncertainty P is low:

The first-order differentiation equation of Lagrange equation is Eq. 9.4.

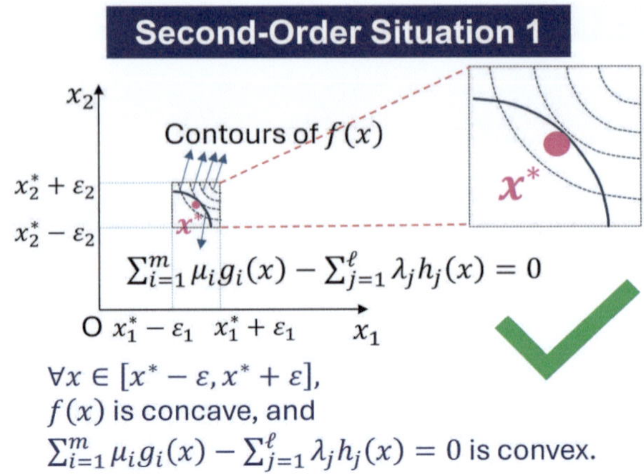

Fig. 9.9 Meeting the Sufficient (Second-Order) KKT Condition—Situation 1: **a** is convex and **b** is concave

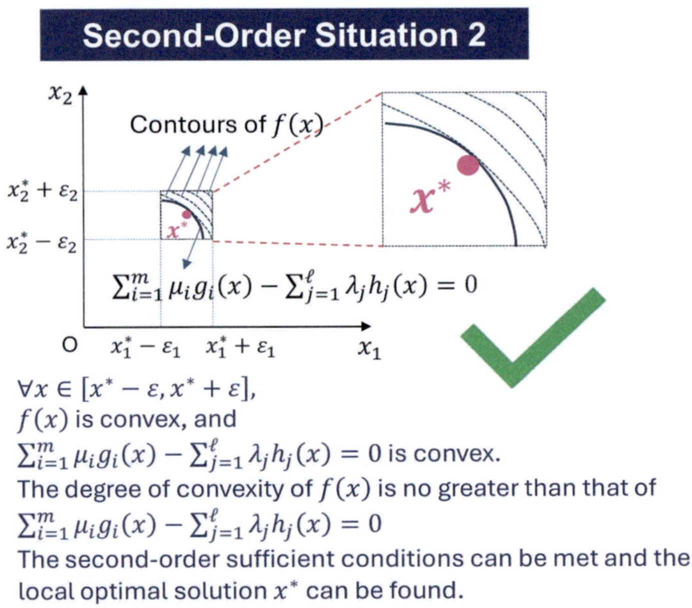

Fig. 9.10 Meeting the Sufficient (Second-Order) KKT Condition—Situation 2: both (**a**) and (**b**) are convex, whereas the convexity of (**a**) is higher than the convexity of (**b**)

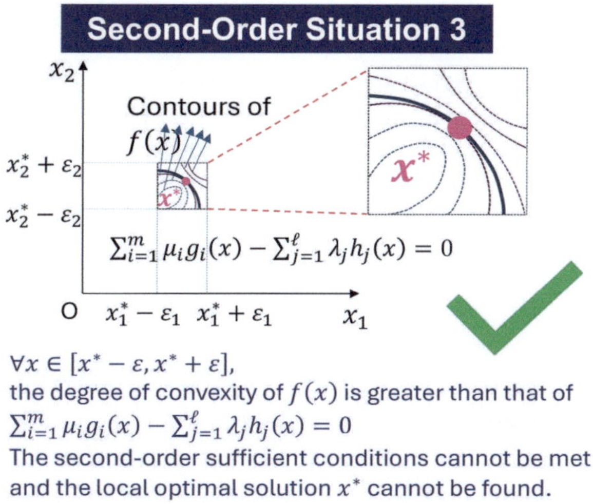

Fig. 9.11 Meeting the Sufficient (Second-Order) KKT Condition—Situation 3: the convexity of (**b**) is higher than the convexity of (**a**) in the feasible space whereas the convexity of (**b**) may be lower than the convexity of (**a**) in the infeasible space, but there is a tangent point of (**a**) and (**b**)

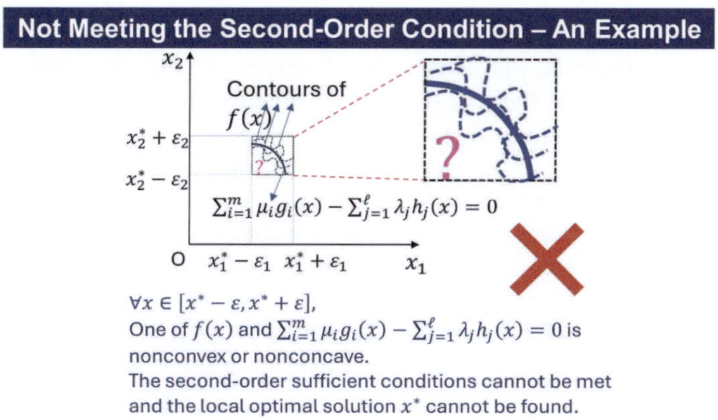

Fig. 9.12 Not Meeting the Sufficient (Second-Order) KKT Condition—an example: at least one of (**a**) and (**b**) is neither convex nor concave

$$\nabla_x L(\boldsymbol{x}, \boldsymbol{\mu}, \boldsymbol{\lambda}) = \mathcal{y}(\mathcal{P}, \boldsymbol{x}, \boldsymbol{\mu}, \boldsymbol{\lambda}, \mathcal{p}), \tag{9.4}$$

where x is the decision variables, μ and λ are Lagrange multipliers, $\mathcal{P}$ are parameters (or coefficients), $\mathcal{p}$ are parameters (for example, weights) to combine the multiple objectives. The second-order differentiation equation of Lagrange equation is Eq. 9.5.

$$\nabla_{xx}^2 L(\boldsymbol{x}, \boldsymbol{\mu}, \boldsymbol{\lambda}) = \nabla_x \mathcal{y}(\mathcal{P}, \boldsymbol{x}, \boldsymbol{\mu}, \boldsymbol{\lambda}, \mathcal{p}) = \mathcal{y}'(\mathcal{P}', \boldsymbol{x}) \tag{9.5}$$

Therefore, the probability of maintaining the optimal solution x^* of a K-objective, N-dimensional, and Q-parameter problem under uncertainty P, represented as Eq. 9.6, can be low, as the right-hand side of Eq. 9.6 consists of many elements, and only none of them is affected by the uncertainty P regarding their value, distribution, etc., can the optimal solution x^* still be optimal under uncertainty P.

$$\boldsymbol{Pr(x^*|P) = \prod_{q=1}^{Q}[1 - Pr(\widetilde{\mathcal{P}_q}|P)]\prod_{n=1}^{N}[1 - Pr(\widetilde{x_n}|P)]\prod_{i=1}^{m}[1 - Pr(\widetilde{\mu_i}|P)]\prod_{j=1}^{\ell}[1 - Pr(\widetilde{\lambda_j}|P)]\prod_{k=1}^{K}[1 - Pr(\widetilde{\wp_k}|P)],} \tag{9.6}$$

where:

- $Pr\left(\widetilde{\mathcal{P}_q}\ |P\right)$ represents the probability that any of the coefficient in the qth objective has a variation under uncertainty P,
- $Pr\left(\widetilde{x_n}\ |P\right)$ represents the probability that the nth decision variable x_n has a variation under uncertainty P,
- $Pr(\widetilde{\mu_i}\ |P)$ and $Pr(\widetilde{\lambda_j}\ |P)$ represent the probabilities that the Lagrange multipliers μ_i and λ_j have any variation under uncertainty P, respectively, and
- $Pr(\widetilde{\wp_k}|P)$ represents the probability that the weight of the kth objective has a variation under uncertainty P.

What are uncertainties? As we discussed in Chaps. 1 and 2 (Sects. 1.2 and 2.4) summarize here, there are four types of uncertainty—they are all considered as the uncertainty P in Eq. 9.6. The four types of uncertainty, their sources, and examples in a supply chain are listed in Table 9.2.

A relatively robust design or complex system should be adaptable to these types of uncertainty, as Charles Darwin said, "It is not the *strongest* of the species that survives, nor the *most intelligent*; it is the one *most adaptable* to change." Therefore, in this monograph, we propose a strategy that is beyond optimization—the satisficing strategy, to manage the three challenges.

"Satisfice" consists of two words—"satisfy" and "suffice." Satisfice is an alternative strategy of the utility maximization in optimization (Kapteyn et al., 1979).

Table 9.2 Four types of uncertainty—sources and examples

Type	Source	Examples in supply chains
Type I: noise factors (Taguchi, 1985)	Uncontrollable parameters causing variations	Fluctuations in market demand
Type II: control factors (1996b; Chen et al., 1996a)	Controllable variables introducing uncertainties	Variability in productivity
Type III: model structure (2006b; Allen et al., 2006a)	Arising from inherent complexity	Supply chain disruptions due to pandemic
Type IV: process chain (Nellippallil et al., 2022a, 2022b, 2022c)	New uncertainties emerging when addressing the first three	Bullwhip effect

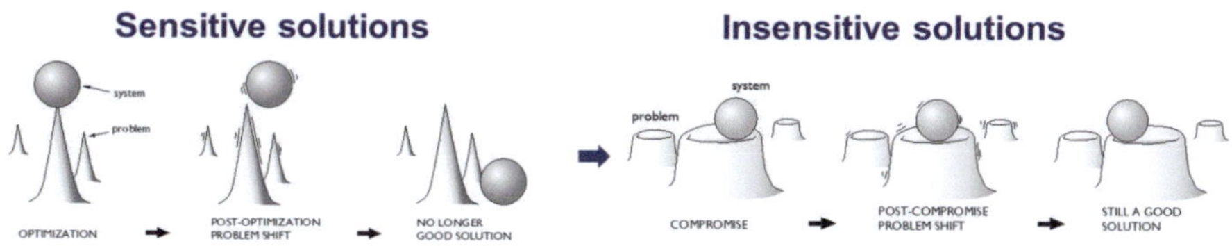

Fig. 9.13 Optimal (Sensitive) solution versus satisficing (Insensitive) solution

Fig. 9.14 The evolution cycle of complex systems realization

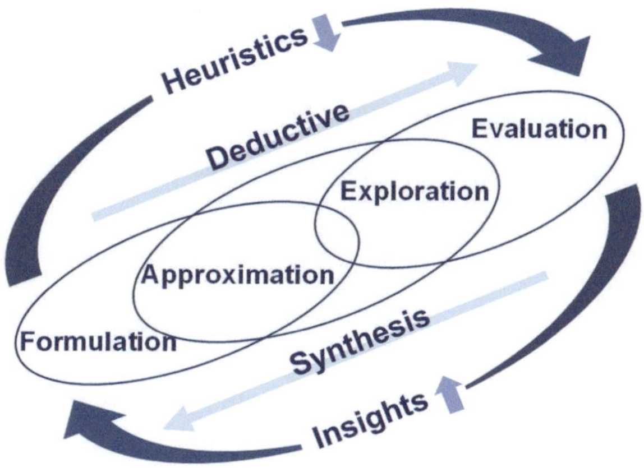

Beyond optimization, we propose using a satisficing strategy that results in solutions that meet an acceptable threshold and are satisfactory, but not necessarily optimal (Simon, 1947, 1956). The purpose is to identify good enough solutions that are relatively insensitive to uncertainties (Fig. 9.13).

To realize the satisficing strategy, we propose the design evolution cycle (Fig. 9.14). We improve model formulation, approximation, exploration, and evaluation through design evolution, and we consider this practice essential for obtaining a solution space that remains relatively insensitive to model errors and inaccuracies. In design evolution, two capabilities of a model are improved:

(1) Model accuracy, as one of the foci of decision methods—the model's capability to capture and incorporate more useful information about the physical world.
(2) Model robustness, as one of the foci of *satisficing* methods—the model's capability to deliver solutions relatively insensitive to uncertainties.

9.2.2 Contributions—Recapping Research Questions and Answers Leading to New Knowledge

We pose four research questions (RQ1-4) and hypotheses (H1-4) to create new knowledge and achieve design evolution.

RQ1: What method is used to evolve the model boundary?

H1: Explore the sensitivity of the segments of the model boundary and improve accordingly.

New Knowledge—M1: Formulation-Exploration framework.

RQ2: What method is used to speed up learning the system's nature?

H2: Learn, evaluate, and update metaheuristics to improve model performance.

New Knowledge—M2: The algorithm of the ALPPL (adaptive linear programming with parameter learning).

RQ3: What method is used to speed up learning the system's structure.

H3: Learn the system's nature, such as the interrelationships among subsystems, and reorganize them based on it.

New Knowledge—M3: The ALWC (adaptive leveling-weighting-clustering) algorithm.

RQ4: What method is used to pass information through multiple system scales?

H4: Capture and quantify emergent properties through scenario planning in simulation.

New Knowledge—M4: Scenario planning workflow for Agent-Based Modeling.

After stating the preceding questions we examine how our verification and validation efforts in Sect. 9.1.3 confirm the knowledge created.

9.2.3 Verification and Validation

Method (theoretical structural validity): To achieve design evolution, we propose the four-stage design evolution cycle—formulation, approximation, exploration, and evaluation. A complex system's model can continuously be improved through evolution by managing complexity and uncertainty among different stages.

Test problems and experiments (empirical structural validity): We use five test problems to illustrate design evolution:

- A dam-network planning problem (T1.1, Sect. 4.2).
- A supply-chain-design problem (T1.2, Sect. 4.3) for design evolution by identifying and removing bottlenecks through evolving the boundary of a continuous and a discrete model.
- A multi-stage manufacturing problem (T2 hot rod rolling process chain, Chap. 5) for design evolution by improving the approximation algorithm through parameter learning.

- A concurrent engineering-design problem (T3 Rankine cycle thermal system, Chap. 6) for design evolution by reformulating the many-objective scalarization function.
- A social system design problem (T4 Learning emergent properties of a social system under interventions, Chap. 7) for design evolution by capturing and incorporating emergent properties in the system.

Results (Empirical Performance Validity): The proposed Formulation-Exploration framework facilitates designers to improve the robustness and achievement of system goals by evolving the model boundary. The proposed ALPPL algorithm allows models to return solutions that are relatively robust to scenario changes; the proposed ALWC algorithm supports design evolution by improving the diversity of the solution space. The proposed multi-scale simulation framework helps decision-makers leverage critical factors to achieve system goals.

Way Forward (Theoretical Performance Validity): The design evolution concept and the derived methods can be used to improve the model of other complex systems, especially when there are more than three system goals, the goals conflict with one another, the changing environment brings complexity and uncertainties to modeling, and information sharing between components is insufficient. Examples include network planning for improving hospital visits, leveraging social drivers in rural development, and knowledge management in designing cyber-physical product-service systems.

The empirical results complete our validation. We next reflect on these findings in Sect. 9.1.4 to highlight their broader significance.

9.2.4 Closing Remarks of Sect. 9.2

In this monograph, we propose a design evolution cycle and derive corresponding methods and algorithms (M1-M4, Chap. 3, Sect. 3.3) for improving model robustness. Applying the proposed methods using five test problems (T1-T5, Chaps. 4–7) demonstrates their internal consistency and general utility. With the explosive development of Artificial Intelligence (AI) in the age of Industry 4.0 and 5.0, the topic of AI-based design evolution will emerge, leading to the creation of more advanced theories and applications for clarity and conciseness. With these findings in place, we next move to Sect. 9.3, where we examine how whole process satisficing builds directly on the design evolution cycle summarized here.

9.3 Whole-Process Satisficing

After we consolidate lessons in Sect. 9.2, we turn in Sect. 9.3 to complex design problems that motivate the whole process satisficing methods summarized later in Sect. 9.3.5. In complex design problems, usually, there are multiple goals with different units, continuous and discrete variables, nonlinear equations, nonconvex equations, and coupled decisions. Ideally, all targets are reached simultaneously within the feasible space. However, the optimal solution may not be available or robust to uncertainties. To deal with all those complexities, a modeling strategy named "satisficing" was proposed in the 1980s. The satisficing strategy allows designers to find "good enough" but may not be optimal solutions. In this section, we provide a broader view by reviewing the publications applying all kinds of satisficing strategies on engineering-design problems and categorize the methods regarding the design stages they manage. **We define the methods dealing with all four design stages—formulation, approximation, solution, and evaluation—as the whole process satisficing methods.** We review the publications using the whole-process satisficing strategy in detail. In the past 30 years, the whole-process satisficing strategy has been improved and applied to a wide variety of engineering-design problems, based on which derived methods, concepts, and platforms are developed. We generalize the specialties, advantages, and scope of applications of the methods in the whole-process satisficing strategy. We hope in this section, we provide information on when and how designers may apply satisficing for their problems.

9.3.1 Special Requirements for Methods in Engineering Design

In short, the special requirements for engineering-design methods include:

- ***Designers should not accept problematic assumptions when applying utility theory.***
- ***Complexities incorporated in engineering design cause convergence failure.***
- ***Multiple sources of variation bring uncertainties.***
- ***The tasks of engineering design include way more than finding the optimal solution.***

Engineering design is a task that involves a group of designers with different interests and knowledge to make decisions that comply with their mutual requirements (Reich et al., 2010). As decision-making requires designers to select the desired solution among several alternatives regarding the output performance, one or more objective functions are usually consciously or subconsciously applied by the designers to measure the output performance. Designers can define an objective function through different methods, for example, through utility theory (Stirling, 2005), game theory (Nash, 1950), analytic hierarchy process (AHP) (Saaty, 1979), Pareto comparisons

(Sen, 2018), etc. However, regardless of the seeming rationale that those methods present, the use of utility theory may result in poor decision-making and may not be appropriate for engineering design (Hazelrigg, 2010). In addition, it is difficult for designers to identify and maintain the optimal solution using optimization methods, given that there are various complexities and uncertainties underlying an engineering-design problem. Moreover, the mission of engineering design is more than obtaining optimal solutions. There are other tasks and foci, such as acquiring knowledge on the output performance in a wide range of solution space (not only the near-optimal area), improving the robustness of the decision model, coupling the continuous and discrete decisions, connecting sub-models with different levels of refinement or fidelity, and so on.

Designers accept problematic assumptions when applying utility theory. For optimization problems that minimize or maximize a mathematical objective, one can define that such problems are based on maximization of expected utility, which enables designers to select the optimal solution among alternatives. Nevertheless, not all design problems can be solved using optimization. Even if the optimal solution to the decision model is acquired, it may not be optimal or even feasible for the physical problem, as the optimality conditions can be easily broken (Guo, 2021). This often results from the incapability of the decision model to capture all information and requirements (Mistree et al., 1993) of the design. One source of the incompleteness and inaccuracy of the model is that by modeling the objective function using utility theory, designers naturally accept the assumption that all decision-makers are rational and every detail in the decision model perfectly maps the physical world to the model abstraction. Such an assumption is often wrong, especially in engineering design.

Complexities incorporated in engineering design cause convergence failure. Engineering-design problems encounter multiple complexities (Guo et al., 2022); therefore, the decision models may have nonlinear, nonconvex equations, multiple objectives with different units and scales, and objectives with different levels of achievability. These complexities sometimes lead to immature convergence or no convergence.

Multiple sources of variation bring uncertainties. How do uncertainties take place? From the view of the source of uncertainties, there are variations (i) in environmental or other noise-noise factors (Taguchi, 1985), (ii) in design (or decision) variables—control factors (Chen et al., 1996a, 1996b), (iii) brought by modeling methods (Allen et al., 2006a, 2006b), and (iv) brought by the process of managing the previous three types of variation (Choi et al., 2005a, 2005b). How do uncertainties affect the design? From the perspective of optimality conditions, uncertainties that break the equilibrium of any Karush–Kuhn–Tucker (KKT) conditions destroy the optimality of a solution, thereby making an optimal solution infeasible or useless (Guo, 2021).

The tasks of engineering design include way more than finding the optimal solution. The demand for case-by-case, general, or reusable knowledge and insight through the designing process entails design methods to enable learning model behaviors (Kumar et al., 2013), exploring the solution space to satisfy a variety of design scenarios (Shukla et al., 2014, 2015), determining the refinement of a model and its sub-models (Messer et al., 2008; Panchal et al., 2008), and choose the proper set of solutions as the input of the previous stage of design (Nellippallil et al., 2018, 2019a, 2019b, 2020a, 2020b). Although optimization methods sometimes work efficiently in returning optimal solutions to mathematical models, they do not support knowledge attainment.

Therefore, we need a design strategy to overcome the shortcomings in optimization methods, including avoiding the assumption in utility theory-based functions, managing the typical complexities and uncertainties in engineering-design problems, and supporting knowledge learning through post-solution analysis.

In Sect. 9.3.2, we summarize the satisficing methods in literature and identify the whole-process satisficing in engineering design. In Sect. 9.3.3, we introduce why the whole process satisficing works as an alternative and effective in managing design problems. We describe a packet of methods that realize the whole process satisficing strategy. In Sect. 9.3.4, we summarize the research topics, derived work (new features and modules, tools, methods, theories, concepts, frameworks, and platforms), applications, and contributions to the whole process satisficing strategy. In Sect. 9.3.5, we anticipate the way forward of whole process satisficing strategy regarding the ideas, functions, and applications. Since these special requirements guide our method selection, we examine in Sect. 9.3.2 how current researchers address them, and we prepare the ground for our rationale in Sect. 9.3.3.

9.3.2 Literatures on Whole-Satisficing Strategies

In this section, we summarize the literature using satisficing strategies to manage engineering-design problems based on four stages—formulation, approximation, exploration, and evaluation. We design the methods that deal with all four stages as "whole-process satisficing" strategies.

An alternative strategy to optimizing is called "satisficing." The satisficing strategy provides "good enough" but not optimal solutions to complex problems. The term satisficing is a portmanteau of satisfy and suffice (Manktelow, 1999). The criteria of "good enough" vary with context. There are a variety of methods and techniques to reach satisficing in literature. To gain a systematic understanding of "what is satisficing," "why and when do designers use satisficing strategy," "how does satisficing strategy work," and "what are the classic satisficing methods," we first generally review the literature using satisficing strategy to manage engineering design problems, and then choose one packet of methods that realize the "whole process satisficing" for a deep review.

Using Web of Science, by searching "satisficing" and "engineering design" in topic, we obtain 189 publications. Through manually selecting the relevant journal articles and research articles that propose new methods, we end up with fifty publications. We review those selected articles regarding the proposed methods, complexities being managed, intellectual merits, foundational theory, application field, and limitations. Among the fifty publications, we select thirty-three representative publications and summarize the focus and technique of the proposed methods In Table 9.3. We align the column headings in Table 9.3 with the four stages of engineering design so that readers can trace how each contribution supports formulation, approximation, exploration, and evaluation. We also note the limitations column so that designers can judge when a published method may fail to meet their needs. By connecting these two resources, we help readers translate the literature survey into actionable guidance for their own projects. "Representative" papers are those that fall into one or more of the following categories.

(i) Articles that manage engineering-design problems with one or more typical features identified in Sect. 9.3.1.
(ii) The baseline methods or foundational theories in those articles are relatively well-known or generally used.
(iii) The most cited articles or the articles with core intellectual merit of an author or a group of authors who have published series of paper in the topic.

During the literature review, we find that so far, there has not been a review article that systematically categorizes the satisficing methods based on different stages in engineering design, specific techniques, and the complexities of their problems. Such categorization is necessary because it may provide guidance to users in selecting appropriate satisficing methods. Therefore, we provide such a categorization in Table 9.3.

We define, in engineering design problems, there are four stages, model formulation or formulation modification (F), model approximation (A), solution searching (S), and evaluation (E). Each method proposed in literature may deal with one or more stages through different means, and we mark the categories of the thirty-three representative papers in Table 9.3. Although researchers reach satisficing using various modeling and measurement methods, they all agree on the idea that "good enough" but not necessarily optimal solutions well balance the performance of their designs and the workload of finding the solutions.

To show the details about the problems, methods, and contributions of the selected publications for readers to further explore each division of the topic, we select thirteen milestone paper from the thirty-three and summarize their information in Table 9.4. The selection of milestone articles is based on our knowledge and experience about the effectiveness of the methods and the typicalness of engineering design problems used in the papers. In Table 9.4, we illustrate the complexities being managed in the thirteen milestone literature, including the proposed methods, foundational theories, design stages, limitations, and application fields. We invite readers to begin with Table 9.3 for a concise overview and then move to Table 9.4 to examine the details that inform method selection. We align the column headings in Table 9.3 with the four stages of engineering design so that readers can trace how each contribution supports formulation, approximation, exploration, and evaluation. We also note the limitations column so that designers can judge when a published method may fail to meet their needs. By connecting these two resources, we help readers translate the literature survey into actionable guidance for their own projects.

Table 9.3 Methods and their focus regarding engineering design stages

Stage	Formulation					Approximation		
Focus or means	Fuzzy formulation	Probability/chance optimization, stochastic programming	Formulation improvement through iterations or gaming	Change the format of objectives or add objectives	Integrate or couple models, concurrent design models	Linearization	Relaxation (change the model)	Use other techniques to approximate the model, (surrogate model, neural network, Bayesian etc.)
Example papers	Hu et al. (2014), Ito and Kogiso (2016), Kiyota and Tsuji et al. (2003), Meng (2014), Rao and Xiong (2005), Sakawa et al. (1998), Wang et al. (2009a, 2009b)	Brown (1990), Goodrich et al. (1999), Li and Wang (2018), Nguyen et al. (2016), Wang and Wang (2014)	Kanno et al. (2017), Li et al. (2019), Rao and Xiong (2005), Stirling and Nokleby (2009), Wang et al. (2009a, 2009b)	Cho et al. (2004), Curtis and Beard (2004), Goodrich et al. (1999), Hu et al. (2014), Kanno et al. (2017), Keyvandarian and Saif (2023), Lewis and Mistree (1999), Ming et al. (2019), Nellippallil et al. (2018), Nellippallil et al. (2020a, 2020b), Salado and Nilchiani (2017), Tao et al. (2006), Toyoda and Kogiso (2015)	Loch et al. (2003), Nellippallil et al. (2018), Nellippallil et al. (2020a, 2020b)	Lewis and Mistree (1999), Ming et al. (2019), Nellippallil et al. (2018), Nellippallil et al. (2020a, 2020b)	Kiyota et al. (2003)	Cho et al. (2004), Li and Wang (2018), Meng (2014), Nellippallil et al. (2018), Nellippallil et al. (2020a, 2020b), Stirling and Nokleby (2009), Wang and Wang (2014)

(continued)

Table 9.3 (continued)

Solution			Evaluation			
Interior-point searching	Forego optimality (relaxation without changing the model)	Combining interior-point searching with simplex algorithm	Tradeoff analysis and discussion	Use information in evaluation to modify formulation	Sensitivity analysis, loss-gain analysis	Develop evaluation functions or judgment parameter
Braha (2001), Kiyota et al. (2003), Meng (2014), Nguyen et al. (2016), Okamoto et al. (2014), Tao et al. (2006)	Loch et al. (2003)	Lewis and Mistree (1999), Ming et al. (2019), Nellippallil et al. (2018), Nellippallil et al. (2020a, 2020b)	Cho et al. (2004), Goodrich et al. (2000), Lounis and Daigle (2013), Okamoto et al. (2014), Wang et al. (2009a, 2009b)	Curtis and Beard (2004), Goodrich et al. (1998), Hu et al. (2014), Kanno et al. (2017), Nellippallil et al. (2018), Nellippallil et al. (2020a, 2020b), Salado and Nilchiani (2017), Stirling and Nokleby (2009)	Goodrich et al. (1999), Meng (2014), Wang and Wang (2014)	Binazadeh and Shafiei (2013), Goodrich et al. (1998), Ito and Kogiso (2016), Keyvandarian and Saif (2023), Lewis and Mistree (1999), Li and Wang (2018), Li et al., (2019), Ming et al. (2019), Nellippallil et al. (2018), Nellippallil et al. (2020a, 2020b), Nguyen et al. (2016), Sakawa et al. (1998), Tao et al. (2006), Wang and Wang (2014)

Table 9.4 Detailed information about the milestone publications

Authors and Year	Problem features and complexities managed	Proposed Method	Method Description	Foundational Theory or Baseline Method	Stage (F, A, S, E)	Limitations	Application Field
Kiyota et al. (2003)	Multi-objective, interdependent, complex systems, and system of systems	Satisficing squared S-2 method	Fuzzy satisficing method using generic algorithms; set aspiration level of each objective, minimizing unsatisfaction rating	Principles of bounded rationality, generic algorithm	F, A, S	May lead to unmature convergence	Multi-objective problems with an aspiration for each objective can be identified but not reached
Toyoda and Kogiso (2015)	Multi-objective, variation in variables and parameters	Satisficing tradeoff method (STOM)	Add mean and variance of the variation as objectives, apply tradeoff analysis, and investigate the effect of uncertainty in Pareto surface	Taguchi's robust design	F	Only investigate the desired space; keen in Pareto solutions	Multi-objective problems with uncertainties, and when designers want to explore how variation may affect Pareto set
Binazadeh and Shafiei (2013)	Nonlinear, time-vary systems with slowly varying parameters	An approach to design a stabilizing control law	Use a parametric version of satisficing control strategy, develop a time-frozen parameter to make a stabilizing control law, and evaluate the maximum admissible rate of change of system dynamics	Satisficing control strategy	F, E	The theory suggesting the stabilizing control law may vary with problems	Slowly varying control systems

(continued)

Table 9.4 (continued)

Authors and Year	Problem features and complexities managed	Proposed Method	Method Description	Foundational Theory or Baseline Method	Stage (F, A, S, E)	Limitations	Application Field
Li et al. (2019)	Conflict resolution among subsystems	A satisficing algorithm to mitigate the conflict resolution problem	Formulate a conflict resolution problem as a game model, use a satisficing game theory to mitigate the model, and develop a judgment parameter to ensure a socially acceptable compromise	Game theory	F, E	Rely on utility theory assuming subsystems' rationality and satisficing level are the same	Systems with multiple subsystems with conflict resolutions, such as unmanned aerial vehicles (UAVs)
Nguyen et al. (2016)	Uncertainty that can be formulated as chance constraints or stochastic model	Satisficing measurement approach (SMA)	Use SMA to mitigate customers' dissatisfaction, and integrate the SMA into tabu search heuristics to solve a set of Solomon instances	Stochastic programming	S, E	Assuming uncertainties can be quantified and represented as stochasticity	Stochastic vehicle routing problems with time windows (VRPTW)

(continued)

Table 9.4 (continued)

Authors and Year	Problem features and complexities managed	Proposed Method	Method Description	Foundational Theory or Baseline Method	Stage (F, A, S, E)	Limitations	Application Field
Stirling and Nokleby (2009)	Robotic systems capable of sophisticated social behavior	Multi-agent satisficing	Use conditional utility considering interest of others, use multi-agent utility aggregation to avoid group interests dominating individual interests, and apply Bayesian network to artificial societies to gain obtain satisficing social welfare	Game theory and Bayesian network theory	F, A, E	Rely on conditional utility assuming individual's rationality and level of consideration of others are the same among individuals	Systems has participants or designers with interaction and gaming nature
Okamoto et al. (2014)	Multi-objective, nonlinear	A solution search method with satisficing trade-off method	Develop a search method to find a preferred solution from Pareto optimal solutions, visualize the Pareto set using a self-organizing map, and perform satisficing trade-off analysis	Pareto optimum	S, E	Assume Pareto front is easy to acquire	Material design

(continued)

Table 9.4 (continued)

Authors and Year	Problem features and complexities managed	Proposed Method	Method Description	Foundational Theory or Baseline Method	Stage (F, A, S, E)	Limitations	Application Field
Ming et al. (2019)	Multi-type decision making—selection, compromise, and coupled	Template-based method configuration and execution of decision workflows	Use template to provide typical decision workflows for designers to execute, identify satisficing solutions, and explore solution space	Bounded rationality realized by compromise decision support problem (cDSP) and adaptive linear programming algorithm (ALP)	F, A, S, E	Did not explain and prove why using cDSP and ALP can obtain satisficing solutions	Product design, system design where coupled decisions are required
Nellippallil et al. (2020a, 2020b)	Multi-objective, nonlinear, multi-stage, multiple ways to integrate models	A goal-oriented inverse design method	Introduce robust design goals and indices, error margin indices (EMIs) and design capability indices (DCIs) to obtain satisficing robust design specifications	Bounded rationality realized by compromise decision support problem (cDSP) and adaptive linear programming algorithm (ALP); robust design realized by EMI, DCI	F, A, S, E	The solution space exploration can be varied from case to case	Multi-objective, nonlinear, and multi-stage design problems, such as the hot rod rolling process chain design

(continued)

Table 9.4 (continued)

Authors and Year	Problem features and complexities managed	Proposed Method	Method Description	Foundational Theory or Baseline Method	Stage (F, A, S, E)	Limitations	Application Field
Meng (2014)	Fuzzy variables	A satisficing data envelopment analysis (DEA) model	For general fuzzy input and output variables, design a hybrid Particle swarm optimization (PSO) algorithm by integrating approximation method, neural network (NN) and PSO to solve the proposed DEA model	PSO, NN, DEA	F, A, S	Based on seeking optimality and utility theory which assume decision-makers have perfect rationality	Problem with fuzzy input and output variables
Lewis and Mistree (1999)	Complex problems that finding a global optimum is difficult, discrete and continuous variables	Foraging-directed adaptive linear programming (ALP)	Hybrid foraging-directed algorithm with ALP—incorporate interior-point searching algorithm with sequential linearization and dual Simplex	Bounded rationality realized by compromise decision support problem (cDSP) and adaptive linear programming algorithm (ALP)	F, A, S, E	Only efficient for problems with multiple local optima and easy to stuck in local optima	Complex problems that finding a global optimum is difficult

(continued)

Table 9.4 (continued)

Authors and Year	Problem features and complexities managed	Proposed Method	Method Description	Foundational Theory or Baseline Method	Stage (F, A, S, E)	Limitations	Application Field
Kanno et al. (2017)	Inherent uncertainty, exogenous uncertainty, vulnerable to certain events	An info-gap robust-satisficing approach	Reverse performance between competing designs is quantified by intersection between the info-gap robustness curves	Info-gap robust-satisficing	F, E	The min–max objective is based on utility theory which assumes the model is complete and accurate	Structure design
Wang and Wang (2014)	Reliability-based problem	A maximum confidence enhancement (MCE)-based sequential sampling approach	Employ surrogate model and define a cumulative confidence level (CCL) to measure reliability estimation accuracy, and select sampling points with largest CCL improvement	Surrogate modeling, reliability-based programming	F, A, E	Only apply to reliability-based design	Reliability-based design, such as vehicle side crash

Among those selected milestone publications, only Mistree's research team proposes methods that deal with all four stages in engineering design. In this paper, we name it "whole-process satisficing," which is also the reviewing focus of Sect. 9.3. We find out the publications in whole process satisficing in engineering design, analyze what specialties in the method that enable designers to acquire satisficing (Sect. 9.3.3), categorize the application fields (Sect. 9.3.4), and anticipate the way forward (Sect. 9.3.5). Having mapped the literature, we now focus in Sect. 9.3.3 on why whole process satisficing uniquely meets the needs identified.

9.3.3 Whole-Process Satisficing and Why

In this sub-section, we introduce the definition, necessity, features, mechanisms of whole-process satisficing and why we recognize it.

Modeling methods for engineering-design problems fall into two categories, i) formulating a complex problem exactly and searching for the optimal solution or near-optimal solutions, and ii) approximating a complex problem and finding a set of solutions that output acceptable performance. We name the first strategy "optimizing" and the second one "satisficing." The major differences between the two strategies are represented in three aspects: model formulation, solution algorithms, and post-solution analysis (Guo, 2021; Mistree et al., 1993). In this section, we describe the typical features of the satisficing strategy regarding those three aspects.

The concept of satisficing solution or satisficing searching is proposed by Simon (1956, 1973, 1976) as "good enough" solutions that make the design output acceptable but maybe not optimal performances. Later, the idea of satisficing is put into practice by several researchers. In the past three decades, Mistree and his research team (Mistree et al., 1981, 1990, 1993) realize the whole-process satisficing strategy in engineering design using the compromise Decision Support Problem (cDSP) as the formulation construct, the Adaptive Linear Programming algorithm (ALP) as the solution algorithm, and the Decision Support in the Design of Engineering Systems (DSIDES) as a platform and solver to formulate and solve a cDSP with the ALP. In this section (Sect. 9.3), when referring to the whole-process satisficing, we specifically mean using the cDSP, ALP, and DSIDES to realize satisficing in each of the four stages of engineering design. For decades, the cDSP-ALP-DSDIES community has systematically developed and derived a packet of design methods along the satisficing strategy, especially for engineering designs (Hajihashemi, 2023). In this chapter, we discuss the whole-process satisficing strategy mainly through reviewing this community's representative works. In the context of using the cDSP-ALP-DSDIES packet, we define satisficing solutions as the solutions only meet the necessary KKT conditions.

The Specialty in the Formulation—Examples of the formulations using the optimizing strategy include mathematical optimization and its variants, such as goal programming. Solutions are usually obtained by identifying the Pareto front consisting of nondominated or near-optimal solutions using optimization solution algorithms. The formulation of design problems using satisficing strategy, namely, the compromise Decision Support Problem (cDSP), has the key features that allow designers to identify satisficing solutions that meet the necessary KKT condition but not the sufficient condition.

The format of a nonlinear optimization problem is like this: For a given objective function $f(x)$, Euler and Lagrange develop the Euler–Lagrange equation forming the second-order ordinary differential equations $\nabla^2_{xx}f(x)$ to facilitate finding the stationary solutions. The value of the variables that maximize $f(x)$ within the feasible set $\mathcal{F}$ is the solution to the optimization problem, where $\mathcal{F}$ is the set bounded by constraints and bounds. The format of an optimization problem $\mathbb{O}$ can be represented as follows. x is the vector of decision variables as real numbers. $g_i(x)$ is the ith inequality constraint. $h_j(x)$ is the ith equality constraint. Any point x that is a local extrema of the set mapped by multiplying active equations with a non-negative vector, is a local optimum of $\mathbb{O}$ (Courant and Hilbert, 1953), denoted as x^*. The elements of such a non-negative vector are Lagrange multipliers, μ and λ.

The format of an optimization problem $\mathbb{O}$:

Given

$$f : \mathbb{R}^n \to \mathbb{R}, \mathcal{F} \subseteq \mathbb{R}^n$$

$$\mathcal{F} = \left\{x \in \mathbb{R}^n | g_i(x) \geq 0, i = 1, \ldots, m, h_j(x) = 0, j = 1, \ldots, \ell\right\}$$

Find

$$x^* : f\left(x^*\right) \succeq f(x), \forall x \in \mathcal{F}$$

One variant of optimization is goal programming. The format of a goal programming problem $\mathbb{O}^{goal}$ is represented as follows. A target value T is predefined for the objective function $f(x)$ as the right-hand side value, so the objective becomes an equation, and we call it a goal. d^- and d^+ are deviation variable measuring the underachievement and over-achievement of the goal towards its target. The problem is solved by minimizing the deviation variables, which is minimizing the difference between $f(x)$ and T. In other words, goal programming is aimed at finding T's closest projection on $\mathcal{F}$.

The format of a goal programming problem $\mathbb{O}^{goal}$:

Given

$$f : \mathbb{R}^n \to \mathbb{R}, \mathcal{F} \subseteq \mathbb{R}^n$$

$$\begin{aligned}\mathcal{F} = \{x \in \mathbb{R}^n | g_i(x) \geq 0, i = 1, \ldots, m, h_j(x) = 0, j = 1, \ldots, \ell, d^- \cdot d^+ \\ = 0, 0 \leq d^{\mp} \leq 1, Goal : f(x) + d^- - d^+ = T\}\end{aligned}$$

Find

$$x^* : \mathbb{P}_{x \in \mathcal{F}}(Goal : f(x) = T)$$

In the cDSP, elements of mathematical programming and goal programming are combined. A cDSP $\mathbb{C}$ is represented as follows. For a nonlinear cDSP, we first linearize the nonlinear equations, including nonlinear constraints and nonlinear goal. Therefore, the nonlinear cDSP first becomes a linear problem with a linear goal $Goal^{linear}$, a linear feasible space $\mathcal{F}^{li}$ bounded by linear constraints $g(x)^{li} \geq 0$ and $h(x)^{li} = 0$. Thus, using a cDSP, we seek the closest projection from the linear goal set onto a linear feasible set. We define the solution as a satisficing solution, and we use x^s to denote it.

The format of a cDSP $\mathbb{C}$:

Given

$$f : \mathbb{R}^n \to \mathbb{R}, \mathcal{F} \subseteq \mathbb{R}^n$$

$$\mathcal{F}^{li} = \Big\{x \in \mathbb{R}^n | g_i(x)^{li} \geq 0, i = 1, \ldots, m, h_j(x)^{li} = 0, j = 1, \ldots, \ell, d^- \cdot d^+ = 0, 0 \leq d^{\mp} \leq 1, Goal^{li} : \frac{f(x)^{li}}{T} + d^- - d^+ = 1\Big\}$$

Find

$$x^s : \mathbb{P}_{x \in \mathcal{F}_{li}}\big(Goal^{li} : f(x)^{li} = T\big)$$

The difference between x^* and x^s is that x^* conforms both the necessary (first-order) and sufficient (second-order) KKT conditions, whereas x^s conforms the necessary KKT condition but may not conforms the sufficient KKT condition. This is because the second derivative of the linear equations, $Goal^{li}$, $g_i(x)^{li} \geq 0$, and $h_j(x)^{li} = 0$, degenerates—as a result, no uncertainty may affect the feasibility of x^s since no uncertainty gets a chance to break the equilibrium of the second-order Lagrange equation. In addition, when the convexity of $\mathcal{F}$ is greater than the convexity of the $f(x)$, x^* may not be identified as the second-order Lagrange equation has no solution, but x^s is obtainable due to its irrelevancy with the second-order KKT condition. The details of the mathematical demonstration and illustration using example problems are in Chap. 2 of Guo (2021).

The Specialty in the Solution Algorithm—There are different ways of categorizing the solution algorithms. Sharma and Kumar (2022) propose that all solution algorithms fall into two categories, classical methods and meta-heuristics algorithms. Classical methods are based on vertex searching or solving equation set to meet the KKT conditions, (e.g., Lagrange's multiplier method, Branching, etc.) Metaheuristics algorithms include deterministic searching methods (Tabu search

Table 9.5 Classifying solution algorithms regarding the optimality of solutions and the searching conditions

	Converge the optimal solution	Type of converged solutions		Strength	Weakness
		Near-optimal	Satisficing		
Vertex search	Simplex, Dual Simplex, Lagrange's multipliers	SLP, SQL	ALP	Requiring relatively low computing power; may give model formulation insight	May not work effective for nonconvex problems
Interior-point search		Meta heuristic interior-point searching algorithms (NSGA, GBO, etc.)		Easy to deal with nonlinear, high-dimension problems	Requiring large computing power; no insight of formulation modification

and local search) and stochastic searching-based methods (Simulated Annealing, Hill Climbing, Ant Colony algorithm, evolutionary algorithm, etc.). Abualigah et al. (2020) classify solution algorithms into two types—metaheuristics algorithms and heuristics algorithms. The former contains three sub-categories, local-based algorithms, evolutionary algorithms, and swarm-based algorithms. The latter specifically refers to vertex searching algorithms such as the Simplex algorithm. Daound et al. (2022) refine the definition by identifying four sub-categories in the metaheuristic algorithms: physics-based, population-based, human-based, and evolutionary-based algorithms; they define the local searching algorithms as heuristic algorithms.

In this monograph, not only do we care about the way of solving a problem, but we also discuss whether the solution is optimal and whether a set of nonoptimal but good enough solutions may provide designers with more options and relevant knowledge in the design space. We use a matrix to categorize the solutions algorithms, as shown in Table 9.5. The two dimensions of the matrix are the optimality of the converged solutions and the searching conditions. No interior-point searching algorithms guarantee the optimality unless KKT conditions are applied to facilitate the examination of optimality. Such an examination is a posteriori supplementary method; therefore, we define all interior-point searching algorithms in the category of converging near-optimal solutions.[4]

The ALP algorithm performs both interior-point searching and vertex searching to maintain a balance between exploitation and exploration. With the module "XPLORE" in DSIDES, a wide-spread interior-point random search is performed

[4] For complex problems that apply an approximation algorithm to solve, the optimal solution may not be found within the available computing power. Near-optimal solutions are found instead. There is not rigid definition of "how close" a near-optimal solution should be to the optimal solution. It depends on the problem and algorithm.

to select a good starting point (see Fig. 9.15), then the problem is linearized around the starting point using the Second-order Sequential linear programming algorithm (SLIP2), and then the linear problem is solved using dual simplex. The linearization can go through a number of iterations to make sure the nonlinear problem is linearized sufficiently and accurately at different local optimums. As there are often more constraints than variables in the engineering-design problems, using the dual simplex is easier than using the simplex. In this way, the solution is on the vertex of the linearized problem and meets the necessary KKT condition to ensure near optimality. The difference between a satisficing solution, a near-optimal solution, and the optimum is illustrated in Fig. 9.16.

With the satisficing solution algorithm—the ALP, why the nonconvex problem can be dealt with, and why is the linearization relatively accurate? Two mechanisms in the ALP make it possible to linearize the nonconvex—using the second-order sequential linearization (SLIP2) algorithm and the accumulation of the linear constraints. The SLIP2 first uses the parabola (the one that passes through the starting point and the two intersection points of the surface and × 1− × 2 plane) to approximate the nonlinear surface, then linearizes the parabola into a plane. And the linearized planes from multiple iterations are accumulated as the linear constraints to replace the nonlinear constraint. In this way, the nonconvex equations can be linearized, and the

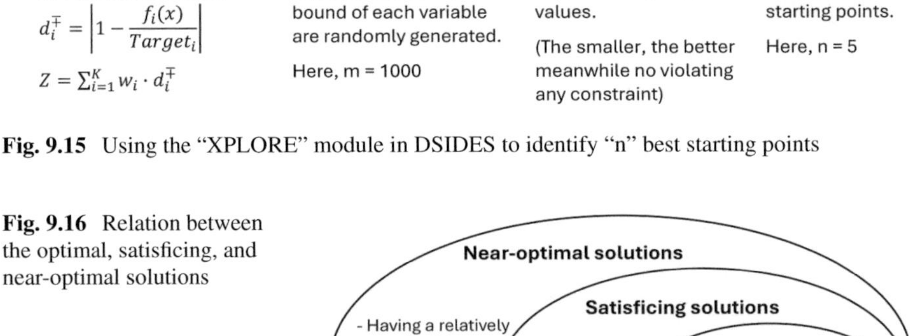

Fig. 9.15 Using the "XPLORE" module in DSIDES to identify "n" best starting points

Fig. 9.16 Relation between the optimal, satisficing, and near-optimal solutions

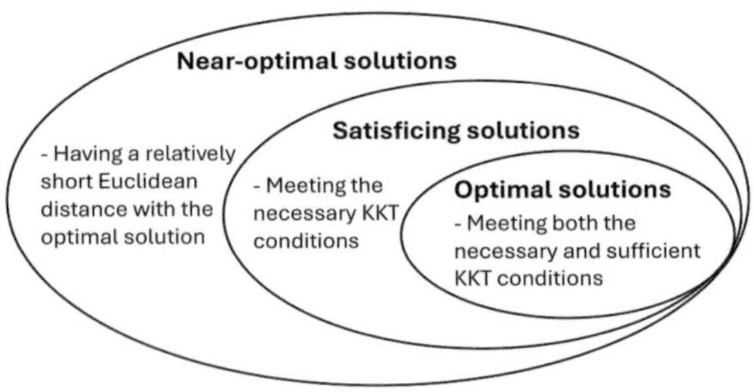

linearized constraints are more accurate. The detailed mathematical descriptions and illustrations are in Nellippallil et al. (2020a, 2020b).

Features that Support Knowledge Attainment and Coupled Decision-Making—Managing engineering-design problems requires designers to find a set of satisficing solutions and way more than that—gain knowledge about the design space, make coupled decisions, and tailor the methods to engineering designs in different fields. The satisficing strategy provides a good construct to complete those tasks. For example, DSIDES provides the formulation construct and solvers for compromise DSP (cDSP), selection DSP (sDSP), and any sequences or combinations of the compromise and selection DSPs, which support the mixture of continuous and discrete decision-making.

We explain how the satisficing strategy realized using cDSP-ALP-DSIDES support various tasks in engineering design by guiding readers through the relevant publications in the past three decades in Sect. 9.3.4.

Scope of Applications and Limitations—Satisficing strategy facilitates designers to deal with engineering problems with multiple goals, goals with various units and achievability, nonlinear features, continuous and discrete variables, evolving design preferences, and coupled decision-making. However, for problems that do not contain the features above, the satisficing strategy may not outperform optimization methods.

In addition, to formulate a problem into a cDSP, a target for each goal should be defined. For an objective that has no target value, or when designers have no idea about the target value, the satisficing strategy may not be the best option.

9.3.4 Application and Evolution of Whole-Process Satisficing

In this sub-section, we review six fields of applications using whole-process satisficing strategy to realize engineering-design and other design problems, and how the whole-process satisficing evolve through the applications over decades.

There are six major contributions (subjects) in the satisficing engineering designs in the past three decades: (i) robust design methods, (ii) multiscale and microstructure design, (iii) multistage design and concurrent design, (iv) coupled decision-making, (v) multigoal (multiobjective) or multidisciplinary design, and (vi) knowledge-based design and platforming. There are overlaps among those subjects, as some applications in one subject are the verifications of the methods in other subjects. In this section, we select the representative work in each subject based on the utility, popularity, and extendibility of the publications and summarize the typical features, contributions, and scope of applications.

1. **Robust Design Managing Four Types of Uncertainty**—The idea of robust design using the satisficing strategy is to make the design relatively insensitive to different types of uncertainty (Allen et al., 2006a, 2006b; Choi et al., 2005a, 2005b). Uncertainties are classified into four types based on their sources (Allen

et al., 2006a, 2006b): Type I—the uncertainty brought by the variation in environmental noise or other noise factors (Taguchi, 1985), Type II—the uncertainty brought by the variation in design variables of control factors (Chen et al., 1996a, 1996b), Type III—the uncertainty introduced by modeling methods (Allen et al., 2006a, 2006b), and Type IV—the uncertainty introduced during the management of the previous three types of uncertainty (Choi et al., 2005a, 2005b). We give an example of the four types of uncertainty in supply chains in Table 9.6. There are four types of robust design using the satisficing strategy, which facilitate the designers to manger the four types of uncertainty.

Taguchi develops the "parameter design concept" to use a two-part orthogonal array for experimental designs using the "signal-to-noise-ratio" as an optimization criterion. This is later defined by Chen et al. (1996a, 1996b). With the identification of one limitation of Taguchi's method—no accurate solution can be yielded for highly nonlinear problems, and given the method is criticized by the statistical community (Box, 1988), Chen et al. propose a variation to Taguchi's method by integrating the Response Surface Methodology (RSM) with the cDSP to manage both Type I and Type II uncertainty. Meanwhile, Simpson et al. (1996) and Chen et al. (1996a, 1996b) propose the Robust Concept Exploration Method (RCEM) and use the principles to determine a range set of top-level design specifications to realize the robust design. Later, Choi et al. (2005a, 2005b) use RCEM with Error Margin Index (RCEM-EMI) to realize the robust design of materials by employing the EMI to indicate the mean and spread of system performance considering the variability in design variables and models. The authors (Choi et al., 2005a, 2005b) also propose RCEM with Design Capability (RCEM-DCI) to determine whether a ranged design specification is capable of satisfying

Table 9.6 The sources of four types of uncertainty in supply chains

Types of robust design	Types of uncertainty	Example in SC	Quantification
I	Uncertainty in parameter	Uncertainty in the demand side, such as unpredictable order	Type I, II: EMI, Monte Carlo simulation, Latin hyper cube, First/ Second moment method, etc.
II	Variable uncertainty	Uncertainty in the supply side, such as variation in productivity	
III	Uncertainty in model structure	Supply chain broken due to unexpected disasters (e.g., the COVID-19 pandemic)	Type III: Design Capability Index (DCI), Variance Function Estimation, Prediction Interval Approach, etc.
IV	Uncertainty created in process chain	The bullwhip effect	Type IV: Satisficing, Exploring the Solution Space (ESS), etc.

a ranged set of design requirements. Choi (2005), Choi et al. (2005a, 2005b) propose an Inductive Design Exploration Method (IDEM) for designing materials and products concurrently and systematically to manage Type IV uncertainty by sequentially identifying a ranged set of feasible specifications and searching the feasible spaces regarding the top-level design requirements.

Given the foundation of four types of robust design—the RCEM and IDEM with indices, the satisficing community later expands the method, derive RCEM methods with new indices, and applies them in different fields of engineering design. Wang et al. (2009a, 2009b) propose a design exploration method for adaptive design systems that include using local regression and inverse IDEM using an example problem in the design of a photonic crystal coupler and waveguide. Sinha et al. (2009, 2011, 2013) propose to use the Inductive Discrete Constraints Evaluation (IDCE), which is to sequentially identify feasible regions in design space using a metric called Hyper Dimensional-Error Margin Index (HD-EMI) indicating the degree of reliability of the model. Kulkarni et al. (2014) and Gautham et al. (2013) later apply HD-EMI in the multistage, inverse design of the heat treatment operation. Samadiani et al. (2010) and Panchal et al. (2005) integrate the Proper Orthogonal Decomposition-based (POD-based), multiscale model with cDSP to realize the robust design of datacenter cell in thermal design. Messer et al. (2006) and Rippel et al. (2010) apply and enrich the RCEM using the early stage of design of a pressure vessel by presenting three alternative methods, namely, second derivative, multiple derivatives, and multiple point method, to measure the robustness through estimating the variance. Goh et al. (2014, 2016, 2018) propose the Integrated Multiscale Robust Design (IMRD) for traversing the Integrated Computational Materials Engineering (ICME) in multiscale heterogeneous internal structure evolution.

In summary, the four types of robust design using satisficing strategy are using the cDSP and ALP as the basic modeling construct and solution algorithm, applying indices to measure and control variations from multiple sources, and sometimes integrating other modeling techniques to identify and work on a range of feasible solutions to ensure the design is insensitive to uncertainties.

2. **Multiscale and Microstructure Design in Material Design**—Material design is a field that often encounters nonlinear equations, variations from multiple sources, multiscale or hierarchical structures, etc., so it requires satisficing as a design strategy. For this topic, the contributions are mainly from the applications of the robust design method in material design.

   Seepersad et al. (2005) present the Type I, II, and III robust design method for designing materials in mesoscopic scales by topologically and parametric tailoring them to achieve desired properties. Allen et al. (2006a, 2006b) use the first three types of the robust design methods on multidisciplinary, multiscale design with multiple sources of uncertainty. Thompson et al. (2006) consider the material properties as uncertain variables and apply the robust design method to provide ranges of acceptable material properties for framing subsequent material design or selection for designing blast-resistant panels. McDowell et al. (2007) give an example for designing a plasticity-related microstructure—a four-phase reactive power metal–metal oxide mixture for initiation of exothermic reactions under shock-wave loading. Later McDowell et al. (2009) integrate the design of multiscale material, multifunctional material, and product design, and the proposed integrated method also works for other systems-based design of materials. Seepersad et al. (2008) propose a two-stage topology design approach that requires customized multifunctional properties for designing cellular materials. Samadiani et al. (2010) use a test problem of datacenter cell internal design to demonstrate the efficacy of the proposed method with the integration of POD-based, multiscale modeling with the cDSP. The method can be used for other simulation-based, multiscale designs. Shukla et al. (2015, 2016) deal with a ladle refining and continuous casting problem by using the satisficing strategy to facilitate producing new grades of steel to meet stringent sets of property specifications by exploring the solution space. Goh et al. (2018) select the satisficing strategy for it allows to find feasible ranged sets of solutions and avails the ICME horizontal Process-Structure–Property-Performance (PSPP) in multiscale heterogeneous internal structure evolution. Beemaraj et al. (2020) propose a method for designing a robust composite structure subjected to different loading conditions.

Why do designers choose the satisficing strategy in material design, especially when designing multiscale, microstructures, or needing to cope with the material properties with other requirements in a process or product? Typical reasons include (i) the product is expected to be robust to the variation in the material property, (ii) the material attributes can be designed to be robust against uncertainties due to random variation of microstructure, (iii) designers only need to determine the value of decision variables that satisfy the rigid requirements (constraints) and achieve the desired properties (goals) as closely as possible, which means they do not have to squeeze to the optimal properties, (iv) material design for a product is often a concurrent design that comprises material design or selection and the dimension or geometry design of the product, (v) robust solutions can be founded to reduce the variability in response for simulation-based design or metamodel-based design, and (vi) the solution space can be explored relatively sufficiently not only to output a solution but to observe and predict model behaviors and in the wide design space, which can be useful information for new material experiment and innovation.

3. **Multistage Design and Concurrent Design**—Multistage design and concurrent design are also applications of robust design methods, specifically RCEM. The applications include engine design, aircraft design, gear transmission, multifunctional material design, process chain design, product and product family design, and steel production.

   Concurrent design: Rangarajan et al. (1998) adopt RCEM as a framework to address tribological considerations in the concurrent design of automobile engine lubricated components. Chen et al. (2000) use RCEM to evaluate design alternatives and develop top-level specifications for High-Speed Civil Transport (HSCT) aircraft. Simpson et al. (2001) propose a product variety tradeoff evaluation method for assessing alternative product platform concepts with varying levels of commonality—the challenge lies in balancing the commonality and performance of products in a product family. The authors verify the method using gear transmission and general aviation aircraft design. McDowell et al. (2009) perform concurrent design by integrating multiscale, multifunctional, and product design at the same time.

Inverse design: Nellippallil and coauthors deal with a series of multistage, inverse design problems in steel manufacturing. Nellippallil et al. (2017) present an inverse design method based on empirical models and response surface models to support the integration that facilitates information flow between stages of the hot rod process chain. The stages are designed by passing the information obtained after exercising the end-stage cDSP to the earlier-stage cDSPs. Then, the authors improve the inverse design method (Nellippallil et al., 2018) to achieve the integrated design exploration of materials, products, and manufacturing processes through the vertical and horizontal integration of models. They demonstrate the efficacy of the method using a hot rod rolling and cooling process chain problem by exploring the processing paths and microstructure in an inverse manner. Nellippallil and the coauthors further improve the method into a goal-oriented Inverse Design method (Nellippallil et al., 2019a, 2019b) to empower the capability to carry out a microstructure-mediated design satisficing specific processing performance of a product. They apply the method to design the thermos-mechanical processing of a steel rod. The authors then summarize their work as an approach (Nellippallil et al., 2020a, 2020b) that facilitates co-design across the materials, products, and manufacturing processes. With this goal-oriented, inverse co-design approach, Nellippallil et al. (2019a, 2019b) design two components of an American football helmet, the composite shell and foam liner. The method can be used in other fields when the multistage, concurrent design of multiple elements (material, product, manufacturing, etc.) is needed and only the end-stage performance or requirements are available.

Designers choose the robust design method under the satisficing strategy for concurrent, multistage, and inverse design projects because they need to manage the uncertainties brought by connecting the stages, concurrent decision-making, and the lack of knowledge on the association between the early-stage decision variable and the end performance of a product or a system.

4. **Coupled Decision-Making**—Coupled-decision-making problems often have common features with the multistage, concurrent design problems—designers need to determine two things while interdependent on each other in a system, for example, the material and the product specification. In this subsection, we focus on the problems that require hierarchical ways of coupling the selection decisions and compromise decisions. Another type of coupled decision-making is making decisions by different teams with multiple backgrounds or fulfilling multiple disciplines and managing interactions.

   Besides the works on steel manufacturing described in 3 Multistage Design and Concurrent Design, which require the design of materials and products at the same time, some applications and derived methods in steel casting and gear design also require material selection and geometric design simultaneously. Kumar et al.

(2013) develop an integrated design framework, Platform for the Realization of Engineered Materials and Products (PREMΛP), based on metamodels and cDSP. The authors utilize the framework to assist in making decisions when an existing configuration for continuous casting is unable to meet the requirements. The approach can be adopted for integrating the host of operations for materials development with specific properties and the coupled design of products and materials. Gautham et al. (2013) envisage PREMΛP as a platform for the purpose of integrating models, knowledge, and data for designing both the material and the product. Kulkarni et al. (2013) use PREMΛP as a platform for the realization of engineered materials and products for gear design by exploiting the synergy between component design, material design, and manufacturing. Later, Kulkarni et al. (2015) propose a method, the concept exploration method, based on the cDSP, which is demonstrated for gear design which requires the simultaneous exploration of geometry, material, and manufacturing spaces to exploit synergies. Sharma et al. (2021) present a method for robust design using coupled decisions to identify design scenarios that are relatively insensitive to uncertainties. Those design scenarios are modeled as coupled decisions to account for influence among decisions. The method is tested using a gearbox design problem—selecting gear material and determining gearbox geometry. Based on this work, Sharma et al. (2022) later propose a decision classification scheme, the Multilevel Decision Scenario Matrix (MDSM), to develop design processes involving decision interactions.

As DSIDES can couple the cDSP and sDSP in different levels, hierarchies, and sequences, when the selection and compromise decisions need to be made at the same time, designers tend to adopt cDSP-ALP-DSIDES and derived design frameworks or methods in coupled decision-making.

5. **Multigoal (Multiobjective) or Multidisciplinary design through Exploring the Solution Space**—Engineering-design problems often have multiple goals (or objectives). Most test problems referred to in 3 Multistage Design and Concurrent Design are multigoal problems. However, in this subsection, we only introduce the publications that focus on the methods of the exploration of the design preferences, tradeoffs among goals, and interrelationship among goals, especially when there are three or more goals and the design encompass multiple disciplines. Since multigoal problems are usually managed through exploring the solution space (consists of vectors of decision variables) and design space (consists of scenarios of formulating and compromising multiple goals), we combine the reviewing of the work in two categories—"(v) exploration of the solution space and design space" and "(vi) multigoal (multiobjective) or multidisciplinary design"—here in one section.

   Xiao et al. (2007) propose the Collaborative Multidisciplinary Decision-making Methodology (CMDM) and use it to design a robot arm with three goals—deformation towards 0.5 mm, von Mises stress around 6 Mpa, and weight around 3.5 g. under working loads around. The CMDM enables designers to integrate the cDSP with Game Theory and apply DCI to obtain robust solutions. The cDSP is used to implement and exchange design scenarios while making coupling decisions. Game Theory is adopted for managing interactions between teams' decisions. DCI is used for maintaining design freedom to accommodate downstream changes. Wang et al. (2009a, 2009b) propose the Design Exploration Method for Adaptive Design Systems (DEM-ADS) to design a photonic crystal coupler and waveguide using the mean response function from simulations as the objective and setting the EMIs between the objective and its upper and lower bounds as goals. The method is a derived method of the IDEM in the robust design. Ahmed et al. (2014) design a multigoal gear blank without prior knowledge of preferences among goals. They also explore different scenarios in the two ways (Archimedean and Preemptive) of goal formulation and gain knowledge on design scenarios associated with the design preferences. Smith et al. (2015) design a thermal plant based on the Rankine cycle with six goals and demonstrate the exploration of the solution space to compromise the goas using both the Archimedean (weighted combination) and Preemptive (Lexicographic) methods. Gautham et al. (2015) and use the IDEM to design a heat treatment process. The five material properties become goals in the decision model. They study the tradeoffs among the properties and find ranged sets of specifications by exploring the design preferences. Anapagaddi et al. (2014) and Sabeghi et al. (2016, 2015) explore the solution space of multigoal design problems in continuous casting and ladle refining by evolving the weights of the goals so as to produce new grades of steels to meet stringent sets of property specifications. Goh et al. (2016), Sabeghi et al. (2015), Nellippallil et al. (2017, 2018, 2020a, 2020b, 2019a, 2019b), and Guo et al. (2021, 2020a, 2020b) apply the Ternary plot to visualize the desired weight range for three-goal problems using applications in steel manufacturing, material design, and supply chain design. Guo et al. (2022) propose a domain-independent algorithm, the Adaptive Leveling-Weighting-Clustering algorithm (ALWC), to explore the design scenarios for multigoal and many-goal (when there are more than three goals) problems. Using ALWC, designers may explore combinations and priorities of the goals based on their interrelationships.

Engineering design problems in nature have multiple goals, and often those goals are formulated using knowledge and requirements from various disciplines. In the early stage of design, designers rely on their domain expertise or intuition to compromise the goals with certain design scenarios and meet different design preferences. However, a robust method should output desired performance without too much domain expertise or assumptions. The methods of the exploration of the solution space and design space can reduce designers' dependence on prior domain knowledge and attain knowledge along with the exploration. The design methods in satisficing strategy well support solution space exploration as sets of satisficing solutions under multiple design scenarios can be identified, and model behaviors associated with those solutions and design scenarios can be recorded and analyzed.

6. **Knowledge-based Design and Platforming**—Theoretically, the awareness, discovery, reconstruction, storage, and reuse of knowledge can take place in any

design problem. Not only the satisficing design strategy supports knowledge-based design, but all kinds of methods should do. However, because the engineering designs using satisficing strategy, especially the RCEM, enforce rich post-solution analyses, designers can obtain relatively more knowledge about robust design, solution space and design space performance, and model behaviors. The beauty of knowledge recognized in design is that it can be reused and guide the next design project, and it can be generalized into rules and packaged into templates or platforms that ease future designs in various fields.

Knowledge and templates/tools about design problems: Chen et al. (2000) propose the RCEM not only as a method for facilitating robust design but also a method enhancing design productivity by increasing design knowledge in the early stages of designs. "Design knowledge is increased by implementing sophisticated integrated systems analysis in the early stages of design and making decisions based on better information." Zha et al. (2008) integrate the cDSP with the Fuzzy Synthetic Decision model (FSD) as a knowledge-intensive collaboration paradigm to support hybrid decisions. Given that collaborative design decisions have objective and subjective features, the proposed method hybrid the selection and compromise decisions and archive knowledge on collaborations. Given the demand for a reusable design process and tool to support interactions in designs, Panchal et al. (2009) derive a method, the modular decision-centric approach. The authors develop generic computational templates to instantiate four foundations to support concurrent design. The four foundations include hierarchical systems, separate information, decision-centric activities, and model interactions. Pederson et al. (2013) propose a domain-independent approach for realizing hierarchical product platforms. The authors focus on how to synthesize numerical taxonomy and technology diffusion in a systematic multiobjective decision-making method.

Knowledge and platforms about design methods: Wang et al. propose an ontology-based uncertainty management approach (Wang et al., 2019) in designing robust decision workflows and a knowledge representation approach (Wang et al., 2021a, 2021b) for designing complex engineered systems. In their two publications (Wang et al., 2019, 2021a, 2021b), they propose a decision-centric design process representation scheme, the Phase-Event-Information X (PEI-X) diagram, for designing workflows. Later, Wang et al. (2021a, 2021b) propose a knowledge-based design guidance system (KBDGS) to support Cyber-Physical-Product-Service Systems (CPPSS) design. Ming et al. (2020, 2018a, 2018b) use an ontology to represent knowledge of decision interactions and realize it as a cloud-based platform (Ming et al., 2018a, 2018b). The authors architecture the cloud-based platform to support a human-cyber-physical view of systems realization ecosystem. The decision support platforms, KBDGS and PDSIDES, can be extended as cloud-based platforms, which lead to possibilities in AI-based design and self-organizing systems design in the age of Industry 5.0. We observe evolving capabilities in these applications. In Sect. 9.3.5 we translate those observations into growth paths that inform the research thrusts presented later in Sect. 9.4.

9.3.5 The Future of Whole-Process Satisficing

In this section, we introduce several future directions of whole-process satisficing strategy.

Industry 5.0, a transformative industrial revolution, with new technologies and applications like cloud-based design and manufacturing, cyber-physical-social systems, and the networked intelligent systems, make it possible for all participants in the whole process chain to be networked, self-organized, and collaborate on decentralized decision-making (Nellippallil et al., 2019a, 2019b). The question is, why satisficing strategies? Especially, why whole-process satisficing?

The answer is: A networked, self-organized system that requires decision-making may have all the features that engineering-design problems have—nonlinear, nonconvex mathematical relations among factors, a number of conflicting goals, variations in controllable and uncontrollable factors, coupled decisions required, multi-stage and inverse decisions required, knowledge is created and can be generalized and reused anywhere all the time, etc. Above all, robust design is still the urged demand in Industry 5.0. We have used the previous three sections to demonstrate that satisficing strategy allows decision-makers to manage the features above relatively well. Some instructive concepts and ideas have already been proposed or foreseen by the satisficing community.

Nellippallil et al. (2019a, 2019b) present the architecture and functionalities of a cloud-based computational platform to support mass collaboration and open innovation. By reviewing the evolution of engineering design in connected products, end-to-end digital integration, mass customization and personalization, data-driven design, digital twins, etc., Jiao et al. (2021) envision a human-cyber-physical view of the systems realization ecosystems. Milisavljevic-Syed et al. (2020) propose to leverage the advantages of digitization and Artificial Intelligence to address smart manufacturing in a networked manufacturing system. Ming et al. (2022a, 2022b, 2022c) identify the requirements of the decision support platform for design Engineering 5.0 and propose an architecture to resolve the challenges. Later they propose a framework for a Cloud-Based Platform for Decision Support in the Design of Engineered Systems (CB-PDSIDES) (Ming et al., 2022a, 2022b, 2022c).

For the whole process satisficing strategy, there are two research directions that may help the community expand its research power—dealing with large-scale engineering design and cyber security. cDSP-ALP-DSIDES and PDSIDES have been proven to be efficient when managing engineering problems with regular size. However, in the new age, as the size of a networked system and system of systems may grow exponentially, the traditional tools may not provide sufficient computing power. Problem partition and parallel computing will be needed.

As CB-PDISEDS can provide cloud-based design support, aware and store knowledge for reuse, learn and record users' preferences, cyber security may become a critical issue. Designing secure protocols and mechanisms for data transmission and storage in the cloud to protect engineering designs from unauthorized access, interception, or tampering will become a necessity. This includes encryption, access control, and secure key management. Investigating strategies to enhance the resilience of cloud-based engineering design systems against cyber-attacks, natural disasters, or system failures is another urgent task.

In summary, future research on engineering design is mainly about i) how to graft new technologies and ideas into the satisficing methodology, ii) how to utilize new technologies and ideas to better derive and innovate design methods under the satisficing strategy, and iii) make the knowledge recognition and knowledge-based design available for all participants in the cyber-physical-social system. In Sect. 9.4 we broaden our future perspective as we group upcoming work into focused research thrusts, cross disciplinary opportunities, and societal connections, each grounded in the design evolution cycle that we introduced earlier.

9.4 Way Forward

Building on the future outlook in Sect. 9.3.5, we now use the design evolution cycle to frame research thrusts, application domains, and interdisciplinary extensions. The design evolution cycle introduced in this monograph provides a foundation for advancing research in complex system modeling, design automation, and decision-making under uncertainty. The methodologies presented here open opportunities for the broader design research community to further develop, refine, and expand these approaches into new domains.

9.4.1 Research Thrusts and Applications

The decision-based methods introduced in this monograph provide a foundation for further exploration in design evolution, AI-driven satisficing strategies, and uncertainty management. Opportunities exist for expanding these methods in the following areas:

- **Expanding Decision Models for Uncertainty Management**: The integration of multiple uncertainty sources can enhance in-situ quality control in smart manufacturing, as discussed in Chap. 7.
- **Advancing Data-Driven Model Training**: Improvements in data curation, learning-based optimization, and predictive modeling can increase efficiency in applications such as recycling Polyether Ether Ketone (PEEK) for in-space 3D printing (Chaps. 5 and 7).
- **Shifting from Metaheuristics to Logic-Based Optimization**: Replacing metaheuristic search techniques with structured mathematical formulations can refine decision workflows in cyber-physical product-service systems (CPPSS) (CPPSS) (Chaps. 5 and 6).

- **Enhancing Multi-Scale Simulations and Information Exchange**: Strengthening cross-scale data sharing can improve the management of emergent properties in self-organizing systems (Chap. 7).
- **Integrating Multi-Criteria Decision Support into Process Optimization**: Embedding decision support algorithms into digital twins can advance lean process design and manufacturing (Chap. 6).

These research thrusts will build upon the **decision-based satisficing strategies** explored in this monograph, enhancing adaptability in **complex design environments**.

9.4.2 Cross-Disciplinary Research Opportunities

The methods introduced in this monograph extend beyond engineering, presenting opportunities for research in economics, healthcare, and policy-making:

- **Data Curation and Knowledge Awareness**: Strengthening data authenticity, validation mechanisms, and provenance tracking in domains such as healthcare networks, supply chains, and CPSS.
- **Satisficing Strategies for Complex Decision-Making**: Applying adaptive satisficing techniques to urban planning, energy optimization, and infrastructure development, expanding upon the multi-objective decision methods outlined in Chap. 6.
- **Lean Process Design for Sustainable Development**: Utilizing adaptive scalarization and Pareto-based optimization to balance economic, social, and environmental factors in sustainable design, as discussed in Chaps. 5 and 7.
- **Mechanism Design for Policy and Governance**: Leveraging game theory, agent-based modeling, and incentive structures to develop sustainable economic policies and environmental strategies, aligning with the decision-support frameworks in Chap. 4.
- **Ontology-Based Decision Support Systems**: Developing knowledge-driven decision platforms to manage uncertainty, variability, and complexity in fields such as engineering, healthcare, and logistics, building upon the structured decision-making models introduced in Chaps. 2 and 3.
- **Impact of Public Events on Education**: Analyzing social media and digital learning trends to assess how major events influence engineering education and adaptive learning models, linking to discussions on knowledge representation in Chap. 8.

The integration of these interdisciplinary approaches will strengthen the decision-based frameworks, AI-driven optimization, and satisficing methods developed in this monograph.

9.4.3 Future Directions for the Design Research Community

Expanding the methods outlined in this monograph requires further exploration in:

- **AI-Driven Design Evolution in Industry 5.0**: AI-assisted predictor–corrector mechanisms and generative design strategies can enhance real-time decision-making in autonomous systems, linking to discussions in Chaps. 6 and 7.
- **Robustness in Self-Organizing Systems**: Strengthening resilient decision frameworks can improve adaptability in cyber-physical-social systems, an area critical to the evolutionary models discussed in Chap. 7.
- **Fail-Safe Network Architectures**: Developing knowledge-based fail-safe systems to enhance decision-making in large-scale networks, expanding on the network modeling approaches presented in Chap. 5.
- **Bridging Design Automation with Policy-Making**: Applying decision-based optimization to sustainability governance and strategic policy development, leveraging the decision-making methods outlined in Chap. 4.

Further research is warranted to refine the decision-based satisficing methods explored throughout this monograph, ensuring continued progress in design automation and uncertainty-aware optimization.

9.4.4 Bridging Design Research and Societal Needs

The methods presented in this monograph align with broader societal challenges, creating opportunities for research that integrates decision-based design evolution with social impact:

- **Human-AI Collaboration for Social Impact**: Exploring how AI-assisted design frameworks can enhance applications such as affordable housing, disaster resilience, and smart infrastructure planning, linking to discussions in Chap. 7.
- **Sustainable Engineering for Circular Economies**: Developing AI-driven waste reduction strategies to support resource-efficient production models, informed by sustainable design concepts in Chap. 5.
- **Knowledge Transfer in AI-Driven Design**: Investigating low-barrier AI interfaces that democratize access to decision-support tools, aligning with knowledge representation strategies in Chap. 8.
- **Ethics and Bias in AI-Augmented Decision Systems**: Addressing algorithmic bias, fairness, and transparency in AI-driven decision models, connecting to discussions on ethical considerations in Chap. 4.
- **Adaptive Learning for Future Engineers**: Creating AI-supported learning environments to foster problem-solving, computational design, adaptability skills, building upon insights from Chap. 8.

By expanding these research directions we prepare for Sect. 9.5, where we articulate a vision that pushes design evolution beyond current disciplinary boundaries.

9.5 A Vision for the Future: Expanding the Boundaries of Design Evolution

As we conclude this monograph, we reflect on the transformative journey from design optimization to design evolution through satisficing strategies. Inspired by pioneers such as Herbert Simon, Albert Einstein, and Genichi Taguchi, we have laid the groundwork for a paradigm shift—one that embraces adaptability, robustness, and uncertainty-aware decision-making in the design of complex systems.

In the Preface we introduced the motivation for satisficing strategies as a pragmatic response to the limitations of traditional optimization, setting the stage for a dynamic, iterative approach to decision-making. In Chapter 9, we expanded this perspective, identifying research thrusts, interdisciplinary applications, and future directions to advance design evolution across engineering and beyond.

Expanding the Impact of Design Evolution: We have demonstrated that the design evolution cycle—comprising formulation, approximation, exploration, and evaluation—is more than a tool for managing complexity. It is a framework for advancing decision-making, AI-driven satisficing strategies, and real-world adaptability. The opportunities outlined in Sects. 9.3.1 through 9.3.4 highlight how satisficing strategies apply to multi-scale modeling, AI-supported decision frameworks, and sustainability, bridging computational design with real-world applications.

By incorporating AI-driven satisficing strategies, designers leverage generative AI, digital twins, and agent based modeling to accelerate decision making while keeping solutions context aware, resilient, and ethically grounded. Using these methods, designers will be able to design future systems that evolve and adapt dynamically to emergent challenges while balancing competing objectives.

The Next Frontier: AI and Human-Centric Design Evolution: Throughout this monograph, we have emphasized that AI enhances satisficing strategies, but human expertise remains essential. Future research must explore hybrid intelligence models, where human intuition and AI-driven analytics work together to create knowledge-aware, uncertainty-resilient design processes. We envision a future where:

- **Cyber-physical-social systems** continuously evolve through AI-augmented satisficing.
- **Decision-support frameworks** integrate human reasoning and AI insights to improve accessibility and impact.
- **Sustainability and equity** are embedded into design evolution cycles, shaping resilient global solutions.

As Industry 5.0 and beyond emerge, satisficing in design evolution will play a crucial role in shaping engineering, business, and policy-making. The concepts presented in this monograph provide a foundation for future advancements, inviting researchers, practitioners, and policymakers to build upon and extend these ideas.

Final Thought: W. Edwards Deming once said, *"It's not enough to do your best; you must know what to do, and then do your best."* We have provided the knowledge of what to do, offering a structured, actionable approach to satisficing and design evolution. Now, the challenge is to put these insights into practice, shaping a future where design is not merely optimized but continuously evolved to meet the changing needs of humanity and technology alike.

Architecting a satisficing strategy for designing complex systems is an ongoing endeavor. As challenges evolve and new uncertainties arise, satisficing remains a powerful framework for balancing trade-offs, fostering innovation, and advancing decision-making. ***The journey of design evolution does not end here—it is continually redefined through every challenge, every iteration, and every breakthrough***.

References

Abualigah, L., Shehab, M., Alshinwan, M., & Alabool, H. (2020). Salp swarm algorithm: A comprehensive survey. *Neural Computing and Applications, 32*, 11195–11215.

Ahmed, S., Goh, C.-H., Allen, J. K., Mistree, F., Zagade, P., & Gautham, B. (2014). Hot forging of automobile steel gear blanks: An exploration of the solution space. In *ASME design automation conference*. Paper Number DETC2014-34157.

Allen, J. K., Seepersad, C., Choi, H., & Mistree, F. (2006a). Robust design for multiscale and multidisciplinary applications. *ASME Journal of Mechanical Design, 128*(4), 832–843.

Allen, J. K., Seepersad, C. C., & Mistree, F. (2006b). A survey of robust design with applications to multidisciplinary and multiscale systems. *ASME Journal of Mechanical Design, 128*(4), 832–843.

Anapagaddi, R., Shukla, R., Goyal, S., Singh, A. K., Allen, J. K., Panchal, J. H., & Mistree, F. (2014). Exploration of the design space in continuous casting tundish. In *ASME design automation conference*. Paper Number DETC2014-34254.

Beemaraj, S. B., Pathan, R. K., Salvi, A. G., Sharma, G., Mistree, F., & Allen, J. K. (2020). Inverse multi-scale robust design of composite structures using design capability indices. In *ASME computers and information in engineering conference* (p. 12). Paper Number DETC2020-22259.

Binazadeh, T., & Shafiei, M. H. (2013). Extending satisficing control strategy to slowly varying nonlinear systems. *Communications in Nonlinear Science and Numerical Simulation, 18*(4), 1071–1078.

Box, G. (1988). Signal-to-noise ratios, performance criteria, and transformations. *Technometrics, 30*(1), 1–17.

Braha, D. (2001). Design-as-stability: A new approach to automated synthesis. *AIEDAM-Artificial Intelligence for Engineering Design Analysis and Manufacturing, 15*(5), 385–399.

Brown, C. B. (1990). *Optimizing and Satisficing Structural Safety, 7*(2–4), 155–163.

Chen, W., Allen, J. K., Mavris, D. N., & Mistree, F. (1996a). A concept exploration method for determining robust top-level specifications. *Engineering Optimization, 26*(2), 137–158.

Chen, W., Allen, J. K., Tsui, K.-L., & Mistree, F. (1996b). A procedure for robust design: Minimizing variations caused by noise factors and control factors. *ASME Journal of Mechanical Design, 118*(4), 478–485.

Chen, W., Allen, J. K., & Mistree, F. (1997). A robust concept exploration method for enhancing productivity in concurrent systems design. *Concurrent Engineering, 5*(3), 203–217.

Chen, W., Allen, J. K., & Mistree, F. (2000). Design knowledge development for productivity enhancement in concurrent systems design. In C. T. Leondes (Ed.), *Knowledge-based systems techniques and applications* (pp. 1037–1060). Academic.

Cho, J. R., Jeong, H. S., Yoo, W. S., & Shin, S. W. (2004). Optimum Tire contour design using systematic STOM and neural network. *KSME International Journal, 18*(8), 1327–1337.

Choi, H.-J., Austin, R., Allen, J. K., Mcdowell, D. L., Mistree, F., & Benson, D. J. (2005b). An approach for robust design of reactive power metal mixtures based on non-deterministic micro-scale shock simulation. *Journal of Computer-Aided Materials Design, 12*, 57–85.

Choi, H.-J., Allen, J. K., Rosen, D., McDowell, D. L., & Mistree, F. (2005). An inductive design exploration method for the integrated design of multi-scale materials and products. In *ASME design automation conference.* Paper Number DETC2005-85335.

Choi, H.-J. (2005). A Robust Design Method for Model and Propagated Uncertainty, Ph.D. Dissertation, THe George W. Woodruff School of Mechanical Engienering, Georgia Institute of Technology.

Courant, R., & Hilbert, D. (1953). *Methods of mathematical physics* (Vol. I). Interscience.

Curtis, J. W., & Beard, R. W. (2004). Satisficing: A new approach to constructive nonlinear control. *IEEE Transactions on Automatic Control, 49*(7), 1090–1102.

Daoud, M. S., Shehab, M., Al-Mimi, H. M., Abualigah, L., Zitar, R. A., & Shambour, M. K. Y. (2022). Gradient-Based Optimizer (GBO): A review, theory, variants, and applications. *Archives of Computational Methods in Engineering, 30*, 2431–2449.

Fonville, T. R., Nellippallil, A. B., Horstemeyer, M., Allen, J. K., & Mistree, F. (2019). A goal-oriented, inverse decision-based method for an American football helmet. In *ASME design automation conference.* Paper Number DETC2019-97388.

Gautham, B., Singh, A. K., Ghaisas, S. S., Reddy, S. S., & Mistree, F. (2013). PREMΛP: a platform for the realization of engineered materials and products. In A. Chakrabarti & R.V. Prakash (Eds.), *Lecture notes in mechanical engineering* (pp. 1301–1313). Springer, India.

Gautham, B., Kulkarni, N., Zagade, P., Allen, J. K., Mistree, F., & Panchal, J. (2015). ICME for the integrated design of an automotive gear considering uncertainty. In *3rd world congress on integrated computational materials engineering* (pp. 323–330) Colorado Springs.

Goh, C.-H., Ahmed, S., Dachowicz, A. P., Allen, J. K., & Mistree, F. (2014). Integrated multiscale robust design considering microstructure evolution and material properties in the hot rolling process. In *ASME design automation conference.* Paper Number DETC2014-34157.

Goh, C.-H., Dachowicz, A. P., Allen, J. K., & Mistree, F. (2016). Exploring the performance-property-structure solution space in friction stir welding. In *Proceedings of the 3rd world congress on integrated computational materials engineering (ICME 2015)* (pp. 347–354). Springer.

Goh, C. H., Dachowicz, A. P., Allen, J. K., & Mistree, F. (2018). A computational method for the design of materials accounting for the Process–Structure–Property–Performance (PSPP) relationship. In M. F. Horstemeyer (Ed.), *Integrated Computational Materials Engineering (ICME) for metals: Concepts and case studies* (pp. 539–572).

Goodrich, M. A., Stirling, W. C., & Frost, R. L. (1998). A theory of satisficing decisions and control. *IEEE Transactions on Systems Man and Cybernetics Part a-Systems and Humans, 28*(6), 763–779.

Goodrich, M. A., Stirling, W. C., & Frost, R. L. (1999). Model predictive satisficing fuzzy logic control. *IEEE Transactions on Fuzzy Systems, 7*(3), 319–332.

Goodrich, M. A., Stirling, W. C., & Boer, E. R. (2000). Satisficing revisited. *Minds and Machines, 10*(1), 79–110.

Guo, L., Chen, S., Allen, J. K., & Mistree, F. (2021). A framework for designing the customer-order decoupling point to facilitate mass customization. *ASME Journal of Mechanical Design, 143*(2), Article 022002.

Guo, L., Milisavljevic-Syed, J., Wang, R., Huang, Y., Allen, J. K., & Mistree, F. (2022). Managing multi-goal design problems using adaptive leveling-weighting-clustering algorithm. *Research in Engineering Design, 34*, 39–60.

Guo, L., Balu Nellippallil, A., Smith, W. F., Allen, J. K., & Mistree, F. (2020). Adaptive linear programming algorithm with parameter learning for managing engineering-design problems. In *ASME design automation conference*. Paper Number DETC2020-22602.

Guo, L. (2021). Model Evolution for the Realization of Complex Systems. School of Industrial and Systems Engineering, University of Oklahoma. Norman, Oklahoma. Ph.D. Dissertation.

Hajihashemi, S. (2023). A user-friendly wrapper for DSIDES (Decision Support in the Design of Engineering Systems). Industrial and Systems Engineering, University of Oklahoma. MS Thesis.

Hazelrigg, G. A. (2010). The pugh controlled convergence method: Model-based evaluation and implications for design theory. *Research in Engineering Design, 21*, 143–144.

Hu, C. F., Zhang, S. K., & Wang, N. (2014). Enhanced interactive satisficing method via alternative tolerance for fuzzy goal programming with progressive preference. *Applied Mathematical Modelling, 38*(19–20), 4673–4685.

Ito, M., & Kogiso, N. (2016). Information uncertainty evaluated by parameter estimation and its effect on reliability-based multiobjective optimization. *Journal of Advanced Mechanical Design, Systems, and Manufacturing, 10*(6), JAMDSM0083–JAMDSM0083

Jiao, R., Commuri, S., Panchal, J., Milisavljevic-Syed, J., Allen, J. K., Mistree, F., & Schaefer, D. (2021). Design engineering in the age of industry 4.0. *ASME Journal of Mechanical Design, 143*(7), 070801.

Kanno, Y., Fujita, S., & Ben-Haim, Y. (2017). Structural design for earthquake resilience: Info-gap management of uncertainty. *Structural Safety, 69*, 23–33.

Kapteyn, A., Wansbeek, T., & Buyze, J. (1979). Maximizing or satisficing. *The Review of Economics and Statistics, 61*(4), 549–563.

Keyvandarian, A., & Saif, A. (2023). Optimal sizing of a reliability-constrained, stand-alone hybrid renewable energy system using robust satisficing. *Renewable Energy, 204*, 569–579.

Kiyota, T., Tsuji, Y., & Kondo, E. (2003). Unsatisfying functions and multiobjective fuzzy satisfaction design using genetic algorithms. *IEEE Transactions on Systems, Man, and Cybernetics, Part B (Cybernetics), 33*(6), 889–897.

Kulkarni, N., Gautham, B., Zagade, P., Panchal, J., Allen, J. K., & Mistree, F. (2015). Exploring the geometry and material space in gear design. *Engineering Optimization, 47*(4), 561–577.

Kulkarni, N., Zagade, P. R., Gautham, B., Panchal, J. H., Allen, J. K., & Mistree, F. (2013). PREMΛP: Exploring the design and materials space for gears. In ICoRD'13: Global product development, lecture notes in mechanical engineering (pp. 759–772). Springer, India.

Kulkarni, N., Gupta, R., Khan, D., Gautham, B., Allen, J. K., Panchal, J., & Mistree, F. (2014). Inverse design of manufacturing process chains. In *ASME design automation conference*. Paper Number DETC2014-34203

Kumar, P., Goyal, S., Singh, A. K., Allen, J. K., Panchal, J. H., & Mistree, F. (2013). PREMΛP: Exploring the design space for continuous casting of steel. In *ICoRD'13: Global product development, lecture notes in mechanical engineering* (pp. 759–772). Springer, India.

Lewis, K., & Mistree, F. (1999). Foraging-directed Adaptive Linear Programming (FALP): A hybrid algorithm for discrete/continuous design problems. *Engineering Optimization, 32*(2), 191–217.

Li, M., & Wang, Z. (2018). Confidence-driven design optimization using gaussian process metamodeling with insufficient ata. *ASME Journal of Mechanical Design, 140*(12), Article 121405.

Li, Y. M., Du, W. B., Yang, P., Wu, T. H., Zhang, J., Wu, D. P., & Perc, M. (2019). A satisficing conflict resolution approach for multiple UAVs. *IEEE Internet of Things Journal, 6*(2), 1866–1878.

Loch, C., Mihm, J., & Huchzermeier, A. (2003). Concurrent engineering and design oscillations in complex engineering projects. *Concurrent Engineering-Research and Applications, 11*(3), 187–199.

Lounis, Z., & Daigle, L. (2013). Multi-objective and probabilistic decision-making approaches to sustainable design and management of highway bridge decks. *Structure and Infrastructure Engineering, 9*(4), 364–383.

Manktelow, K. I. (1999). *Reasoning and thinking*. Psychology Press.

McDowell, D. L., Panchal, J., Choi, H.-J., Seepersad, C., Allen, J. K., & Mistree, F. (2009). *Integrated design of multiscale*. Butterworth-Heinemann.

McDowell, D. L., Choi, H.-J., Panchal, J., Austin, R., Allen, J. K., & Mistree, F. (2007). Plasticity-related Microstructure-property relations for materials design. *Key Engineering Materials, 340*, 21–30.

Meng, M. Q. (2014). A hybrid particle swarm optimization algorithm for satisficing data envelopment analysis under fuzzy chance constraints. *Expert Systems with Applications, 41*(4), 2074–2082.

Messer, M., Pedersen, K., Allen, J. K., & Mistree, F. (2006). Domain independent approach to designing hierarchical platforms on multiple levels of abstraction and scales. In *ASME design automation conference*. Paper Number: DETC2006-99474.

Messer, M., Panchal, J., Allen, J., Mistree, F., Krishnamurthy, V., Klein, B., & Yoder, P. (2008). Designing embodiment design processes using a value-of-information-based approach with applications for integrated product and materials design. In *ASME design automation conference*. Paper Number: DETC2008-49395.

Milisavljevic-Syed, J., Allen, J. K., Commuri, S., Mistree, F., Milisavljevic-Syed, J. (2020). Decision-based design of networked manufacturing systems (NMS). In J. Milisavljevic-Syed, J. K. Allen, S. Commuri & F. Mistree (Eds.), *Architecting networked engineered systems: Manufacturing systems design for industry 4.0* (pp. 41–70).

Ming, Z., Nellippallil, A. B., Yan, Y., Wang, G., Goh, C. H., Allen, J. K., & Mistree, F. (2018a). PDSIDES—a knowledge-based platform for decision support in the design of engineering systems. *ASME Journal of Computing and Information Science in Engineering, 18*(4), Article 041001.

Ming, Z., Wang, G., Yan, Y., Panchal, J. H., Goh, C. H., Allen, J. K., & Mistree, F. (2018b). Ontology-based representation of design decision hierarchies. *ASME Journal of Computing and Information Science in Engineering, 18*(1), Article 011001.

Ming, Z., Sharma, G., Allen, J. K., & Mistree, F. (2019). Template-based configuration and execution of decision workflows in design of complex engineered systems. *Advanced Engineering Informatics, 42*, Article 100985.

Ming, Z., Sharma, G., Allen, J. K., & Mistree, F. (2020). An ontology for representing knowledge of decision interactions in decision-based design. *Computers in Industry, 114*, Article 103145.

Ming, Z., Nellippallil, A.B., Wang, R., Allen, J. K., Wang, G., Yan, Y., & Mistree, F. (2022). Extending PDSIDES to CB-PDSIDES: New opportunities in design engineering 4.0. In Z. Ming, A. B. Nellippallil, R. Wang, J. K. Allen, G. Wang, Y. Yan & F. Mistree (Eds.), *Architecting a knowledge-based platform for design engineering 4.0* (pp. 213–237). Springer.

Ming, Z., Nellippallil, A. B., Wang, R., Allen, J. K., Wang, G., Yan, Y., & Mistree, F. (2022). Foundations for design decision support in model-based complex engineered systems realization. In Z. Ming, A. B. Nellippallil, R. Wang, J. K. Allen, G. Wang, Y. Yan & F. Mistree (Eds.), *Architecting a knowledge-based platform for design engineering 4.0* (pp. 23–46). Springer.

Ming, Z., Nellippallil, A. B., Wang, R., Allen, J. K., Wang, G., Yan, Y., & Mistree, F. (2022). Requirements and architecture of the decision support platform for design engineering 4.0. In Z. Ming, A. B. Nellippallil, R. Wang, J. K. Allen, G. Wang, Y. Yan & F. Mistree (Eds.), *Architecting a knowledge-based platform for design engineering 4.0* (pp. 1–22). Springer.

Mistree, F., Hughes, O., & Phuoc, H. (1981). An optimization method for the design of large, highly constrained complex systems. *Engineering Optimization, 5*(3), 179–197.

Mistree, F., Smith, W., Bras, B., Allen, J. K., & Muster, D. (1990). Decision-based design: A contemporary paradigm for ship design. *Transactions, Society of Naval Architects and Marine Engineers, 98*, 565–597.

Mistree, F., Hughes, O. F., & Bras, B. (1993). The compromise decision support problem and the adaptive linear programming algorithm. In M. P. Kamat (Ed.), *Structural optimization: Status and promise* (pp. 247–286). AIAA.

Nash Jr., J. F. (1950). The bargaining problem. *Econometrica: Journal of the Econometric Society*, 155–162.

Nellippallil, A. B., Song, K. N., Goh, C.-H., Zagade, P., Gautham, B., Allen, J. K., & Mistree, F. (2017). A goal-oriented, sequential, inverse design method for the horizontal integration of a multistage hot rod rolling system. *ASME Journal of Mechanical Design, 139*(3), Article 031403.

Nellippallil, A. B., Rangaraj, V., Gautham, B. P., Singh, A. K., Allen, J. K., & Mistree, F. (2018). An inverse, decision-based design method for integrated design exploration of materials, products, and manufacturing processes. *ASME Journal of Mechanical Design, 140*(11), Article 111403.

Nellippallil, A. B., Ming, Z., Allen, J. K., & Mistree, F. (2019a). Cloud-based materials and product realization—fostering ICME Via Industry 4.0. *Integrating Materials and Manufacturing Innovation, 8*, 107–121.

Nellippallil, A. B., Allen, J. K., Gautham, B., Singh, A. K., & Mistree, F. (2020a). *Architecting robust co-design of materials, products, and manufacturing processes.* Springer.

Nellippallil, A. B., Mohan, P., Allen, J. K., & Mistree, F. (2020b). An inverse, decision-based design method for robust concept exploration. *ASME Journal of Mechanical Design, 142*(8), Article 081703.

Nellippallil, A. B., Mohan, P., Allen, J. K., & Mistree, F. (2019). Inverse Thermo-mechanical Processing (ITMP) design of a steel rod during hot rolling process. In *ASME design automation conference.* Paper Number DETC 2019-97390.

Nguyen, V. A., Jiang, J., Ng, K. M., & Teo, K. M. (2016). Satisficing measure approach for vehicle routing problem with time windows under uncertainty. *European Journal of Operational Research, 248*(2), 404–414.

Okamoto, T., Hanaoka, Y., Aiyoshi, E., & Kobayashi, Y. (2014). Optimal design of buffer material in the geological disposal of radioactive wastes using the satisficing trade-off method and a self-organizing map. *Electrical Engineering in Japan, 187*(2), 17–32.

Panchal, J. H., Paredis, C. J., Allen, J. K., & Mistree, F. (2008). A value-of-information based approach to simulation model refinement. *Engineering Optimization, 40*(3), 223–251.

Panchal, J. H., Gero Fernández, M., Paredis, C. J., Allen, J. K., & Mistree, F. (2009). A modular decision-centric approach for reusable design processes. *Concurrent Engineering, 17*(1), 5–19.

Panchal, J. H., Choi, H.-J., Shephard, J., Allen, J. K., McDowell, D. L., & Mistree, F. (2005). A strategy for simulation-based multiscale, multi-functional products and associated design processes. In *ASME design automation conference.* DETC2005-85316

Pedersen, K., Messer, M., Allen, J. K., & Mistree, F. (2013). Hierarchical product platform design: A domain-independent approach. *Ships and Offshore Structures, 8*(3–4), 367–382.

Rangarajan, B., Mistree, F., & Sorab, J. (1998). Incorporating tribological considerations in the robust concurrent design of automobile engine lubricated components. In *ASME advances in design automation.* DETC98/DAC-5594.

Rao, S. S., & Xiong, Y. (2005). Mixed-discrete fuzzy multiobjective programming for engineering optimization using hybrid genetic algorithm. *AIAA Journal, 43*(7), 1580–1590.

Reich, D., Green, R. E., Kircher, M., Krause, J., Patterson, N., Durand, E. Y., Viola, B., Briggs, A. W., Stenzel, U., & Johnson, P. L. (2010). Genetic history of an Archaic Hominin group from Denisova Cave in Siberia. *Nature, 468*(7327), 1053–1060.

Rippel, M., Choi, S., Mistree, F., Allen, J. K. (2010). Alternatives to taylor series approximation for the variance estimation in robust design. In *13th AIAA/ISSMO multidisciplinary analysis optimization conference.* Paper Number AIAA-2101-816659.

Saaty, T. L. (1979). Optimization by the Analytic Hierarchy Process, Air Force Office of Scientific ResearchBolling AFB DC, Technical Report, Accession Number: ADA214804.

Sabeghi, M., Smith, W., Allen, J. K., Mistree, F. (2015). Solution space exploration in model-based realization of engineered systems. In *ASME design automation conference.* Paper Number DETC2015-46457.

Sabeghi, M., Shukla, R., Allen, J. K., & Mistree, F. (2016). Solution space exploration of the process design for continuous casting of steel. In *ASME design automation conference.* Paper Number DETC2016-59399.

Sakawa, M., Shibano, T., & Obata, H. (1998). An interactive fuzzy method for multiobjective 0–1 programming problems with fuzzy number criteria using genetic algorithms. *Electronics and Communications in Japan Part III-Fundamental Electronic Science, 81*(8), 64–72.

Salado, A., & Nilchiani, R. (2017). Reducing excess requirements through orthogonal categorizations during problem formulation: Results of a factorial experiment. *IEEE Transactions on Systems Man Cybernetics-Systems, 47*(3), 405–415.

Samadiani, E., Joshi, Y., Allen, J. K., & Mistree, F. (2010). Adaptable robust design of multi-scale convective systems applied to energy efficient data centers. *Numerical Heat Transfer, Part a: Applications, 57*(2), 69–100.

Seepersad, C. C., Allen, J. K., McDowell, D. L., & Mistree, F. (2008). Multifunctional topology design of cellular material structures. *ASME Journal of Mechanical Design, 130*(3), 1285–1297.

Seepersad, C. C., Allen, J. K., McDowell, D. L., & Mistree, F. (2005). Robust design of cellular materials with topological and dimensional imperfections. In *ASME design automation conference*. Paper Number DETC2005-85061.

Sen, A. (2018). *Collective choice and social welfare*. Harvard University Press.

Sharma, S., & Kumar, V. (2022). A comprehensive review on multi-objective optimization techniques: past, present and future. *Archives of Computational Methods in Engineering, 29*(7), 5605–5633.

Sharma, G., Allen, J. K., & Mistree, F. (2021). A method for robust design in a coupled decision environment. *Design Science, 7*, Article e23.

Sharma, G., Allen, J. K., & Mistree, F. (2022). Designing concurrently and hierarchically coupled engineered systems. *Engineering Optimization, 55*(9), 1556–1576.

Shukla, R., Anapagaddi, R., Singh, A. K., Panchal, J. H., Mistree, F., & Allen, J. K. (2016). Design exploration to determine process parameters of ladle refining for an industrial application. *Steel Research International, 87*(10), 1333–1343.

Shukla, R., Goyal, S., Singh, A. K., Panchal, J. H., Allen, J. K., & Mistree, F. (2014). An approach to robust process design for continuous casting of slab. In *ASME design automation conference*. Paper Number DETC2014-34208.

Shukla, R., Kulkarni, N. H., Gautham, B., Singh, A. K., Mistree, F., Allen, J. K., Panchal., J. H. (2015). Design exploration of engineered materials, products, and associated manufacturing processes. *JOM: The Journal of the Minerals, Metals and Materials Society, 67*(1), 94–107.

Simon, H. A. (1947). *Administrative behavior*. Simon and Schuster.

Simon, H. A. (1956). Rational choice and the structure of the environment. *Psychological Review, 63*(2), 129.

Simon, H. A. (1973). Organization man: Rational or self-actualizing? *Public Administration Review, 33*(4), 346–353.

Simon, H. A. (1976). From substantive to procedural rationality. In T. J. Kastelein, S. K. Kuipers, W. A. Nijenhuis & G. R. Wagenaar (Eds.), *25 Years of economic theory* (pp. 65–86). Springer.

Simpson, T. W., Seepersad, C. C., & Mistree, F. (2001). Balancing commonality and performance within the concurrent design of multiple products in a product family. *Concurrent Engineering, 9*(3), 177–190.

Simpson, T., Allen, J., Chen, W., Mistree, F. (1996). Conceptual design of a family of products through the use of the robust concept extrapolation method. In *AIAA/NASA/USAF/ISSMO symposium on multidisciplinary analysis and optimization* (pp. 1535–1545). AIAA Paper Number 96-4161.

Sinha, A., Bera, N., Allen, J. K., Panchal, J. H., & Mistree, F. (2013). Uncertainty management in the design of multiscale systems. *ASME Journal of Mechanical Design, 135*(1), 0011008.

Sinha, A., Chakraborty, M., Ghosh, S., Kumar, C., Panchal, J. H., Allen, J. K., McDowell, D. L., & Mistree, F. (2009). Microstructure-mediated integration of material and product design: Undersea submersible. In *ASME design automation conference*. Paper Number: DETC2009/DAC-87276.

Sinha, A., Panchal, J. H., Allen, J. K., Mistree, F. (2011). Managing uncertainty in multiscale systems via simulation model refinement. In *ASME design automation conference*. Paper Number. DETC2011-47655.

Smith, W. F., Milisavljevic, J., Sabeghi, M., Allen, J. K., & Mistree, F. (2015).The realization of engineered systems with considerations of complexity. In *ASME design automation conference*. Paper Number DETC2015-46211.

Stirling, W. C. (2005). Social utility functions—Part I: Theory. *IEEE Transactions on Systems Man and Cybernetics Part C-Applications and Reviews, 35*(4), 522–532.

Stirling, W. C., & Nokleby, M. S. (2009). Satisficing coordination and social welfare for robotic societies. *International Journal of Social Robotics, 1*(1), 53–69.

Taguchi, G. (1985). Quality engineering in Japan. *Communications in Statistics-Theory and Methods, 14*(11), 2785–2801.

Tao, Y., Huang, H. Z., & Yang, B. (2006). An interactive preference-weight genetic algorithm for multi-criterion satisficing optimization. In L. Jiao, L. Wang, X. Gao, J. Liu, & F. Wu (Eds.), *Advances in natural computation* (pp. 643–652). Springer.

Thompson, S., Muchnick, H., Choi, H.-J., McDowell, D., Allen, J. K., & Mistree, F. (2006). Robust materials design of blast resistant panels. In *11th AIAA/ISSMO multidisciplinary analysis and optimization conference*. Paper No: AIAA-2006-7005.

Toyoda, M., & Kogiso, N. (2015). Robust multiobjective optimization method using satisficing trade-off method. *Journal of Mechanical Science and Technology, 29*(4), 1361–1367.

Wang, Z., & Wang, P. (2014). A maximum confidence enhancement based sequential sampling scheme for simulation-based design. *ASME Journal of Mechanical Design, 136*(2), Article 021006.

Wang, J., Si, W. J., & Li, H. Y. (2009b). Robust ISS-satisficing variable universe indirect fuzzy control for chaotic systems. *Chaos Solitons and Fractals, 39*(1), 28–38.

Wang, R., Nellippallil, A. B., Wang, G., Yan, Y., Allen, J. K., & Mistree, F. (2019). Ontology-based uncertainty management approach in designing of robust decision workflows. *Journal of Engineering Design, 30*(10–12), 726–757.

Wang, R., Nellippallil, A. B., Wang, G., Yan, Y., Allen, J. K., & Mistree, F. (2021b). A process knowledge representation approach for decision support in design of complex engineered systems. *Advanced Engineering Informatics, 48*, Article 101257.

Wang, C., Krishnamurthy, V., Klein, B., Choi, S.-K., Allen, J. K., & Mistree, F. (2009). A design exploration method for adaptive design systems. In *ASME design automation conference*. Paper DETC2009-87217.

Wang, R., Milisavljevic-Syed, J., Guo, L., Huang, Y., & Wang, G. (2021). Knowledge-based design guidance system for cloud-based decision support in the design of complex engineered systems. *ASME Journal of Mechanical Design, 143*(7), Article 072001.

Xiao, A., Seepersad, C. C., Allen, J. K., Rosen, D. W., & Mistree, F. (2007). Design for manufacturing: Application of collaborative multidisciplinary decision-making methodology. *Engineering Optimization, 39*(4), 22.

Zha, X. F., Sriram, R. D., Fernandez, M. G., & Mistree, F. (2008). Knowledge-intensive collaborative decision support for design processes: A hybrid decision support model and agent. *Computers in Industry, 59*(9), 905–922.

Index

L. Guo et al., *Architecting a Satisficing Strategy for Designing Complex Systems*,
https://doi.org/10.1007/978-3-032-04577-5

MIX
Papier aus verantwortungsvollen Quellen
Paper from responsible sources
FSC® C105338

If you have any concerns about our products,
you can contact us on
ProductSafety@springernature.com

In case Publisher is established outside the EU,
the EU authorized representative is:
Springer Nature Customer Service Center GmbH
Europaplatz 3, 69115 Heidelberg, Germany

Printed by Libri Plureos GmbH
in Hamburg, Germany